DARK EAGLES

DARK EAGLES

A History of Top Secret U.S. Aircraft Programs

CURTIS PEEBLES

PRESIDIO

Published by Presidio Press
505 B San Marin Dr., Suite 300
Novato, CA 94945-1340

Library of Congress Cataloging-in-Publication Data

Peebles, Curtis.
Dark Eagles : a history of top secret U.S. aircraft programs / by Curtis Peebles.
p. cm.
Includes bibliographical references and index.
ISBN 0-89141-535-1 (hardcover)
ISBN 0-89141-623-4 (paperback)
1. Aeronautics, Military—Research—United States—History. 2. Airplanes, Military—United States—History. 3. Defense information, Classified—United States. I. Title.
UG643.P44 1995
623.7'46'0973—dc20 95-11086
CIP

Printed in the United States of America

Contents

Introduction

On February 3, 1964, Lockheed test pilot James D. Eastham reached a speed of nearly Mach 3.3 at an altitude of 83,000 feet during the test flight of a new aircraft. This was a world's record for a jet-powered aircraft. For ten minutes, the plane held this speed. This epic flight was the culmination of five years of effort, frustration, and, finally, success. There was not one word about this singular achievement in that evening's newspapers. There was no mention of the event on the television news. No articles were published about the flight in the technical press. As far as the larger world knew, it had never happened.

This was because the airplane did not officially exist.

For the past five decades, some of the most significant advances in aerospace technology were made by airplanes that the larger world knew nothing about. Since 1941, the United States has produced a series of "Black" airplanes—planes developed, tested, and operated in deep secrecy. Years, even a decade or more, would pass before their existence was made public. Some remain secret still.

The impact of these Dark Eagles has been profound. The first introduced America to the jet age. The next revolutionized the way intelligence was gathered. Another pushed aviation technology to its farthest limits. A series of unmanned reconnaissance drones would venture to places too dangerous for conventional aircraft. One group would change U.S. aerial combat techniques and training. The latest series would fundamentally alter the role of airpower and strategic bombing, leading the way to the Coalition victory in the Gulf War. Each would do things most engineers thought impossible.

The names given to these Dark Eagles were meant to conceal—"Aquatone," "Oxcart," "Tagboard," "Have Blue," "YF-110," "YF-113," "F-117A," "HALSOL," "Amber," and "GNAT-750." There was no hint of the wonders that lay behind those bland titles. Similarly, the place where so many of these planes would make their first flights was called "The Ranch," "Area 51," "Red Square," or the sinister and evocative "Dreamland." The true name of this place was never spoken, never mentioned, not even in classified documents.

The Dark Eagles would first come to public attention in the 1980s. The secret reality and public speculation would combine into a dark mixture to form a shadow called Aurora.

It is only now, with the passage of the years and the end of the Cold War, that the story can be told of these remarkable aircraft, the people who designed and flew them, and the secret place where they were tested. Their story is the history of our time.

To understand how it came to be, we must first return to the beginnings of the Black airplane program—to a place under a blue desert sky, in the midst of a terrible war—and to an airplane that was one of America's most closely guarded secrets.

CHAPTER 1

The First Black Airplane
The XP-59A Airacomet

. . . come like the wind, go like the thunder.

Sun Tzu
ca. 400 B.C.

On the cool morning of October 1, 1942, a group of Bell Aircraft Company engineers prepared their new plane for its first flight. Finally, shortly after noon, the XP-59A stood ready. The aircraft had a midposition straight wing and tricycle landing gear. The tail was on a raised boom, while the center section of the fuselage seemed to bulge. The plane was painted dark olive green with dark gray undersides and had the U.S. insignia of a white star in a blue circle. It had no serial number. The XP-59A's design owed much to Bell's earlier P-39 and P-63. But in one aspect, this first Dark Eagle had nothing in common with any aircraft of the previous four decades of American aviation technology. It was the plane that separated all that was from all that would follow.

The XP-59A had no propeller.

BIRTH OF THE JET AGE

The events that set in motion development of the first U.S. Black airplane had begun more than a decade before. In 1928 Royal Air Force (RAF) Pilot Officer Frank Whittle, then only twenty, realized that the conventional propeller engine was nearing its performance limit. To fly faster, a larger, more powerful engine was needed. Such an engine would burn more fuel, thus requiring a larger, heavier airframe and canceling out any gain. As a plane flew higher, it flew into thinner air. This resulted in a loss of engine power. Propellers, as they approached supersonic speed, also lost efficiency. The high-speed planes of the 1930s, such as the GeeBee racer, were little more than the biggest possible engine attached to the smallest possible airframe. They flew fast, but, like the GeeBee, often proved lethal.[1]

Whittle proposed the idea of using a gas turbine to power an aircraft. Incoming air would be compressed, then mixed with fuel and ignited. The hot gas would be vented out an exhaust pipe to produce thrust. This offered speeds and altitudes far beyond the reach of propeller-driven fighters and bombers. Whittle submitted his idea to the British Air Ministry, which promptly rejected it as unattainable. For the next seven years, Whittle struggled to find money to build his "Whittle Unit." It was not until 1937 that the Air Ministry agreed to a small study contract, and it was another year before the money was actually provided.

In March 1939, the first Whittle jet engine was making test-bed runs. In the meantime, Nazi Germany had absorbed Austria and taken over Czechoslovakia. With war clouds looming over Europe, a few far-sighted individuals realized the strategic advantages of jet aircraft. In July 1939, Whittle was given a contract to develop the W.1 jet engine, which would power the experimental Gloster E28/39 Pioneer aircraft. Two months later, Germany invaded Poland and World War II began. By the following summer, Hitler was the master of Europe, and England stood alone. In the sky above London, the RAF and the Luftwaffe fought the Battle of Britain to decide the fate of Western Civilization.

As these monumental events were being played out, Whittle and a small group of engineers were working in an empty factory near Coventry, England. The engine that was built was unlike any power plant ever flown before. A conventional aircraft engine operated at 2,000 rpm. The W.1's turbine spun at 17,750 rpm. The temperatures inside the combustion chambers were also far higher than those of piston engines.

Equally daunting was the political situation. With London in flames and England needing every Spitfire it could produce, Lord Beaverbrook, head of the Ministry of Aircraft Production, stripped the W.1 of its priority order. By this time, however, Whittle had gained powerful supporters who were able to convince Beaverbrook to restore the W.1's priority.[2]

By the spring of 1941, the first E28/39 aircraft was finished and Whittle delivered a "lash-up" prototype engine, the W.1X. (The *X* indicated it was not to be flown, but only used for taxi tests.) The taxi tests were made on April 7 and 8, 1941, by Flight Lieutenant P. E. G. Sayer, Gloster's chief test pilot. In the final series, the aircraft lifted off on three short hops. On May 15, the E28/39 was ready for its first flight. Due to poor weather, it was delayed. Finally, at 7:35 P.M., Sayer took off for a seventeen-minute flight. It was the culmination of more than a decade of efforts by Whittle. The Air Ministry did not bother to send an official photographer to record the event.[3]

THE JET ENGINE COMES TO AMERICA

The Battle of Britain had ended in victory for England, but it was clear that the country lacked the industrial capacity to defeat Nazi Germany on its own. The only option was to share military technology with America, including jet engines. On April 11, 1941, U.S. Army Air Corps Chief Maj. Gen. Harold "Hap" Arnold arrived in England to examine jet propulsion projects. Arnold quickly realized what Whittle had achieved—the E28/39 could outfly a Spitfire, then the fastest British aircraft. Every aircraft the army air corps and navy were building or planning was about to be made obsolete.

In late May, General Arnold formally requested access to jet engine technology. Initially, the British provided only a nine-page secret memo describing the engine. On July 15, the British agreed to release the Whittle engine to the United States "subject to special care being taken to safeguard its secrecy." To meet this requirement, the concept of the "Black airplane" would be developed.[4]

On September 4, 1941, General Arnold met with senior Army Air Forces (AAF) and War Department officials. Also on hand were four General Electric representatives. Arnold opened a safe and pulled out several reports. After discussing the recommendation that the United States embark on a crash program to mass-produce the engine, General Arnold said, "Gentlemen, I give you the Whittle engine—consult all you wish and arrive at any decision you please—just so long as General Electric accepts a contract to build 15 of them."

It was also agreed that Bell Aircraft would build three prototype jet fighters. Because of the low thrust of the Whittle engine, it would have to be a twin-engine design. Unlike the experimental E28/39, it would be intended as an operational fighter. Bell was picked for several reasons. Bell's Buffalo, New York, plant was near General Electric's Schenectady and Lynn, Massachuetts engine plants. Bell's engineering and design staff were not overloaded with existing contracts. This was important, as General Arnold had imposed a one-year deadline. Larry Bell, president of the company, had a reputation for undertaking unusual projects and could be counted on to keep close watch on the effort. That evening, Arnold's office contacted Larry Bell and asked that he and his chief engineer, Harland M. Poyer, come to Washington, D.C. On September 5, they were briefed on the jet engine and were asked to build the airplane. They agreed.[5]

On September 22, the British Air Commission told the U.S. Secretary of War, Henry L. Stimson, that all information on jet engines would be released. The British provided one engine (the old W.1X used in the taxi tests) and a set of manufacturing drawings for the W.2B, an advanced version of the original W.1 engine. On October 1, 1941, the engine and drawings were loaded on a B-24 at Prestwick, Scotland, and flown west across the Atlantic. The plane arrived the next day at Bolling Field in Washington, D.C. Then began a standoff; customs agents demanded to inspect the cargo. It took two days before they relented and agreed only to count the three crates, which were sent on, finally, to the General Electric plant in Boston.[6]

THE SECRET SIX DESIGN A BLACK AIRPLANE

When Poyer returned from the September 5 meeting, he selected a small group of engineers and called them to Larry Bell's office. They were sworn to secrecy, then briefed on the project. Larry Bell told them they would design the first U.S. jet fighter. The group, quickly dubbed the "Secret Six," were Poyer, Robert Wolf, E. P. Rhodes, Jim Limage, H. L. Bowers, and Brian Sparks. This established the pattern for later Black airplanes—they were developed by a very small group, using streamlined procedures and working on a tight schedule.

The project was protected by layers of secrecy and deception, far beyond the normal secrecy involved in building a new aircraft. All information on jet technology was classified "Special Secret." This was the predecessor to today's "Top Secret (Codeword)" and "Special Access" classifications. The designation XP-59A was an example of such "cover." The original XP-59 was a single-seat, twin-boom pusher fighter. Preliminary design work had been done and a wooden mock-up had been built. By reusing the designation and adding an *A*, it was made to seem to be only a revised version of the old plane.

The Secret Six had a preliminary proposal and a one-twentieth scale model of the aircraft ready in two weeks. General Arnold approved the design. On September 30, 1941, an eight-month, fixed-fee contract was signed for three XP-59A aircraft, a wind-tunnel model, and data. The total price was $1,644,431. The XP-59As were described only as "twin engine, single place interceptor pursuit models." The contract required that General Electric engines be used, but otherwise Bell had a free hand in determining the plane's configuration.

It was clear from the start that the project would have to be done outside Bell's existing development-production facilities. The first drawings were done at an old Pierce-Arrow factory on Elmwood Avenue in Buffalo. Before long, the work was moved to a four-story building owned by the Ford Motor Company at Main and Rodney Streets. As the Secret Six moved into the second floor, a Ford dealership was still selling cars on the first floor. The dealership was soon eased out, and the machine shop and storage areas were set up on the first floor. To ensure security, all entrances to the "Main Street Plant" were guarded, and special passes were required to enter the building.

The metal window frames were welded shut, and the first- and second-floor window panes were painted over.[7]

Once again, the XP-59A introduced an aspect of later Black airplane development. The contractor was now split into a "White" half (which conducted normal production) and a "Black" half, for secret work. The Black company was a duplicate of the larger White part, with its own design and production facilities. These facilities were isolated—both physically and in terms of secrecy—from the main company.

The Secret Six were embarking on an unknown sea. They were about to reinvent the airplane, yet the only information they had initially on the jet engine was a single, freehand one-twentieth scale sketch. There was nothing about the specific dimensions, weight, thrust, attachment points, accessories, cooling, inflow and outflow—just a drawing the size of a cigarette pack. Another difficulty was the expectation that the XP-59A could be directly converted into an operational fighter, skipping the test aircraft step. This was made more difficult by the low thrust of the jet engines. The XP-59A's thrust-to-aircraft-weight ratio was lower than contemporary fighters.

The secrecy of the program also complicated development. Outside wind tunnels could not be used. (The one exception was the use of the Wright Field low-speed tunnel to refine the engine inlet design.) The Secret Six could not consult with outside technical experts or contractors. They had to either build equipment in-house or use off-the-shelf hardware.

The secrecy problems became more complex once fabrication of the first XP-59A began on January 9, 1942. Much of the work was done in the machine shop at the Main Street Plant. However, large parts had to be made at the main Bell plant. The drawings were purposely mislabeled—the engine exhaust pipes, for example, were "heater ducts" (a full fourteen inches in diameter).

As construction of the prototype continued in March and April 1942, more man power was needed. People began to "disappear." Desks and drawing boards were now empty. When the "lost ones" met their ex-coworkers at social events and meetings, they were asked what they were doing. It reached the point that the XP-59A personnel were discouraged from attending such outside activities.[8]

As the Secret Six worked on the prototype, General Electric was producing the engine. Once the W.1X was delivered to Boston, General Electric constructed a special test cell in Building 34 North at the River Works Plant. Dubbed "Fort Knox," it was constructed of reinforced concrete and had a heavy steel door. The engine was viewed through a small slit. The exhaust was vented out a sixty-four-foot unused chimney. The W.1X was ignited for the first time on October 16, 1941.

General Electric then began building the production W.2B engines. As with the XP-59A, cover designations were used. The engine was called "I-A turbosupercharger." General Electric was then producing aircraft superchargers in the *A* through *F* series; calling it the "I-A" made it seem to be the eighth in this series. General Electric found that the drawings were not complete. They also suggested changes in the gear train and accessories, a new alloy, and modified compressor blades. Even so, it was still a copy of the W.2B. The first test I-A engine made a brief run on March 18, 1942.[9] It was another thirty days before the problems were ironed out and the engine reached 1,250 pounds of thrust. To help work out the problems, Whittle, now an RAF wing commander, came to the United States in early June and remained until the first week of August. By that time, the first two production I-A jet engines were shipped to Bell. They were installed in the prototype XP-59A and final assembly began.[10]

MUROC—A NEST FOR THE DARK EAGLE

Having gone to these extraordinary lengths to keep the XP-59A secret, it was clear the plane could not be test flown from the Bell plant in Buffalo. An isolated site would be needed to ensure secrecy during the highly visible test flights. In early 1942, Lt. Col. Benjamin W. Chidlaw and Maj. Ralph R. Swofford Jr. made a tour of possible sites. They selected the Muroc Bombing and Gunnery Range on Rogers Dry Lake, in the Mojave Desert of California.

Rogers Dry Lake is a flat expanse some sixty-five square miles in area. The site was originally settled by Clifford and Effie Corum in 1910. The little desert community that soon grew on the edge of the lake bed was named Muroc (Corum spelled backward). In September 1933, the army air corps set up a gunnery range on the lake bed. With the attack on Pearl Harbor, a base was built at the south end of the lake bed to train B-24, B-25, and P-38 crews. Out on the lake bed the "Muroc Maru" was built—a false-front mock-up of a Japanese *Mogambi*-class heavy cruiser—to act as a bombing target.[11] The site was isolated, far from any major city.

In mid-May 1942, Bell was told Muroc would be the test site. The flight test facility was constructed on the north end of the lake bed, about five miles from Muroc Field. In later years, the two areas would become known as "North Base" and "South Base." North Base consisted of a large portable hangar (which had lost many parts in its travels), a two-story barracks, and a mess hall. Water came from a two-hundred-foot-deep well and was stored in a wooden tank. The ground control for the test flights was a two-way radio and an old recorder set up on the ramp. Transport was provided by two Bell-owned station wagons.[12]

Larry Bell told Robert M. Stanley, Bell Aircraft's chief test pilot, that the company was building a jet aircraft and that he would fly it. This was the first time Stanley had heard of the project. Stanley arrived at North Base on August 20, 1942. He found that progress on the buildings had been slow, and the contractor said there was little chance of completing the work by mid-September. As it turned out, the barracks was completed by the deadline, while the hangar lacked only the floor and electrical wiring. With the prototype aircraft about to arrive, the civilian contractors were sent away, and Stanley and the Bell crew finished the work. They built to last—a half century later, the original XP-59A hangar is still in use at North Base, and is still used for Black airplanes.

The prototype XP-59A was ready to ship in September 1942. It was decided to send the fuselage to Muroc with the two I-A engines in place. This meant there would be no time lost removing the engines, then reinstalling them at Muroc. The problem was that jolts during the long train trip could damage the engines' bearings. It was decided to slowly spin the engines for the whole trip.

The fuselage and wings were wrapped in fabric for the journey. A hole was knocked through the second-story wall, and the packages were lowered by a crane. They were then loaded in two boxcars at 2:00 A.M. on September 12, while army guards patrolled the rail yard. The train set out with three General Electric engineers and five army guards to watch over the plane. A gasoline-powered air compressor was used to keep the engines turning. The compressor's gasoline tank had to be refilled constantly—a difficult job on a moving railroad car. On the second night out, the compressor repeatedly failed due to contaminated gasoline, but the General Electric engineers were able to keep restarting it before the jet engines spun down. Finally, at 8:00 A.M., September 19, a full six days after leaving Buffalo, the XP-59A arrived at Muroc.

The next week was spent getting the aircraft ready for its planned first flight on October 2. The first engine ground test runs were made on September 26. Both engines made three five-minute runs. The plane was judged ready for taxi tests. On September 30, Stanley made several high-speed taxi runs to check out the handling of the aircraft. Several times, the XP-59A lifted off the lake bed. Based on this, Stanley wanted to press on and make the first flight. It was late in the afternoon, however, and Larry Bell said it would be better to wait until the next morning.

FIRST FLIGHT

On October 1, 1942, a year to the day after the nonflyable W.1X engine and an incomplete set of drawings were sent to the United States, the

XP-59A stood ready to try its wings. Given the technological unknowns, this was a remarkable achievement. It was also an indication of what Black development procedures could accomplish.

On this morning there were the usual last-minute problems. The ignition wires on both engines had to be changed before they could be started. Once they were running, Stanley taxied about three miles downwind onto the lake bed. He then turned the XP-59A and ran up the engines. The first flight reached an altitude of approximately 25 feet, and landing was made using partial power without flaps.

In all, four flights were made. In each case, the landing gear was left down and altitude did not exceed 100 feet.[13] For those who had worked on the project, who knew the secret and understood what had been accomplished, it was a remarkable experience. Ted Rogers, a General Electric engineer wrote, "What a strange feeling this seemingly giant bird gave us as it approached. There was dead silence as it passed overhead—then a low rumbling like a blowtorch—and it was gone, leaving a smell of kerosene in the air."[14]

The following day, a second series of test flights was made. Stanley made the first two flights, reaching 6,000 and 10,000 feet. The day's third flight was made by Col. Laurence C. Craigie, chief of the Aircraft Project Section at Wright Field. Stanley told Craigie that the engines had only about a half hour left before they would have to be overhauled, then asked if he would like to fly the plane. Craigie was a program manager and was not even a test pilot. He had come to North Base only as an observer, but Craigie did not have to be asked twice. Later, he recalled, "I didn't get very high. I didn't go very fast. The most vivid impression I received, after a very long takeoff run, occurred at the moment we broke contact with the ground—it was so quiet."

Thus, quite by chance, Craigie became the first U.S. military pilot to make a jet flight.[15] Stanley made the day's final flight.

The two days of flights indicated the igniter wires, landing gear, and oil pressure gauges all needed modification. The two I-A engines were also replaced. All early jet engines had very low operating lifetimes—in the case of the I-A, a mere five hours.[16]

The test procedures did not match XP-59A's sophistication. The test pilot would radio instrument readings to the ground or jot down notes on a knee board. Control stick forces were measured with a fish scale. The engine thrust was measured with an industrial spring scale attached to the landing gear and anchored to the ground. Testing the pressurized cockpit (the first on a U.S. fighter) was a constant problem. The cabin seals had to be checked and replaced frequently. To check them, Angus McEahem, a General Electric technician, would close the canopy, start up the engine, and pressurize the cockpit. He would then light up a cigar. The smoke would show any leaks.

It was clear from the start that the XP-59A required a new level of flight test data. As an interim solution, an observer's position was fitted into the nose section. A twenty-inch hole was cut in the upper fuselage, and a seat and instrument panel were fitted into the empty gun compartment. It resembled a World War I biplane cockpit. The XP-59A thus became the first two-seat jet (and the first open-cockpit jet aircraft). When test flights resumed on October 30, the observer's position proved highly successful. The first observer was E. P. Rhodes, Bell project engineer for the XP-59A.[17]

Test flights of the XP-59A continued at a slow pace, due, in part, to the maintenance and modifications required of all new aircraft. The main problem was the I-A engines. They needed constant inspection and troubleshooting. This was aggravated by slow engine production at General Electric. Delays in engine deliveries were a constant problem. Because of the short lifetime of each engine, the shortage interfered with early flight operations. What test flights were made indicated the engine bearings were overheating.

The engine delivery problems also affected the second and third XP-59A prototypes. The second aircraft was sent to Muroc without engines. The wings arrived on December 27, 1942, with the fuselage following on January 4, 1943. Delays in the engine shipments pushed back the first flight until February 15. It was flown by Bell test pilot Frank H. "Bud" Kelly Jr., who had replaced Stanley in November. At takeoff, the cabin defroster failed, filling the cockpit with smoke. Kelly made a tight turn, cut the engines, and made a dead-stick landing.

The third XP-59A arrived at Muroc on February 21. Again, the engines were not ready, so it was shipped without them. Due to the lack of engines and the press of modifications on the first two aircraft, it was not assembled until April. Adding to the engine delivery problems was the weather. In late January 1943, heavy winter rains flooded Rogers "Dry" Lake. While waiting for the lake bed to dry, the Bell and General Electric engineers worked on the bearings problem. They found it was caused by excessive tolerances.

With more rain expected, it was decided to shift operations away from North Base temporarily. Hawes Field, near Victorville Field (later George Air Force Base) would be used. On March 10, the second XP-59A was towed thirty-five miles by road to the new site. The XP-59A was still classified Special Secret, however. The solution would create the most lasting image of the first Black airplane. To hide the telltale intakes and exhausts, the fuselage, from the nose to behind the wing's trailing edge, was covered in fabric during the move. Fitted to the nose was a four-bladed "prop" made by Joe Brown. Although crude, from a distance it would fool a witness. To make sure no one came close enough to see through the camouflage, the road was temporarily closed.

Only one flight was made from Hawes Field before it was decided that the facility had inadequate security. About March 15, the XP-59A was again moved, this time to Harpers Lake. The site was some forty-two miles from North Base, and it soon proved difficult to transport personnel, supplies, and food to the area. By April 7, Rogers Dry Lake was again usable. The plane was flown back to North Base.[18]

FLIGHT TESTS ACCELERATE

April 1943 marked a turning point in the XP-59A program. Up to April 11, the first aircraft had made only thirty flights for a total of fifteen hours fifteen minutes of flight time. The second aircraft totaled twenty-four flights and thirteen hours forty-five minutes in the air, while the third aircraft had yet to fly. During April and May, the pace of flight testing picked up. The third aircraft was flown, and the trio made sixty-seven flights to conduct glide tests, speed-power calibrations, landing gear tests, and performance checks.

All those who flew the XP-59A noted its smooth and quiet ride. In fact the instruments often stuck due to the lack of vibration. To solve that problem, a two-dollar doorbell ringer was mounted on the instrument panel to provide the necessary vibration.

The circle of those with jet flight experience was expanding. On April 21, Capt. Frederick M. Trapnell, chief of flight test for the Bureau of Aeronautics, became the first navy jet pilot. Trapnell, who retired as an admiral, had mixed feelings about his flight in the first XP-59A. Many years later he recalled,

> In ground run-ups the jet was very impressive for its unusual nose and the "blow-torch" slipstream, but the aircraft was obviously a very gentle type of high-altitude fighter with low wing-loading. It was a great surprise to find that the thing was very quiet and smooth from the pilot's point of view. During takeoff the rattling of the landing gear was audible and the general impression was that of a glider. The XP-59A was comparatively low-powered and this was apparent from the shallow climb-out. Its performance was, at first, distinctly unimpressive—long takeoff and slow rate of climb.

The Bell test pilots also underwent changes—Kelly left North Base and was replaced by Jack Woolams as chief pilot. Woolams set an altitude record of 45,765 feet on July 14, 1943. He broke his own record on December 15, reaching 47,600 feet. In September 1943, Alvin M. "Tex" Johnston joined the program. Tex Johnston took over as chief pilot at the end of the year. Soon after, R. J. O'Gorman was added to the flight test effort. One famous pilot who did not get a chance to fly the plane was Howard Hughes. He came to North Base to fly the XP-59A, but the crew faked an engine problem—they did not want him flying "their" plane just for fun.

The number of aircraft was also growing. On March 26, 1942, a contract had been approved to deliver thirteen YP-59A service test aircraft. These were preproduction aircraft, more similar to operational aircraft. Unlike the three XP-59As, these aircraft would be armed with either two 37mm cannons or one 37mm cannon and three .50-caliber machine guns. The first two YP-59As arrived at North Base in June 1943, but problems delayed their first flights until August and September. Initially, they had to use the original I-A engines, as the more powerful I-16 engines were delayed. By the end of the year, more YP-59As had been delivered, and the airplane had been given its official name of "Airacomet," which had been selected from crew suggestions.

LIFE WITH A BLACK AIRPLANE

To enter the world of Black airplanes is to embark on a strange adventure. Tex Johnston was asked only if he wanted to be project test pilot on a secret airplane. He drove from Buffalo to North Base, arriving at lunchtime. He was about to sit down in the mess hall (called the "Desert Rat Hotel") when "there was a sudden swish and a roar overhead." He asked, "What the hell was that?" He went outside and, as he wrote later: "I spotted the plane coming in for another pass. As it swooshed by, I understood. No prop. I had just witnessed my first jet-propelled airplane."

The XP-59A personnel (and their counterparts on later Black airplanes) were doing things no others had the chance to do or would even dream possible. But they could not tell any one about it. Captain Trapnell later gave a firsthand example of this: "I found myself in a group discussing rumors then emanating from Europe, of a weird and wonderful means of propulsion—without a propeller. The discussion became quite intense and very inaccurate, to say the least. I was supposed to be the most knowledgeable of those present but I had to sit silent and act dumb. I couldn't say that I not only knew about it but had flown one. I was forbidden to say a word."

Life at North Base was rugged—the hours were long, living quarters spartan, and the weather ranged from extreme heat during the day to freezing cold at night. Such shared hardship creates a unity that people in nine to five jobs can never know. Such a brotherhood of experience finds expression in symbols. Woolams returned from a trip to Hollywood with several dozen black derby hats and some fake mustaches. He gave them out to the Bell personnel. The "Bell Bowlers" would wear the hats as symbols of jet service while drinking in bars such as Juanita's in nearby Rosamond. The army air forces personnel removed the propeller from their collar insignia. To those who were part of the group, the meaning was understood. To those outside the secret club, the symbols were meaningless.

The airspace over North Base was restricted, and pilots training at South Base were told never to approach it. Being human, they sometimes tried to sneak a peek. In mid-June 1943, Lt. Royal D. Frey was flying near North Base when he saw a plane take off. It was silhouetted against the lake bed, and he noticed the shadow of a smoke trail from the aircraft. A few minutes later, the plane passed his P-38 in a steep rolling climb. During the brief "sighting," he saw it had no propeller. When he landed, Frey told the other student pilots but was disbelieved. After all, an airplane could not fly without a propeller.

Frey was more fortunate than another group of Muroc pilots. They were flying in formation when an XP-59 pulled up alongside. Their shock at seeing an airplane flying without a propeller was considerable. A bigger shock came when they saw the pilot was a gorilla wearing a black derby hat and waving a cigar! It was Jack Woolams in a Halloween mask and the Bell trademark hat. The "pilot" then tipped his hat and peeled off. It is reported that throttles were bent and vows of abstinence taken by several pilots in that fall of 1943.[19]

But sometimes the secrecy of a Black airplane asked a great deal. On September 24, 1943, Woolams was flying a photo mission with another airplane. After they took off, a sandstorm swept in, covering North Base with a blanket of blowing dust. The chase plane flew on to Burbank for a landing. Woolams did not have that option—he *had* to land at North Base. Whatever the circumstances, the XP-59A could not be seen. Woolams made a risky instrument landing in the midst of the storm.[20]

THE END OF THE P-59

It was not until January 6, 1944, that the existence of the P–59 program was revealed. The joint U.S. Army Air Forces–Royal Air Force announcement gave a brief history of jet propulsion and limited details such as the date of the first flight. It did not say where the test flights were made, the name of the aircraft, or did it include a picture. This set the pattern for later announcements.

The lack of official information did not stop the press from speculating, however. Typical quotes included, "Speed of the plane was placed at between 500 and 600 mph," and "Its top speed has been estimated by ground observers to exceed 500 mph." This, too, would become typical of later Dark Eagles.

In February 1944, operational tests were conducted at Muroc by army air forces test pilots to determine the production YP-59's tactical suitability. Three YP-59As with the more powerful I-16 engines were used. The results were disappointing—in mock dogfights with P-47Ds and P-38Js, the YP-59As were outclassed in both performance and maneuverability. The P-47s and P-38s could break off combat at will by either diving away or going into a full-power climb.

As the suitability tests were being conducted, Bell's North Base operations were being brought to an end. The final days were spent giving rides to Bell mechanics in the observer's cockpit of the first XP-59A. By February 18, 1944, the aircraft and equipment were turned over to the army air forces. On February 27, Bell flight operations at North Base were formally closed. The three XP-59As and six YP-59As had put in 242 hours and 30 minutes of flight time without a mishap.

The shortcomings of the P-59 were reinforced in April 1944 when a YP-59A underwent gunnery tests. Using its three .50-caliber machine guns, the plane made firing runs at speeds between 220 and 340 mph. The tests showed poor directional stability at speeds above 290 mph. The army air forces concluded ". . . it is not believed that the P-59 airplane is operationally or tactically suited for combat nor is it believed that any modification to this aircraft, short of a completely new design, would improve its combat suitability . . ."

It was still felt the P-59A had a useful role:

> . . . although the aircraft is not suitable for combat, there is a requirement for a limited number of subject aircraft to be utilized for jet training and for general Air Force familiarization. The Army Air Forces Board is further of the opinion that use of jet propelled aircraft will become widespread in the immediate future and that the P-59 . . . is an excellent aircraft for purposes of conducting research on jet power plants and pressure cabins. The P-59 will also make an excellent training ship in that its low wing-loading makes the airplane very safe for transition flying and the fact that it has two engines is an added safety factor.

Due to its planned role as a jet trainer, only limited P-59 production was undertaken. The operational aircraft incorporated a number of modifications first tested on the YP-59As. The wing tips and rudder were reduced in size to improve maneuverability, and a vertical fin was added to improve spin recovery. The aft fuselage was strengthened, metal flaps and ailerons replaced the original fabric-covered ones, and the main landing gear was modified. After twenty P-59As were delivered, fuel tanks were added to the outer wings. These final thirty aircraft were redesignated P-59Bs. Most of the aircraft were operated by the 412th Fighter Group. In July 1946, less than a year after the last P-59 was delivered, they were retired. One YP-59A and three P-59As were also provided to the U.S. Navy. They were operated for several years in a test role, introducing that service to the jet age.[21]

The Bell P-59 Airacomet was a ground-breaking aircraft in many ways. For American aviation, it ushered in the jet age and a half-century dominance of both military and civilian aerospace technology. Although unsuccessful as a fighter, the P-59 provided valuable experience. It underlined the kind of change jet engines brought to aviation. Although its propulsion was revolutionary, the P-59 was limited by outmoded aerodynamics. With its broad, straight wings and teardrop-shaped fuselage, the P-59 was very much a late-1930's design. The fake prop did not look at all out of place. Its top speed was limited to 389 mph at 35,000 feet—inferior to that of prop fighters.

In contrast, the German Me 262, with sweep wings and more refined aerodynamics, had a top speed of 580 mph. Clearly, it was not enough to simply stick jet engines on a propeller-driven airplane. (One early XP-59A design was a P-39 with two jet engines hung under the wings.) The revolutionary engines had to be matched with an equally revolutionary airframe.

Bell and the XP-59A created the modern concept of the Black airplane. All the elements—the secret task, small design group, tight schedule, separate facilities, and the isolated test site—were present. Yet the plane marked the decline of Bell's role in fixed-wing aviation. The formal end came with another Black airplane, also unsuccessful. The heritage of the first Dark Eagle would be carried by another company, and at another place.

On January 8, 1944, the Lockheed XP-80 Shooting Star jet fighter made its first flight at Muroc. At the controls was Milo Burcham. The plane soon proved capable of reaching over 500 mph. Tex Johnston knew what it meant for the P-59. After seeing the first flight, he telegraphed Bob Stanley: "Witnessed Lockheed XP-80 initial flight STOP Very impressive STOP Back to the drawing board."[22] Later, a mock dogfight was held between a P-80 and a Grumman F8F Bearcat, the navy's latest prop fighter. Unlike the YP-59A, the P-80 held the initiative, controlling the fight. The F8F was never able to catch the jet in its sights long enough to get a shot. The era of the prop fighter was over.[23]

The XP-80 contract specified that the prototype was to be delivered in 180 days. Clarence L. "Kelly" Johnson, Lockheed's chief designer, went to company chairman Robert Gross. Gross told Johnson, "Go ahead and do it. But you've got to rake up your own engineering department and your own production people and figure out where to put this project."

For some time, Johnson had been asking Lockheed management to set up an experimental department where there would be direct links between designer, engineer, and manufacturing. Johnson decided to run the XP-80 program on this basis. The only place for the new section was next to the wind tunnel. The tools came from a small machine shop Lockheed bought out. The walls were wooden engine boxes, while the roof was a rented circus tent. Johnson assembled a group of twenty-two engineers; the new group had its own purchasing department and could function independently of the main plant. Working ten hours a day, six days a week, they had the XP-80 ready in 163 days.

Part of the secrecy surrounding the project was that Johnson's new section had no name. Soon after the makeshift shop was finished, Lockheed engineer Irving H. Culver was at the phone desk. The phone rang, Culver was alone, and he had not been told how to answer the phone. Culver was a fan of Al Capp's comic strip "L'l Abner." In the strip, "Hairless Joe" brewed up "Kickapoo Joy Juice" using old shoes, dead skunks, and other ingredients. On impulse, Culver answered the phone with the name of that brewery.[24]

It was called "the Skunk Works."

Chapter 2

The Angel of Paradise Ranch
The U-2 Aquatone

Now the reason the enlightened prince and the wise general conquer the enemy whenever they move and their achievements surpass those of ordinary men is foreknowledge.

Sun Tzu
ca. 400 B.C.

With the end of World War II, the shaky alliance between the Soviet Union and the Western powers unraveled before the reality of Stalinism. Events during 1948 and 1949, such as the Berlin Blockade and the testing of the first Soviet A-bomb, underlined the need for information on the Soviet Union. The CIA and the British Secret Intelligence Service attempted to parachute agents into the Soviet Union between 1949 and 1953. The West also attempted to support resistance groups in the Ukraine, the Baltic States, Albania, and Poland. The efforts ended in failure. The agents were captured as soon as they landed, while the resistance groups were ruthlessly hunted down.[1]

The intelligence would have to be gathered from the air.

COLD WAR OVERFLIGHTS

With the start of the Cold War, overflights of the Soviet Union began. In the late-1940s, the British used de Havilland Mosquito PR.34s to photograph northern ports such as Murmansk and Archangel. The armor was removed to raise the maximum altitude to 43,000 feet, above that of Soviet propeller fighters. The Mosquito overflights continued into 1949, until the introduction of the MiG 15 jet fighter made them too dangerous.[2]

With the start of the Korean War in June 1950, overflights began in earnest. In the fall of 1950, President Harry S Truman authorized a program to cover Soviet ports, islands, and coastal areas.[3] Initially, two different aircraft were used—the RB-36D Peacemaker and the RB-45C Tornado. The

RB-36s were stripped of unnecessary equipment, including all the guns except the twin 20mm cannons in the tail turret. These featherweight RB-36s could reach altitudes of 58,000 feet, which gave them virtual immunity from Soviet MiG 15s.

The RB-45s were light jet reconnaissance bombers, which relied on speed and the brief duration of the overflight, rather than altitude, to escape detection. (Its performance was less than that of the MiG 15.) In 1952 and 1954, RB-45Cs were painted in RAF markings and made overflights of the western Soviet Union.[4] U.S. Air Force RB-45Cs, based in Japan, also overflew the Pacific coast of the Soviet Union.[5]

In 1953, overflight missions were taken over by RB-47 Stratojets, medium jet bombers with much better speed than the RB-36s or 45s. Their most spectacular mission was a mass overflight of Vladivostok at high noon by the entire RB-47 force. Each target was photographed by two or three aircraft. Only two planes saw MiGs, but no interceptions were made.[6]

These overflights were brief. The planes crossed the border, covered their targets, and were back across before Soviet air defenses could react. If the Soviets protested, the incident could be described as an "off-course training flight." These flights could not cover the Soviet interior, where the bulk of Soviet industrial and military facilities were located. In the Ural Mountains whole cities had been built that no Westerner had ever seen. Filling the blank spots would take a Dark Eagle.

ORIGINS OF THE U-2

The U-2 project was set in motion by Maj. John Seaberg, an air force reservist recalled to duty as assistant chief of the New Developments Office at Wright-Patterson Air Force Base. Seaberg, who had worked as an aeronautical engineer at Chance Vought, realized the new generation of jet engines being developed in the early 1950s had an inherent high-altitude capability. If matched with extremely efficient wings, the resulting aircraft would have a maximum altitude far above the reach of any interceptor. It would be ideal for reconnaissance.

By March 1953, Seaberg had written a formal design study. It envisioned an aircraft with a maximum altitude of 70,000 feet, a range of 3,000 miles, subsonic speed, up to 700 pounds of reconnaissance equipment, a one-man crew, and the use of existing engines. Two requirements would have a long-term impact on both this and later Black airplanes. Under "Detectability," the requirement stated: "Consideration will be given in the design of the vehicle to minimize the detectability by enemy radar." Under the category of "Vulnerability," it stated: "It is anticipated that the enemy will have limited means of detection and/or interception of a vehicle of the required performance. The greatest opposition to the operation of this aircraft can be expected to be encountered from guided missiles."

Major Seaberg went to possible contractors. It was decided to bypass large prime contractors such as Boeing, Convair, North American, Douglas, and Lockheed. The aircraft was highly specialized, and the number produced would be small. A smaller company would give it both a higher priority and a more innovative design. Bell Aircraft and Fairchild were asked to submit designs, while Martin was asked to modify its B-57 Canberra light bomber.

THE X-16 BALD EAGLE

By January 1954, the three proposals were finished. Fairchild's M-195 design featured an intake behind the cockpit and a tail mounted on a short boom. The landing gear was a tail wheel and two main wheels in the wing. The Bell Model 67 was a large, twin-engine design of a more conventional appearance. The Martin RB-57D proposal had extended outer wings and new engines. All three designs used Pratt and Whitney J57 engines, which were the best then available for such extreme altitudes.

In early May 1954, the evaluation was completed and approval was given to build two of the designs. The Martin RB-57D was to be the interim aircraft, able to provide a limited high-altitude capability in a short time. It would not meet the full requirement, however.[7] To do this, the Bell design was selected. To hide its reconnaissance mission, it was designated the "X-16." *X* (for experimental) normally signified such research aircraft as the rocket-powered X-1, X-2, and X-15. The program was given the code name "Bald Eagle." A total of twenty-eight were to be built.

The X-16 was a very large, yet delicate-looking aircraft. It was 60.85 feet long, with a wingspan of 114.85 feet, but the cylindrical fuselage was only 4 feet in diameter. It used "bicycle" landing gear—front and back wheels with two retractable outriggers in the middle of the wings for balance (similar to the B-47). There were three sensor bays—one in the nose, and one in front of and one behind the rear landing gear bay. The cameras could cover an area 50 miles wide and 795 miles long. Weight was the primary concern. The two J57 engines produced only 743 pounds of thrust each at 65,000 feet. Above this, the thrust dropped off sharply. The X-16 weighed 36,200 pounds fully loaded, and there was no margin for added weight or drag.

The X-16 mock-up was finished in mid-1954. It consisted of the fuselage and included the cockpit, camera bays, a stub wing, and landing gear. A separate mock-up of the J57 engine was also built. The mock-ups were housed in a special tent, with access limited to project personnel. By October 1955, the prototype X-16 was 80 percent completed. But it was already too late.[8]

THE CL-282

Kelly Johnson also had heard about the request for a special high-altitude aircraft. Although Lockheed had not been asked to submit a proposal, he did so anyway. The CL-282 was an XF-104 fuselage fitted with long-span wings. It was much smaller than the X-16—44 feet long with a 70.67-foot wingspan. The emphasis was on weight savings. The airplane had no landing gear; the CL-282 was to take off from a wheeled dolly, then land on two skids. The cockpit was unpressurized, and there was no ejector seat. The camera bay was behind the cockpit.

Johnson's proposal arrived on Seaberg's desk on May 18, 1954, some two weeks after the go-ahead for the RB-57D and X-16. In June, Seaberg recommended the CL-282 be rejected. The primary reason was its use of the General Electric J73 engine. This was an unproven (and ultimately unsuccessful) design. The XF-104 fuselage could not be easily enlarged to accommodate the preferred J57 engine. The air force agreed with Seaberg's recommendations. But Johnson had already submitted the proposal elsewhere.[9]

The CIA was keeping in close touch with the air force on the emerging special reconnaissance program. The CIA's assessment of the CL-282 was very different from Seaberg's. It was "considered to be the best." It had a maximum altitude of 73,000 feet over the target and a speed between 450 and 500 knots. The prototype could be ready in a year, with five planes delivered in two years. The X-16, it was noted, could only reach 69,500 feet and had inferior speed and range.[10]

The CL-282 proposal came at a critical time. Although the Soviets exploded their first A-bomb in 1949, they had limited production facilities. In the late 1940s and early 1950s, Soviet delivery systems were similarly limited. They had only Tu-4 bombers—copies of the B-29 propeller bombers. They were slow and capable of only a one-way attack on the United States. By the early 1950s, this situation started to change. The first operational Soviet A-bombs were being deployed. Test flights of the Soviet turboprop Tu-95 and Mya-4 jet bombers were under way. The Soviets were also working on long-range ballistic missiles. For the first time, a surprise nuclear attack on the United States seemed possible.

On March 27, 1954, President Dwight D. Eisenhower told the Science Advisory Committee of the Office of Defense Mobilization, "Modern weapons had made it easier for a hostile nation with a closed society to plan an attack in secrecy and thus gain an advantage denied to the nation with an open society." The United States was vulnerable to both a surprise attack and political blackmail. These fears also drove up military spending, which Eisenhower saw as a danger to the U.S. economy. On July 26, Eisenhower asked Dr. James R. Killian, the president of Massachusetts Institute of Technology, "to direct a study of the country's technological capabilities to meet some of its current problems." The group was to look at ways of overcoming the lack of strategic intelligence on the Soviet Union.

One member of the panel Dr. Killian assembled was Dr. Edwin H. Land, a noted photo scientist. Dr. Land was named to head Project 3 of the Technological Capabilities Panel, often called the "Land Panel." Dr. Land once said that discoveries were made by people who had freed themselves from conventional thinking and had the ability to take a new look at old data. Dr. Land realized that technological advances promised a revolution in photo reconnaissance—thin plastic film bases, lens designs which increased resolution 500 percent, computer custom grinding of lenses, and innovative camera designs that allowed a single camera to provide both high-resolution and horizon-to-horizon coverage. Use of a camera incorporating these advances would save weight, increase performance, and provide the intelligence the United States needed.[11]

The Land Panel was briefed by Seaberg on the Fairchild M-195, the Bell X-16, the Martin RB-57D, and the Lockheed CL-282. Aware of the growing support for the Lockheed plane, Seaberg showed the panel a graph that demonstrated that all three designs were aerodynamically similar. If the CL-282 was fitted with a J57 engine, it would be competitive. The following day, Johnson underwent intensive questioning about the CL-282. Johnson said he would use a J57 engine. He also promised to have the prototype ready within eight months after the go-ahead. This was an extraordinary schedule—the RB-57D and X-16 projects had been under way for several months yet the RB-57D would not fly until November 1955, while the X-16 would not go aloft until early 1956. Johnson said he could beat them both, although all Lockheed had was a "paper airplane."[12]

On November 5, 1954, Dr. Land wrote a memo to CIA Director Allen Dulles. Entitled, "A Unique Opportunity for Comprehensive Intelligence," it stated about the CL-282: "No proposal or program that we have seen in intelligence planning can so quickly bring so much vital information at so little risk and at so little cost."

Land noted that, "We have been forced to imagine what [the Soviet's] program is, and it could well be argued that peace is always in danger when one great power is essentially ignorant of the major economic, military, and political activities . . . of another great power. . . . We cannot fulfill our responsibility for maintaining the peace if we are left in ignorance of Russian activities."

The memo recommended that the CIA develop the CL-282, then set up a task force to operate the aircraft, make the overflights, and analyze the photos. Cost for six aircraft, training, and operations would be $22 million. The memo stressed time was of the essence—soon the Soviets would develop radars, interceptors, and guided missiles able to reach 70,000 feet.[13]

On November 24, 1954, a meeting was held with President Eisenhower. Present were Allen Dulles, Secretary of State John Foster Dulles, Defense Secretary Charles Wilson, Air Force Secretary Harold Talbott, and Air Force Generals Nathan Twining, Donald Putt, and C. P. Cabell. Eisenhower agreed to production of thirty "special high-performance aircraft." Initial funding would come from a special CIA fund. The air force would buy the special high-altitude version of the J57 engines. Total cost was $35 million. Once the aircraft was ready, he would decide on the overflight program.[14]

Eisenhower was very specific about how the program was to be run. The aircraft "should be handled in an unconventional way so that it would not become entangled in the bureaucracy of the Department of Defense or troubled by rivalries among the services." This meant the air force would provide support, but the CIA would have control of the program. By having the CIA make the overflights and analyze the photos, the intelligence would not become entangled with the internal politics of the air force. Additionally, as a "civilian" aircraft, it could be "disowned" if shot down.

The initial contract for twenty airplanes at a cost of $22 million was signed on December 9, 1954. Unlike an air force contract, there was not a long list of technical specifications. The CIA listed only performance specifications. The project was given the initial code name "Aquatone." Later, this was changed to "Idealist." To the Skunk Works, the aircraft was "the Article" or "the Angel."

THE ANGEL

Johnson quickly assembled a group of twenty-nine engineers to develop the aircraft. They were warned that the project was so secret that their employment record might have a two-year gap. Dick Boehme was named chief engineer, and Art Viereck was head of manufacturing. The Skunk Works engineers were crammed together in "slumlike conditions," but were only a few steps from the production floor. They began a punishing six-day, 70 hours a week development schedule, but could tell no one what they were doing.

The Angel was a much more refined aircraft than the original CL-282 design. The fuselage was lengthened and widened to accommodate the J57 engine. Dimensions were now 49.72 feet long with an 80.17-foot wingspan. The XF-104's "T" tail was replaced by a conventional unit. The emphasis was on weight control—its empty weight was only 12,000 pounds. (This was equivalent to the X-16's fuel load!) The aluminum skin was only 0.02 inches thick and lacked the structural stiffeners of conventional aircraft. Johnson said at one point that he would "trade his grandma" for several pounds of weight reduction. (After this, every pound saved was a "grandma.") The tail was held on with three five-eighth-inch bolts. The Angel was stressed for only plus-1.8 gs and negative-0.8 gs in some flight conditions.

The cockpit, unlike the original CL-282 design, was pressurized. It was very cramped, especially as the pilot had to wear a partial-pressure suit for protection in case pressurization was lost. Rather than a stick, it had a large control yoke, like that on a transport. On the instrument panel was a driftsight-sextant. This allowed views of ground landmarks (and any fighters trying to intercept the plane), and of the sun and stars for navigation. There was no ejector seat.

Behind the cockpit was the pressurized "Q-bay" which held the camera. Three camera systems were originally developed for use on the Angel. The "A camera" was a set of three twenty-four-inch focal-length cameras, one vertical and two oblique. The "B camera" had a thirty-six-inch focal-length lens. The lens assembly pivoted to provide panoramic coverage. The camera was loaded with two rolls of film, each nine inches wide and five thousand feet long. Both rolls were exposed during each shot, forming an eighteen-by-eighteen-inch frame. As each shot was taken, the B camera moved forward slightly to compensate for the aircraft's angular motion over the ground. The resolution of the camera was two and a half feet from 70,000 feet. The B camera was the embodiment of Dr. Land's vision. The "C camera" used a sixty-six-inch focal-length lens and was to be used for high-resolution technical intelligence.[15]

Unlike the original CL-282, this plane was fitted with bicycle landing gear. Two "pogos" kept the wings level during taxi and takeoff. When the plane left the ground, the pogos fell out. When the plane landed, the pilot would have to keep the wings level through touchdown and rollout. When it came to a stop, the plane would tip and come to rest on one wing-tip skid.

The long narrow wings were the key to the Angel's high-altitude capability. Between its high-aspect ratio, very high camber, and very low wing loading, the aircraft was given the best possible lift-drag ratio for cruise efficiency. Because the wings were shorter than those of the RB-57D or X-16, they were not affected by "aeroelastic divergence," a twisting of the wings caused by aerodynamic forces. (The RB-57's operational life was cut short by structural failures caused by this problem.)

The long wings did create a particular problem—they generated a strong pitch force, which had to be counteracted by the tail. This was particularly evident at high speeds and in turbulence. Rather than beefing up the tail structure (and adding weight), the ailerons and flaps would be raised slightly. This moved the wing's center of lift slightly and reduced wing and tail loading. (The procedure, called "gust control," was later used on airliners.)

The fuel carried in the wing tanks was also special. The Angel would be exposed to negative-95-degree Fahrenheit (F) temperatures for eight hours or more. Normal JP-4 jet fuel would freeze. Shell Oil developed a special kerosene that would not freeze or evaporate in the extreme cold and low pressure at 70,000-plus feet. The military called it JP-TS (for thermally stable), while Lockheed referred to it as LF-1A. The *lf* stood for "lighter fluid," since it smelled very similar to that found in a cigarette lighter.

By the end of 1954, the aircraft's design was set and construction of two prototypes could begin. Johnson selected Lockheed chief test pilot Anthony W. LeVier to make the initial flights. LeVier had worked on earlier Lockheed projects such as the P-38, P-80, and XF-104. In one harrowing accident, he had bailed out of a P-80 that was cut in half by a disintegrating engine. LeVier was called into Johnson's office and asked if he wanted to fly a new airplane. LeVier asked, "What plane?" Johnson responded, "I can't tell you unless you agree to fly it!" LeVier agreed and was told his first job was to find a secret test site for the plane.[16]

THE RANCH

With the extreme secrecy enveloping the project, the flight test and pilot training programs could not be conducted at Edwards Air Force Base or Lockheed's Palmdale facility. LeVier spent several days plotting a route to visit potential test sites in the deserts of southern California, Nevada, and Arizona. Scattered throughout the area are dry lake beds, ranging from less than a mile to several miles in diameter. Johnson asked him to look for a site that was "remote, but not *too* remote."

The search was conducted under the same extreme security as the rest of the project. LeVier and Dorsey Kammerer, the Skunk Works foreman, told everyone they were going on a hunting trip to Mexico; they even dressed the part when they took off in the Lockheed Flight Test Department's Beech V-tail Bonanza. Once out of sight of the factory, they changed course and headed toward the desert. For the next two weeks, LeVier and Kammerer spent their "vacation" photographing and mapping possible sites.[17]

In all, fifty possible sites were looked at. When Richard M. Bissell Jr., the CIA official selected to direct the program, and his air force liaison, Col. Osmond J. "Ozzie" Ritland, reviewed the list, they felt none of them met the security requirements. Then Ritland recalled "a little *X*-shaped field" in Nevada he had flown over many times while involved with U.S. nuclear testing. He offered to show it to Bissell and Johnson.

Soon after, LeVier flew Johnson, Ritland, and Bissell out for an on-site inspection. They did not have a clearance, so flew in at low altitude. Ritland said later, "We flew over it and within thirty seconds, you knew that was the place . . . it was right by a [dry] lake. Man alive, we looked at that lake, and we all looked at each other. It was another Edwards, so we wheeled around, landed on that lake, taxied up to one end of it, and Kelly Johnson said, 'We'll put it right here, that's the hangar.'"[18]

Bissell recalled later that it was "a perfect natural landing field . . . as smooth as a billiard table without anything being done to it."[19] Johnson used a compass to lay out the direction of the first runway, kicking away spent shell cases as he walked.

The place was called "Groom Lake."

Groom Lake is square-shaped, about three by four miles in size. It is on the floor of Emigrant Valley in Lincoln County, Nevada. Like all such dry lakes (including Edwards Air Force Base), Groom Lake was formed by water runoff. (Yearly rainfall was only four and a third inches.) The sediment flows to low areas, where it settles. The 100-degree F heat of summer dries the mud, leaving a flat, hard surface. In winter, temperatures drop to below freezing and light snowfall can dust the area. Strong afternoon winds often hit the area, although thunderstorms are rare. (One such storm would have an important part in Groom Lake's history, however.)

During World War II, Groom Lake was used as a gunnery range. The lake bed was littered with empty shell cases and debris from target practice. An airstrip was built on the east side of the lake bed. With the end of the war, the site was abandoned. By early 1955, the runway had reverted to sand and was unusable. Ritland said it "had got hummocks and sagebrush that wouldn't quit."

Groom Lake is cut off from the surrounding desert by the Timphute Range to the west, the Groom Mountains to the east, and the Papoose Range to the south. A few miles to the north is the 9,380-foot summit of Bald Mountain. The mountains loom like walls above the lake bed. The only nearby towns are "wide spots in the road" such as Rachel, Nevada. Las Vegas is nearly 100 miles to the southwest. To the west, just over the surrounding hills from Groom Lake, is Nellis Air Force Base and the Atomic Energy Commission's (AEC) Nuclear Test Site. It was the perfect place to hide a secret. The only access to the site was by air. The AEC's security restrictions would cut off both ground and air access, effectively protecting the site and its secrets. The Groom Lake site was approved, and the restricted area around the nuclear test site was extended to encompass it.

A small but complete flight test center would have to be created out in the desert. To hide Lockheed's involvement, "CLJ" (Johnson's initials) became its company name. The facility plans were given to a contractor who had the special license needed to build at the nuclear test site. This led to a problem—when the contractor asked for bids, he was told to watch out for "this CLJ outfit" because it had no Dun and Bradstreet credit rating.[20]

Throughout the summer of 1955, with temperatures over 100 degrees F, the crews worked to build the test center. They had no idea what the facility would be used for. The site included a 5,000-foot tarmac runway, two hangars, a small tower, several water wells, fuel storage tanks, a mess hall, a road, plus some temporary buildings and trailers for living quarters. These were located on the southwestern edge of the lake bed. Total cost was $800,000. The site was isolated, rugged, barren, and lacking in personal comfort. This was more than made up for by a pioneering spirit.

In early July 1955, LeVier was told to fly out to the site. This was his first visit since the first survey with Johnson, Ritland, and Bissell. He was stunned by the changes. His first action was to get the lake bed ready. As at Edwards Air Force Base, the lake bed would be used for takeoffs and landings. LeVier and fellow Lockheed test pilot Bob Matye spent nearly a month driving around the lake bed in a pickup truck cleaning up spent shell cases, rocks, brush, and even half a steamroller.[21] Flying over a flat surface like the lake bed, it was very difficult to judge height, so LeVier also wanted to paint markings for four three-mile runways on the lake bed. Johnson turned down the proposal when told it would cost $450. The money was not in the budget.[22]

By late July 1955, the facility was completed. In order to recruit people, Johnson dubbed the site "Paradise Ranch." Years later, he admitted, "It was kind of a dirty trick since Paradise Ranch was a dry lake where quarter-inch rock blew around every afternoon." Soon, the name was shortened to "the Ranch."

THE ANGEL TESTS ITS WINGS

By this time, the first prototype was ready. "Article 341," as it was designated, was disassembled, and the fuselage and wings were wrapped in fabric and loaded on two carts. At 4:30 A.M. on July 24, 1955, they were loaded on a C-124 transport for the flight to Groom Lake. The Skunk Works crew would follow in a C-47. There was a delay—the local commander refused permission for the C-124 to land on the runway at Groom Lake, because the wheels of the heavily loaded plane would break through the thin surface. He wanted it to land at another base, then have the prototype moved to Groom Lake over bad dirt roads. This would delay the first flight by a week, however. Johnson argued that they could let most of the air out of the C-124's tires, reducing the surface pressure. When the local commander refused, Johnson called Washington to get approval to override him. Permission was given, the tire pressure was reduced, and Article 341 was successfully flown to Groom Lake.[23]

Once it was reassembled, Article 341 was towed out of the hangar by a pickup truck and underwent engine run-up tests. It was in a bare-metal finish—no U.S. star and bar insignia, no "USAF," not even a civilian "N-number" registration.

Article 341 was ready for its first taxi tests on August 1, 1955. The first run, to a speed of 50 knots, was successful, even though the brakes were found to be ineffective. The second taxi run reached 70 knots. LeVier cut the throttle to idle, then realized he was some twenty feet in the air. Article 341 continued to fly for over a quarter of a mile. LeVier tried to land the plane, but it was impossible to judge his height above the lake bed. The plane contacted the lake bed in a 10-degree bank—the left wing-tip skid hit first, then the left pogo, main gear, and finally, the tail wheel. The landing was hard, and the plane bounced back into the air. The second landing was much smoother, and LeVier was able to regain control. As the plane rolled to a stop, the right tire blew and caught fire. This was extinguished in short order. Despite the mishap, no major damage was done, and repairs were completed the next day. LeVier, in his pilot report, said, "The lake bed during this run was absolutely unsatisfactory from the standpoint of being able to distinguish distance or height."

While Article 341 was being repaired, LeVier and Matye put crude markings on the lake bed to make a north-south runway. The following day, August 2, two more taxi runs were made. LeVier pushed the control wheel forward to keep the plane on the ground. The runs uncovered a few minor problems: poor braking, reflections on the windshield, and the need for a sunshade to keep the cockpit from becoming too hot. LeVier wrote in his pilot report, "I believe the aircraft is ready for flight."

Article 341's first flight was set for August 4. It was planned for a maximum speed of 150 knots and an altitude of 8,000 feet. The aircraft's low-speed control would be checked. The plane would stay close to the lake bed. The weather for the first flight was threatening, with thunderstorms near Groom Lake. The C-47 made a weather check. At 2:28 P.M. the C-47 landed and the flight was allowed to proceed. At 2:57 P.M. the T-33 chase plane took off and preparations began to start Article 341's engine.

Then began a series of events that turned the first flight into a cliff-hanger. At 3:06 P.M. LeVier twice tried to start the plane's engine, but his attempts failed. At 3:12 P.M. the T-33 landed for refueling. The fuel was not immediately available, and the T-33 did not take off again until 3:46 P.M. At 3:51 P.M. LeVier was finally able to start the engine. During the delay, the wind had shifted and LeVier had to reposition the aircraft.

Finally, at 3:55 P.M., nearly an hour late, Article 341 began its takeoff roll. It lifted off the lake bed thirty seconds later. LeVier made a circle of the lake bed while the landing gear retracted. He operated the speed brakes, then made six stall checks. LeVier was very satisfied, radioing at one point, "Flies like a baby buggy." LeVier then started his descent for the landing at 4:10 P.M. At this point, as he wrote in his postflight comments, "It wasn't difficult to realize that this was no ordinary aircraft. With the power lever in almost idle, the wing flaps partially down and dive brakes extended, the aircraft had a very flat glide and a long float on flaring out."

LeVier and Johnson had earlier discussed the best landing technique. Johnson thought the forward landing gear should touch down first, to avoid stalling the wings. LeVier believed he should make a two-point landing. He had talked with B-47 pilots who warned that the aircraft would "porpoise" if it landed nose wheel first. At 4:20 P.M. LeVier made his first landing try, but he said, "attempting to touch the main wheels first while pushing on the control wheel to lower the nose only served to produce a most erratic and uncontrollable porpoise. I immediately applied more power and took off." Over the next few minutes, LeVier made three more attempts to land nose gear first. Each time, the attempt failed.

Another factor was the weather. A few minutes after takeoff, the thunderstorms moved into the area and light rain began to fall. As LeVier lined up for his first attempt, he radioed, "Hardly enough speed to take water off the windshield." The rain squalls were getting closer as LeVier made his fourth landing attempt. This time, LeVier stalled the aircraft just above the ground, and it touched down on both gears in a perfect landing. As the plane rolled out at 55–60 knots, the pogos, which had been locked in place, were still off the ground. LeVier used the gust control to reduce lift. Article 341 came to a stop at 4:34 P.M.[24]

As LeVier climbed out of Article 341, he saw Johnson, who had been flying as a passenger in the T-33. LeVier jokingly "saluted" him with an obscene gesture and accused Johnson of trying to kill him. Johnson responded with the same gesture and a loud, "You too," which was heard by the ground crew. LeVier answered back, "You did." So was born the "U-2" name. Ten minutes later, the rain squalls flooded Groom Lake with two inches of water. The Lockheed personnel celebrated that evening with beer-drinking and arm-wrestling contests.

The following day, LeVier made a second, short flight to check out the landing technique. The plane's official first flight took place on August 8. On hand were Bissell and other government officials. LeVier made a low pass, then zoomed up to 30,000 feet. The T-33 chase plane, with Matye at the controls, struggled to follow. At the end of the hour-long flight, LeVier made another low pass and landed.[25]

LeVier made a total of twenty flights, which completed the Phase 1 testing. These flights took the aircraft to its maximum speed of Mach 0.84, an altitude of 50,000 feet, and a successful dead-stick landing. LeVier said the plane "went up like a homesick angel." With the Phase 1 testing completed, LeVier left to join the F-104 program.[26]

Lockheed test pilots Bob Matye and Ray Goudy replaced LeVier. They expanded the altitude envelope to 74,500 feet. On three occasions, Matye broke the world altitude record of 65,890 feet set on August 29, 1955, by Wing Commander Walter Gibb in an English Electric Canberra. The Canberra record had made headlines; there was no announcement from Groom Lake. On the third flight, Matye suffered an engine flameout. This qualified the pressure suit emergency oxygen system and emergency descent procedures.[27]

Despite these successes, Matye's flameout indicated a major problem with the J57 engine. When the engine flamed out, the aircraft would have to descend to 35,000 feet before the pilot could attempt a relight. On test flights, this was no problem. On an overflight, however, the plane would be helpless against MiGs. Bissell said later, "Plainly, unless this problem could be licked, it would be altogether too hazardous to fly this aircraft over unfriendly territory." The early J57-37 engines also dumped oil into the cockpit pressurization system. This left an oily film on the windshield. The test pilots had to carry a swab on a stick to clean it. Pratt and Whitney made a number of small fixes, but with only limited success. It would require a new version, the J57-31, before the flameout problem was solved. And this would not be accomplished until early 1956.

By November 1955, there were four or five U-2s in the test program at Groom Lake. Robert Sieker and Robert Schumacher were added to the flight test staff. The initial flight tests were of airframe and engine, followed later by tests of subsystems, such as the autopilot. Finally, with the arrival of the cameras, these would be tested on simulated operational missions.[28] The initial test flights did not venture more than two hundred miles from the Ranch. From 70,000 feet, the U-2 could glide back to Groom Lake. As confidence in the aircraft grew, the Lockheed pilots began flying triangular patterns up to one thousand miles away from the Ranch. These flights could last up to nine and a half hours.

If the triumphs of Groom Lake were secret, so too were its tragedies. At 7:00 A.M. on Wednesday, November 17, 1955, the daily air force flight to Groom Lake took off from Burbank. Aboard the C-54 transport were ten Lockheed and CIA personnel and five crewmen. There would have been more passengers, but a party at the Flight Test Division had left some people with hangovers. The weather was poor and the C-54 hit the peak of

Mount Charleston near Las Vegas, killing all fifteen. It took three days to reach the wreckage, which was only thirty feet from the eleven-thousand-foot summit. An air force colonel accompanied the rescue party to recover briefcases and classified documents from the bodies.

The air force issued a statement saying they were civilian technicians and consultants. It was assumed by the press that they had been scientists connected with the AEC's nuclear tests. They would not be the last to meet secret deaths. In the wake of the tragedy, Johnson insisted Lockheed take over the daily flights to the Ranch. A company-owned C-47 was used.

THE END OF THE X-16

During this time, Bell had continued work on the X-16. In early October 1955 (two months after the first U-2 test flight), Bell signed a contract with the air force for twenty-two aircraft. Then, a few hours later, Bell was notified that the project had been terminated. It had been realized that the U-2, even with the engine problems, was a vastly superior aircraft. Loss of the X-16 was a major blow to Bell; it was one of the few contracts the company had.[29] The loss meant the end of Bell's involvement with fixed-wing aircraft. Ironically, the X-16 would remain secret for another decade—it was not until 1976 that photos of the aircraft would be released.

Following the cancellation of the X-16, Lockheed received contracts for a total of fifty U-2s. Lockheed gave back some $2 million on the initial contract. Later, an additional five U-2s would be assembled from spare parts.[30]

CIA PILOT RECRUITMENT

In late 1955 and early 1956, recruitment of the CIA U-2 pilots began. They were all F-84 pilots from two Strategic Air Command (SAC) bases, Turner Air Force Base, Georgia, and Bergstrom Air Force Base, Texas. The Strategic Fighter Wings at these bases were being phased out. The "disappearance" of a few pilots would not be noticed.[31] The pilots approached were all reserve officers with indefinite service tours, Top Secret clearances, exceptional pilot ratings, and more than the required flight time in single-seat, single-engine aircraft.

The pilots initially were told only that a flying job was available. If they were interested, an interview would be arranged. These interviews were held at night, at nearby motels. The pilots were not told much more—simply that they had been picked to be part of a group that would carry out a special mission. It would be risky, but they would be doing something important for the United States. They would be well paid but would have to be overseas for eighteen months without their families. If they were interested, they should call the motel the next day and arrange another interview.

Several pilots refused because of the separation from their families. The remainder were highly curious. There was wild speculation on what the job offer was really about. Marty Knutson thought they were going to be astronauts.[32] Francis Gary Powers thought it sounded like the Flying Tigers.[33]

It was not until the third interview that the tantalizing mysteries were made clear. The pilots were told they would be working for the CIA and that they would be flying a new airplane that could go higher than any other. Their pay, during training, would be $1,500 per month; overseas it would be raised to $2,500 per month. This was almost as much as an airline captain's salary. Their time with the CIA would count toward air force retirement and rank. Part of their job would be to fly along the Soviet border to record radio and radar signals. Their main job, they were stunned to learn, would be to overfly the Soviet Union. They were given a day to think it over.

Those pilots who agreed underwent several months of briefings, lie-detector tests, and medical checks at the Lovelace Clinic in Albuquerque, New Mexico. The examination lasted a week and involved tests developed specifically for the prospective U-2 pilots. (They were later used for the Mercury astronauts.) One series tested for claustrophobia—a necessity given the cramped cockpit and restrictive partial-pressure suit. A handful of pilots washed out. The rest, about twenty-five in all, resigned from the air force (a process called "sheep dipping") and signed eighteen-month contracts with the CIA. They were then sent to the Ranch.

MEANWHILE, BACK AT THE RANCH . . .

The training program at the Ranch was a joint CIA–air force operation. The group was commanded by Col. Bill Yancey and included four experienced instructor pilots. Because there were no two-seat U-2s at this time (or even a ground simulator), the instructor pilots were limited to conducting the ground school.[34] The CIA pilots underwent training in three groups, starting in early 1956 and continuing through the year. While at the Ranch, the pilots used cover names. Francis Gary Powers became "Francis G. Palmer" (same initials and similar last name).[35] The pilots' gray green flight suits had no name tags nor squadron patches. They did wear film badges that measured radiation exposure, because of the nearby nuclear test site.[36]

The pilots first underwent ground school, which included training in use of the pressure suit. Then flight training began. The first two flights were landing practice in a T-33. The technique used to land the U-2 was directly counter to that used in conventional aircraft.

Once this was completed, the pilots could begin flying the U-2. The initial flights would again be landing training. This was followed by high-altitude flights. Then the pilots would begin flying long-range simulated missions, lasting up to eight hours.[37] These training flights went from Groom Lake to the Allegheny Mountains and back—a flight of some 4,000 miles.[38] In all, some sixteen flights were made. As the training progressed, each pilot was evaluated.[39]

The U-2 flight training was much more extensive than that for other air force planes. The U-2 was a very demanding aircraft. The takeoff roll was only a few hundred feet. The U-2 would then go into a spectacular climb at better than a 45-degree angle. The first few times the pilots thought the U-2 would continue right over on its back. The U-2 would continue up to 60,000 feet before leveling off. Then, as fuel was used, the plane slowly climbed. The peak altitude was about 75,000 feet. This depended on both the fuel and equipment load, and on the air temperature. Between 55,000 and 60,000 feet, the air temperature could vary widely. This could cause the aircraft to climb more rapidly or even force it to descend.

Above 68,000 feet, the difference between the U-2's stall speed and its maximum speed was only 10 knots. This was called the "coffin corner." The aircraft could easily exceed these limits due to control inputs or pilot inattention. The result would be the plane tearing itself apart within seconds. The pilot would have to maintain this balancing act for hours on end, plus navigate, operate the camera, and monitor fuel consumption. Therefore the autopilot was critical in controlling the plane.

Coming down from this lofty perch was difficult. The pilot could not simply point the nose down—the aircraft would overspeed and break up. Rather, the throttle was eased back to idle, then the landing gear and speed brakes were deployed. Even so, the descent was very slow—a striking contrast to the rocketlike climb. As the U-2 descended, the margin between the stall and maximum speed would widen.[40]

As the aircraft approached for the landing, fuel had to be transferred to balance the wings. If one wing became too heavy, the plane could go into an uncontrollable spin. Unlike at high altitudes, where the U-2 had to be flown with a light touch, at lower altitudes, the pilot had to manhandle the plane. Even the touchdown was critical. The pilot had to hold the long wings level, a difficult task in a crosswind. If the plane was stalled too high, it would hit the runway, bounce into the air, stall, and crash before the engine could come to full power.

The U-2 was a plane that required the pilot's complete attention every second. There was no margin for error. The flights were so exhausting that a pilot would not be allowed to fly again for two days.[41]

But, as Powers later noted, the rewards of a U-2 flight were far greater. From altitude, above Arizona, the pilot could see from the Monterey Peninsula to midway down Baja California. Above was the blue black of space. Powers wrote, "Being so high gave you a unique satisfaction. Not a feeling of superiority or omnipotence, but a special aloneness." He added, "There was only one thing wrong with flying higher than any other man had flown. You couldn't brag about it."[42]

The first group of pilots had the roughest time. The U-2 was barely out of the test phase and was still plagued with engine flameout problems. In one incident, Bissell received a call that a U-2 flying over the Mississippi River had suffered a flameout. The engine had apparently been damaged as it was vibrating and could not be restarted. The pilot radioed he would land at Kirtland Air Force Base at Albuquerque, New Mexico. Bissell called the base commander at Kirtland and told him a U-2 would be landing in about fifteen minutes. He asked the commander to have the base air police at the runway when it landed. They should cover the plane with tarps to hide its configuration. A half hour later, Bissell received a call from Kirtland. The base commander reported that the plane had landed safely and that he was talking with the pilot in his office.[43] In another case, a U-2 flamed out and landed at the Palm Springs Airport. A C-124 transport and recovery crew took off within an hour to pick it up. The incident was reported in the local newspaper but attracted little attention.

The first group also suffered a fatal crash. Wilbur Rose took off on a training flight when one of the pogos failed to fall out. He flew low over the field trying to shake it free. He misjudged, and the plane, heavy with fuel, stalled. Rose died in the crash.[44]

The second group, which included Powers, went through the Ranch between May and August 1956. They suffered no crashes or washouts. Powers recalled that he was nervous before making his first high-altitude flight and forgot to retract the landing gear after takeoff. As he flew above California and Nevada, his first impression of the U-2 was disappointment—the plane was not capable of the altitude that had been promised. When time came to begin the descent, Powers started to lower the landing gear, only to realize it had been down the whole time. His impression improved considerably. He had broken the world altitude record with the gear down.[45]

At the same time, a special group of pilots was undergoing training. President Eisenhower was worried about the possibility of an American citizen being killed or captured during an overflight. This would generate tremendous political problems. Eisenhower told CIA director Allen Dulles, "It would seem that you could be able to recruit some Russians or pilots of other nationalities." Eventually, one Polish and four Greek pilots were recruited. The Greek pilots underwent training at the Ranch, but all washed out. The Polish pilot was never allowed to fly the U-2.[46]

The third group underwent training in late 1956. The group suffered two crashes, one fatal. In December, Bob Ericson was flying at 35,000 feet when his oxygen ran out. As he began to lose consciousness, the aircraft began to overspeed and go out of control. Ericson fought his way out of the cockpit and parachuted to a landing in Arizona. Less fortunate was Frank Grace. He took off on a night training mission, became disoriented, and flew into a telephone pole at the end of the runway. Grace died in the crash.[47]

Training operations followed a pattern. The pilots arrived at Groom Lake on the Monday morning flight. They turned in their IDs, which gave their true names and described them as pilots with Lockheed, then assumed the cover names. Each pilot would make two or three U-2 flights per week. Then, on Friday afternoon, the pilots left the site to spend the weekend in Los Angeles.

While at the Ranch, the pilots lived in trailers, four in each. Powers called "Watertown Strip," which was the pilots' name for the site, "one of those 'you can't get there from here' places." The population had grown from about 20, at the time of the first flight, to around 150 air force personnel, Lockheed maintenance crews, and CIA guards. A third hangar had been added, as had more trailers. The Ranch was still a remote desert airstrip. The growing numbers of U-2s were parked on the hard-packed dirt on the edge of the lake bed; there was no concrete apron. U-2 takeoffs and landings were made from the lake bed. The whole facility was temporary; it was never built to last.[48]

Amusements were limited. There was no PX or Officers' Club. The mess hall, however, was likened to a first-class civilian cafeteria. The food was excellent and second helpings were available. The mess hall also had several pool tables. A sixteen-millimeter projector provided nightly movies. Given the isolation of the site, the pilots were forced to create their own entertainment. Alcohol was freely available and consumed in abundance. Marathon poker games were also organized by the pilots. The first group of pilots scrounged up gunpowder, woodshavings, and cigar tubes to build small rockets. They made a satisfying "woosh" when launched, but the fun ended when one nearly hit a C-131 transport in the landing pattern.[49]

From time to time, official visitors would come to Groom Lake. In December 1955, Defense Secretary Charles Wilson was shown around the Skunk Works and the Ranch. Allen Dulles also visited the Ranch and met with the first group of pilots.

The only "outsiders" allowed into Groom Lake were the C-124 transport crews, and they did not know where they were. The production U-2s were not built at Burbank, but in the small town of Oildale, California, near Bakersfield. The factory was a tin-roofed warehouse called "Unit 80." During 1956 and 1957, the aircraft were completed, then disassembled, covered, and taken to a local airport, where they were loaded on the C-124s.

It was important that no one know the Ranch's location, so the flights were made at night. The crew was instructed to fly to a point on the California-Nevada border, then contact "Sage Control." The radio voice would tell them not to acknowledge further transmissions. The C-124 would then be given new headings and altitudes. Soon the crew would be contacted by "Delta," who would tell them to start descending into the black desert night. The voice would then tell the transport's crew to lower their flaps and landing gear. Yet their maps showed no civilian or military airports in the area, only empty desert. Then the runway lights would come on, and Delta would clear them to land. Following the landing, the runway lights would be turned off and a "follow me" truck would direct them to a parking spot. The buildings were visible only as lights in the distance. A group of tight-lipped men with names like "Smith" would unload the U-2.[50]

Once delivered to Groom Lake, the U-2s would be reassembled and test flown. The process would be reversed when the time came to send the U-2s to their overseas bases.

OVERFLIGHTS BEGIN

The need for intelligence on the Soviet Union had grown since the start of the U-2 program. The pace of Soviet nuclear testing was picking up. The Soviets had also staged mass flybys of Mya-4 and Tu-95 bombers. Estimates began to appear that the Soviets would soon have upwards of five hundred to eight hundred Mya-4s. So began the "bomber gap" controversy. The problem was that these estimates were based on fragmentary data; they were little better than guesses. There was no way to *know*.

Eisenhower made two efforts—one political, the other clandestine—to gain intelligence. At the July 1955 Geneva Summit, he made the "Open Skies" proposal. The United States and Soviets would be allowed to overfly each others' territory as a guard against surprise attack. Eisenhower also believed such an effort would be a step toward disarmament. The Soviets, relying on secrecy to hide their military strengths and weaknesses, rejected the proposal.[51]

The other effort was the Genetrix reconnaissance balloon program. The plan, which had been in development since 1950, envisioned the launch of some twenty-five hundred Skyhook balloons, carrying camera gondolas from England, Norway, West Germany, and Turkey. The balloons would drift across the Soviet Union on the winter jet stream. The large number would cover nearly all of the Soviet land mass. The randomly drifting balloons could not cover specific targets, but this did not matter. The Soviet Union was a huge blank. Once clear of Soviet airspace, the gondolas would be cut free of the balloons by radio signals. As the gondolas descended by parachutes, they would be caught in midair by specially equipped C-119 transports.

The Genetrix launches began on January 10, 1956. For the first two weeks, the loss rate of the balloons was acceptable and the Soviets made no protest. By late January and early February, however, the balloons were no longer making it through. Soviet air defenses were able to stop the high-flying intruders. On February 6, following a Soviet protest, Eisenhower ordered the balloon launches halted. In all, only 448 balloons were launched; of these, 44 gondolas were successfully recovered. These provided 13,813 photos covering 1,116,449 square miles of the USSR and China (8 percent of their total land mass).[52] This daring and desperate attempt to cover the Soviet interior had ended in disappointment. It was now the U-2's turn.

By early April 1956, flight training of the first group of CIA pilots was completed and the new J57-31 engine had proven itself virtually immune to flameouts. The U-2s, pilots, and ground crews were sent to Lakenheath, England. The unusual looking plane soon attracted attention. The June 1, 1956, issue of *Flight* carried a report of a sighting over Lakenheath. It stated, "In the sky, it looks like the war-time Horsa glider. He believes it to have one jet engine and reports a high tailplane and unswept wings of high-aspect ratio."

The U-2's time in England was brief. At the same time the unit was being set up, a British frogman died while investigating the hull of a Soviet cruiser in an English harbor. The resulting press furor caused Prime Minister Anthony Eden to withdraw permission for the U-2 to operate from Lakenheath. Bissell and General Cabell then went to see West German Chancellor Konrad Adenauer. He said, "This is a wonderful idea. It's just what ought to be done." He gave permission to use an old Luftwaffe base fifty miles east of Wiesbaden. The four U-2s and the seven pilots moved to the new base. The unit was called "Detachment A."[53]

The first public word of the U-2's existence came with a May 7, 1956, press release from the National Advisory Committee for Aeronautics (NACA). It announced: "Start of a new research program [using a] new airplane, the Lockheed U-2 . . . expected to reach 10-mile-high altitudes as a matter of routine . . . The availability of a new type of airplane . . . helps to obtain the needed data . . . about gust-meteorological conditions to be found at high altitude . . . in an economical and expeditious manner." Specific areas of research included clear air turbulence, convective clouds, wind sheer, the jet stream, ozone, and water vapor. "The first data, covering conditions in the Rocky Mountain area," the press release said, "are being obtained from flights from Watertown Strip, Nevada."

On July 9, NACA issued a second press release titled, "High Altitude Research Program Proves Valuable." It stated:

> Initial data about gust-meteorological conditions to be found at 10-mile altitudes which have been obtained to date by the relatively few flights of the Lockheed U-2 airplane have already proven the value of the aircraft for this purpose . . .

"Within recent weeks, preliminary data-gathering flights have been made from an Air Force base at Lakenheath, England . . . As the program continues, flights will be made in other parts of the world."[54]

Indeed, the U-2 had been making flights that provided highly valuable data. But the data was not about the weather.

In the early summer, the CIA sought Eisenhower's agreement to begin overflights. At this point, the project was seen as a short-term, high-risk operation.[55] The U-2s were also considered too delicate to have a long-operating lifetime. The desperate need for intelligence outweighed the risks.[56]

Eisenhower initially authorized two test overflights of Eastern Europe.[57] The first was made on June 20, 1956. Carl Overstreet was selected to be the first pilot to take the U-2 into "denied" airspace. The route went to Warsaw, Poland, then over Berlin and Potsdam, East Germany. Following the two overflights, Eisenhower was shown a number of photo briefing boards from the missions. At the same time, the Soviets put on another mass flyby of bombers.

On July 2, Bissell sent a request to begin Soviet overflights to Eisenhower's personal assistant Gen. Andrew Goodpaster. The following day, Goodpaster sent word that Eisenhower had authorized overflights of the USSR for a ten-day period. Bissell asked if this meant ten days of good weather. Goodpaster replied, "It means ten days from when you start."[58]

At 6:00 A.M. on July 4, 1956, Hervey Stockman took off in U-2 Article 347. The plane carried the A camera and was in a bare-metal finish with no national markings. Stockman headed over East Berlin and northern Poland via Poznan, then crossed the Soviet border. The overflight covered a number of bomber bases in the western USSR, as far east as Minsk. The Soviets made more than twenty intercept attempts. The camera photographed MiG fighters trying to reach the U-2's altitude, only to have their engines flame out. Stockman then turned north, toward Leningrad. Once he reached the city, he turned west and flew along the Baltic coast. The U-2 landed back at Wiesbaden after an eight-hour-forty-five-minute flight.

The next overflight would go directly to Moscow. When asked to justify such a dangerous target, the mission planners told Bissell, "Let's go for the big one straight away. We're safer the first time than we'll ever be again."

Article 347 took off at 5:00 A.M. on July 5. The pilot was Carmen Vito. The flight path was farther south than the first mission—over Kracow, Poland, then due east to Kiev, then north to Minsk. There was heavy cloud cover, which started to clear as Vito turned toward Moscow. Again, MiGs tried to reach the U-2. Several crashed when they were unable to recover after flaming out. Over Moscow, a new danger loomed—the SA-1 Guild surface-to-air missile (SAM) sites that ringed the capital. Vito could see several "herring bone" shaped sites, but no missiles were fired. Vito flew back along the Baltic coast to Wiesbaden.[59]

In all, five overflights were made during the first series—one on July 4, and two each on July 5 and 9. Their photos were highly illuminating. The bomber airfields in the western USSR had been equipped with nuclear weapons loading pits, but no Mya-4 bombers were spotted. Within weeks, the bomber gap controversy was over. The U-2 photos had proven the Soviets did not have a large bomber force. The photos brought a revolution in intelligence. It was now possible to know, not to estimate, not guess, but to *know* the military capabilities of an enemy. That was the most important accomplishment of the Angel of Paradise Ranch.

On July 10, the Soviets protested the overflights, and Eisenhower decided to halt the missions for the time being. He was very impressed, however, with the photos of bomber bases and the shipyards around Leningrad.[60]

In September 1956, the second group of U-2 pilots completed training at the Ranch. The seven pilots of Detachment B were based at Incirlik, Turkey. From there, the southern Soviet Union, as well as targets throughout the Mideast, could be covered. It was not until November that the first overflights were made from Turkey, a pair of short overflights to examine Soviet air defenses. The first was flown by Powers.

The third group, Detachment C, was established in early 1957 at Atsugi, Japan. The unit made overflights of targets in the eastern USSR, such as Vladivostok and Sakhalin Island, as well as flying missions over China, North Korea, North Vietnam, and Indonesia.[61]

The year 1957 saw a step-up in overflight activities. This represented a change in attitude toward the operation. Eisenhower had come to rely on the U-2 photos, comparing other intelligence data to them. Soon they were providing 90 percent of the intelligence on the Soviet Union. Instead of a short-term project, it had become an open-ended one. Fears about the U-2's fragile structure had eased and early problems with the B camera had also been overcome. Detachment A in Germany was closed down and combined with the Turkish-based Detachment B.

The risks were also clearer; a year after the first overflight, Bissell asked for a special estimate of the U-2's vulnerability. The Soviets had begun deployment of a new SAM, the SA-2 Guideline, in late 1956. The study concluded the SA-2 could reach the U-2's altitude, but they had been designed to hit much lower-flying B-47s and B-52s. Above 60,000 feet, the SAM's accuracy was so poor only a lucky hit could be made. The risk was not a serious one.[62]

A primary target of the overflights was Soviet ballistic missile activities. Since World War II, the Soviets had undertaken an aggressive development program. The SS-3 and SS-4 medium range ballistic missiles, then in the final stages of testing, could threaten U.S. bases throughout Western Europe, North Africa, and Asia. It was the U-2 that provided the first good photos of the Kapustin Yar test site on the Volga River.

A new missile threat was also emerging. Development work was under way on the R-7 intercontinental ballistic missile (ICBM). A new test site, in Soviet central Asia, was completed in late 1956. In March 1957, the prototype R-7 was undergoing checkout. The missile was fired on May 15 but exploded fifty seconds after launch. Two more launch attempts were made in the spring and summer, also ending in failure.

These activities were detected, and Eisenhower authorized a series of overflights to find the launch site. These overflights were along the main railroad lines. During one of them, the pilot spotted construction in the distance and altered course to photograph it. When the photos were developed, they showed the launch pad. Within days, the analysis was completed. The site was named "Tyuratam," after the rail stop at the end of the fifteen-mile spur that connected the site to the main Moscow-Tashkent line.

The string of R-7 failures ended on August 21, 1957, when the fourth attempt made a successful 3,500-nautical-mile flight. A second successful R-7 launch followed on September 7. Emboldened by the twin flights, Communist Party Secretary Nikita Khrushchev authorized the launching of an earth satellite by an R-7. This was *Sputnik 1,* orbited on October 4, 1957. *Sputnik 2,* which carried a dog named Laika, followed on November 3.[63]

The R-7 and *Sputnik* launches showed the Soviets had achieved a breakthrough in rocket technology. Estimates began to appear that the Soviets would deploy their ICBMs, which had been given the NATO code name SS-6 Sapwood, in huge numbers. If true, the United States would be vulnerable to a surprise nuclear attack. So began the "missile gap" controversy. Unlike the bomber gap, this new intelligence question was not so easily answered. The Soviet Union was vast. Even with the use of bases in Iran, Pakistan, and Norway, many areas were out of the U-2's range. Another factor was Eisenhower's growing reluctance to authorize overflights. He feared that large numbers of such flights would provoke the Soviets, possibly starting World War III.[64]

What overflights were authorized concentrated on Soviet rail lines. Because of the SS-6's huge size, it could only be moved by rail, and any operational sites would also be located near rail lines. The problem was the U-2s were failing to bring back any photos of deployment. No ICBMs were spotted in transit, nor were any operational sites found. Despite this lack of evidence, the air force continued to insist that the Soviets would deploy large numbers of SS-6s.

Eisenhower was increasingly frustrated. From the U-2 photos, he knew Soviet nuclear forces were a pale shadow of those of the United States, but without evidence of the true SS-6 deployment rate, he was attacked as downplaying the Soviet threat in order to balance the budget. To aggravate matters, Khrushchev was using the missile gap and Soviet successes in space to promote an image of superiority. This, in turn, was used to put pressure on the West over Berlin.[65]

THE DIRTY BIRDS

While the overflights were under way, the Ranch housed the headquarters squadron, called Detachment D, and the training unit for the first group of air force U-2 pilots.[66] The Ranch also served as Lockheed's U-2 flight test center. Starting in late 1956, work was under way on a program that would influence the design of every Dark Eagle to follow and that would see final success two decades later.

When the U-2 was first developed, it had been hoped that the aircraft would fly so high the Soviets would have only fragmentary tracking data and would not be sure what was going on. During training flights, this theory had been borne out: only one or two radar sites would detect the aircraft. When the overflights started, however, the Soviets were not only able to track the U-2s but vector fighters toward them.[67]

President Eisenhower was extremely disturbed by the ease with which the Soviets were detecting the U-2 overflights. He directed that work be undertaken to reduce the U-2's "radar cross section." Called Project Rainbow, it had the highest priority and the attention of all the Skunk Works engineers. Eisenhower threatened to end the overflight program should Rainbow fail. Johnson asked advice from two radar experts, Dr. Frank Rogers and Ed Purcell.

They suggested stringing wires of varying lengths from the nose and tail to the wings. The idea was to scatter the radar signals away from the receiver, which would weaken the radar echo. The modification worked, but with major shortcomings—the U-2's range was cut and its maximum altitude was reduced by seven thousand feet. The wires also whistled and sometimes broke, flapping against the cockpit and fuselage.

The other attempt was more elaborate. The U-2's underside was covered with a metallic grid, called a Salisbury Screen, and then overlayed with a black foam rubber called Echosorb. The grid would deflect the radar signal into the absorber. The modifications proved to have only limited usefulness. At some radar frequencies, they did reduce the U-2's radar cross section. At others, however, the plane's radar echo was made worse. The coating also prevented the engine's heat from dissipating out the skin.

This latter problem caused the loss of Article 341, the U-2 prototype. On April 4, 1957, Lockheed test pilot Bob Sieker was making a flight at 72,000 feet. This involved flying the aircraft up and down a radar range for hours on end. The heat buildup caused the plane's engine to flame out. When this happened, cockpit pressurization was lost and Sieker's suit inflated. As it did, the clip holding the bottom of his faceplate failed and it popped open. The suit lost pressure, and Sieker passed out within ten seconds. The U-2 went into a flat spin and crashed.

A search was launched, but the wreckage could not be found. It was Lockheed test pilot Herman "Fish" Salmon who discovered the crash site. He rented a twin-engine Cessna from Las Vegas, and three days after the crash, Salmon found the U-2 in a valley near Pioche, Nevada, about ninety miles from the Ranch. The faceplate was still in the cockpit. Sieker's body was fifty feet away. This suggested that he had revived at the last moment and jumped from the plane but was too low for his parachute to open. In the wake of this and other accidents, an ejector seat was added to the U-2. (In June 1957, following the crash, Lockheed moved its test operations from the Ranch to North Base at Edwards Air Force Base.)[68]

Finally, Johnson decided it was more practical to cover the aircraft with a paint that contained iron ferrite. Later called "Iron Ball" paint, it absorbed some of the radar signals, which reduced the cross section by an order of magnitude.

The first "Dirty Bird" U-2, as the modified aircraft was called, was sent to Turkey in July 1957. It had wires strung from the nose to poles on the wings, as well as the radar absorbing paint. On July 7, CIA pilot James Cherbonneaux made a Dirty Bird flight along the Black Sea coast to probe Soviet air defenses. Intercepted communications indicated the wires and coatings worked well but that the Soviets were able to pick up radar returns from the cockpit and tailpipe. Two weeks later, he made a Dirty Bird overflight of the central Soviet Union from Pakistan. It covered Omsk and the Tyuratam launch site, before landing back in Pakistan.[69]

In the end, the loss of altitude caused by the wires was too great and they were removed. The Iron Ball paint continued to be used. At first, it was a light color. This was soon changed to midnight blue, which matched the color of the sky at 70,000 feet. This would make it harder for MiG pilots to spot the plane and gave the U-2 a sinister appearance.

With the failure of the Dirty Bird U-2, it was now clear to Johnson that a reduced radar cross section would have to designed into a plane from the start, not added on later. Ironically, when final success was achieved, Johnson was one of those who doubted it could work.

MAY DAY

U-2 overflights of the Soviet Union remained sporadic throughout 1958 and 1959. Months would pass without one. Eisenhower continued to express fears that the overflights would provoke a Soviet response, even World War III. The U-2 pilots had a feeling time was running out. Starting in 1959, SA-2 SAMs had been fired at the planes. Some had come dangerously close. What overflights were being made could not settle the question of the Soviet ICBM force size.

Then, in early 1960, information was received that an ICBM site was being built at Plesetsk in the northwest Soviet Union. From this site, SS-6 missiles could reach the northeast United States, including New York, Boston, and Washington, D.C. Although there had been a number of false alarms before, this report seemed solid.

After some four months without an overflight, Eisenhower approved two in succession for April 1960. One complication was the upcoming Paris Summit Conference set for May 16, 1960, to be followed in June by a visit by Eisenhower to the USSR. Eisenhower observed that the one asset he had at a summit meeting was his reputation for honesty. If a U-2 was lost during the Summit, it could be put on display in Moscow; a disclosure like that would ruin his effectiveness.[70]

Accordingly, when Eisenhower approved the two missions, he added an April 25 cutoff date. Detachment B commander Col. William Shelton selected Bob Ericson and Francis Powers for the overflights. Ericson was a member of the third group who had been originally stationed with Detachment C in Japan. He had later been transferred to Turkey. Powers was the only original member of Detachment B still with the group. He and another pilot had each made at least three overflights.

Both of the overflights were made from Peshawar, Pakistan. On April 9,

Ericson took off. He crossed the Hindu Kush Mountain Range and crossed into Soviet airspace. The primary target was Sary-Shagen, the Soviet's test site for both SAMs and antiballistic missiles, as well as long-range radars. This was not the first visit to the site, but earlier photos had been poor. This time the results were good. Ericson headed for the nuclear test site at Semipalatinsk. At this time, both the United States and Soviets were observing a nuclear test moratorium. The U-2 then headed west and photographed Tyuratam before landing at the Zahedan airstrip in Iran.[71]

The second April overflight would be different. It was to go all the way across the Soviet Union. After takeoff from Peshawar, the route went from Stalinabad, Tyuratam, Chelyabinsk, the Soviet's main nuclear weapons production facility at Sverdlovsk, suspected ICBM sites at Yurya and Plesetsk, then submarine shipyards at Severodvinsk, and naval bases at Murmansk, before landing at Bodo, Norway.[72] The flight demanded the most of both plane and pilot.

Detachment B was alerted for the overflight, but weather was bad. This mission required the whole of the USSR to be clear. By this time, the April 25 deadline had expired, and Bissell required an extension. Due to lighting conditions caused by its northerly location, Plesetsk could only be covered between April and early September, and during this period, only a few days per month were clear. If not covered now, the Summit and Eisenhower's visit could delay the flight beyond the weather-lighting window. Eisenhower agreed, with May 1, 1960, as the final allowable date.

Finally, on April 27, the weather looked good, and Powers and the support crew headed for Peshawar. The overflight was to begin at 6:00 A.M. the next morning. Powers and the backup pilot were awakened at 2:00 A.M., but almost immediately weather forced a scrub. Powers went through the same routine the next morning, but again weather forced a scrub, this time for forty-eight hours. It was not until Sunday, May 1, the last authorized day, that the weather cleared enough to allow the flight to be made. A last-minute communications problem delayed the takeoff until 6:26 A.M. This invalidated the precomputed navigation data.

As Powers crossed the Soviet border, he found the weather was worse than expected. A solid cloud cover extended below him. An hour and a half into the flight, Powers spotted the first break in the clouds. The plane was slightly off course and Powers corrected his heading. Far below, Powers could see the contrail of a Soviet fighter. He knew the U-2 was being tracked.

The clouds cleared again when the U-2 reached Tyuratam. Several large thunderheads hid the pad area, but the surrounding area was clear. The clouds closed in again until about three hours into the overflight. As they began to clear, Powers could see a town. Using the plane's radio compass, Powers took a bearing on a Soviet radio station and corrected his course again. About fifty miles south of Chelyabinsk, the clouds finally broke and Powers could see the snowcapped Urals.[73]

At this point, Powers's plane, Article 360, suffered an autopilot failure. The aircraft's nose pitched up. Powers disconnected the autopilot, retrimmed the aircraft, and flew it manually for several minutes. He then reengaged the autopilot, and the plane flew normally. After ten or fifteen minutes, the pitch control again went full up. This could not continue, so Powers left the autopilot disconnected. He now faced the daunting task of hand flying the plane. The weather was now clear, however, and the plane was nearing the halfway point. Powers decided to press on rather than turning back.[74]

The U-2 was approaching Sverdlovsk at an altitude of 72,000 feet when it was picked up on Soviet radar. A prototype Su-9 fighter, still in testing and not even armed, was ordered to ram the U-2. The pilot was unable to spot the U-2, however, and flew far past it. Two MiG 19s were also sent up, but with a maximum altitude of 66,000 feet, they could not reach the U-2.

As yet, Powers was unaware of these intercept attempts. He had just completed a 90-degree turn and was lining up for the next photo run. As he wrote entries in his logbook, an SA-2 battery opened fire.[75] One of the missiles exploded below and behind the U-2. Powers saw an orange flash. The shock wave damaged the right stabilizer. The U-2 held steady for a moment, then the stabilizer broke off, the U-2 flipped over on its back, and the wings broke off.[76] Powers struggled to escape from the tumbling forward fuselage. He was unable to trigger the plane's destruct system. At 15,000 feet, he was able to escape and parachute to a landing. Powers was captured almost immediately.

The Soviets did not realize they had shot down the U-2. The MiG 19 pilots saw the explosion, but thought the SAM had self-destructed after a miss. On the ground, the fluttering debris from the U-2 filled the radar screens with echoes, but the Soviets thought it was chaff being ejected from the U-2 to confuse the radar. At least three SAM sites continued to fire—reportedly fourteen SA-2s in all. An SA-2 hit one of the MiG 19s, killing its pilot, Sergei Safronov. Soon after the MiG was hit, the destruction of the U-2 was confirmed.[77]

The confusion of the Soviet air defenses was echoed by that of U.S. intelligence. The Soviet radio transmissions had been intercepted. They were interpreted as indicating the U-2 had gradually descended for a half hour before being shot down. It was assumed the U-2 had flamed out. A cover story was issued that an unarmed civilian weather plane had crossed the Soviet border after the pilot had reported problems with his oxygen system.

Several days later, Khrushchev revealed that Powers had been captured and had confessed to spying. As Eisenhower feared, the U-2 wreckage was put on display in Moscow. Eisenhower made the unprecedented admission that he had personally authorized the overflights. No head of state had ever before admitted that his country spied in peacetime. The Paris Summit ended when Khrushchev demanded Eisenhower apologize for the overflights. Eisenhower would only give a promise that no future overflights would be made. Powers underwent a show trial and was sentenced to ten years. He was exchanged in February 1962 for a Soviet spy. He later worked for Lockheed as a U-2 test pilot.[78]

The U-2 detachments were brought home following the loss of Powers's aircraft. The number of CIA U-2 pilots was cut from about twenty-five to only seven.[79] The Detachment D headquarters squadron moved from the Ranch to North Base at Edwards Air Force Base in June 1960. The Lockheed test operation was moved to Burbank.[80] Groom Lake was about to become home for the greatest Dark Eagle ever built.

Chapter 3

The Archangel from Area 51
The A-12 Oxcart

What is called "foreknowledge" cannot be elicited from spirits, nor from gods, nor by analogy with past events, nor by calculations. It must be obtained from men who know the enemy situation.

Sun Tzu
ca. 400 B.C.

The December 24, 1962, issue of *Aviation Week and Space Technology* carried an editorial titled "Laurels for 1962." It was a listing of significant accomplishments for the previous year. The sixth item was one of the magazine's most significant scoops. It read: "Clarence (Kelly) Johnson of Lockheed Aircraft for his continued ingenuity in the 'Skunk Works.'"[1] Behind those bland words was the greatest achievement of aeronautical technology. The program had already been under way for six years. The full dimension of the achievement of this greatest of the Dark Eagles would not be revealed for another three decades.

Despite the success of the U-2, its top speed of just over 400 knots was slower than that of some World War II prop fighters. It could only survive through height. With development of the SA-2 SAM, this was no longer enough. Well before Powers was shot down, it was clear any U-2 successor would have to fly both higher and faster. Much faster.

SUN TAN

There had been early, pre-U-2 studies of high-speed reconnaissance aircraft. One was by Bell aircraft of the "RX-1," a second-generation X-1 rocket-powered research aircraft with camera equipment. In the early 1950s, the X-1A reached a speed of Mach 2.44 and an altitude of 90,440 feet, both world records. The RX-1 would be carried to the target area by a bomber; it would then be released, make the overflight, and be retrieved. It does not appear the idea progressed beyond the concept stage.[2]

A somewhat more practical idea was the air force–AVRO Canada's Project Y, also called WS-606A. This was a vertical-takeoff and landing (VTOL) aircraft that used six Armstrong-Siddeley Viper jet engines, a CF-105 fuselage, and a disk-shaped wing. It was 37 feet long, with a dish-span of 29 feet. The top speed was Mach 3 to 4, with a maximum altitude of 95,000-plus feet. The combat radius was a mere 800 nautical miles in the VTOL mode. Although WS-606A had 1-A priority for a time in the mid-1950s, technical problems soon ended the effort. Of the various air force–CIA Black airplanes, WS-606A remains unique in that it was the only one to involve a foreign contractor.

What proved to be the most serious of these early attempts grew out of early 1950s work on aircraft fueled by liquid hydrogen. In early 1956, Johnson proposed to the air force a study design for a hydrogen-powered reconnaissance aircraft called the CL-400. It had a top speed of Mach 2.5, an altitude of 100,000 feet, and a range of 2,200 nautical miles. Johnson said he could have the prototype ready in eighteen months.

The CL-400 would be a huge aircraft—164.8 feet long with a wingspan of 83.8 feet. It used a T-tail and a retractable vertical fin that spanned nearly 30 feet. The fuselage was nearly 10 feet in diameter. The plane's two engines were located on the wingtips. It used a bicycle-type landing gear with the outriggers retracting into the engine pods. In shape, the CL-400 resembled a scaled-up F-104. The plane's insulated tanks held 21,440 pounds of liquid hydrogen. It had a crew of two and 1,500 pounds of reconnaissance equipment.

Lieutenant General Donald Putt, the deputy chief of staff for development, was very impressed with the CL-400 proposal and indicated that the air force wanted such a high-speed aircraft within two or three years (the expected operating lifetime of the U-2). In February 1956, Pratt and Whitney was selected to build the engines, and Lockheed was given a contract for two prototypes. This was soon followed by a contract for six production CL-400s. By April, a full three months before the first U-2 overflight of the USSR, work on the project was under way. Lieutenant Colonel John Seaberg, who had set in motion the U-2, was named to manage the liquid hydrogen tanks, airframe, and systems. Major Alfred J. Gardner was to manage the engines, while Capt. Jay R. Brill would work on the logistical problems of producing, transporting, and storing liquid hydrogen.

The CL-400 was to be a Black airplane, due to the advanced technology and the need for rapid development. It was classified Top Secret (Codeword) and only twenty-five people had full access to the project. To speed development, near complete power to issue contracts was given to the managers. The project number was changed regularly and some contracts were written by other air force offices to hide their connection with the CL-400. At contractors' plants, CL-400 personnel were isolated from other employees. The project was given the code name "Sun Tan."[3]

Johnson saw the development of Sun Tan as more than aeronautical; the plane would require the routine production and transport of huge quantities of liquid hydrogen. Ben R. Rich, the Skunk Works engineer with dual responsibility for propulsion and hydrogen handling, liked to talk about "acre-feet" of liquid hydrogen (code named "SF-1" fuel). This was at a time when the *Mechanical Engineering Handbook* described it as only a laboratory curiosity.

A major concern was the danger of hydrogen fire and explosion. The vivid images of the destruction of the *Hindenberg* were very much in mind. The tests were done at "Fort Robertson," a converted bomb shelter near the Skunk Works. Surprisingly, in many cases, the liquid hydrogen simply escaped without igniting. In sixty-one attempts to cause an explosion, only two succeeded. When a fire did occur, the fireball quickly dissipated. In contrast, gasoline fires did much more damage. Clearly, with proper care, liquid hydrogen was a practical fuel.

Despite the high level of security that enveloped Sun Tan, several incidents occurred, funny in retrospect, that threatened to expose the project. All of these related to the use of liquid hydrogen. The first such hydrogen "leak" occurred when a female Skunk Works engineer (a rarity in the 1950s) attended a conference on hydrogen. Another engineer recognized her and began to wonder why Lockheed was interested in liquid hydrogen.

Another problem was the semitrailer used to transport liquid hydrogen. Because of the light weight of liquid hydrogen (one gallon weighed one pound), the vehicle had only a single axle instead of the two a trailer of this size normally required. The single-axle arrangement attracted undue attention every time it went through state weighing stations. At one weigh station, a trailer was found to be 100 pounds overweight, and the driver was ordered to unload the excess. The air force had to go to the governor to get the load released. The Sun Tan group thought about painting on a second axle but quickly realized this would be too obvious. When the new trailer was built, it had two axles, the second purely for cover.

A third incident occurred during construction of a liquid hydrogen plant near Pratt and Whitney's Florida test facility. Its cover was as a "fertilizer plant," but word soon spread that the facility produced hydrogen. A local civil defense official became alarmed that a hydrogen bomb was being built in the area. It took a delegation of security officials to convince him to keep quiet.

Use of liquid hydrogen affected every part of the CL-400. It boiled at negative 423 degrees F, yet, at Mach 2.5, the plane's skin would reach 746 degrees F. The liquid hydrogen would have to be protected from this heat. The fuel lines, which would have to pass through the hot wing structure before reaching the engines, had a vacuum-jacketed insulation. Tests of the insulation were done at Fort Robertson using five ovens. Heat tests were also run on the engines, booster pumps, valves, controls, and other components.[4]

While the Skunk Works was designing the CL-400, Pratt and Whitney was conducting tests on the hydrogen-fueled engine. The initial work, code-named "Shamrock," was to convert a J57 engine to burn hydrogen. The modifications worked very well; the engine could be throttled down until the fan blades were spinning slowly enough to be counted. The throttle could then be smoothly opened to full power.

The success of the modified J57 encouraged development of the Model 304 engine that would power the CL-400. On a normal jet engine, fuel is sprayed directly into the combustion chamber. With the Model 304 engine, the liquid hydrogen first passed through a heat exchanger. This contained nearly five miles of stainless steel tubing. The liquid hydrogen was heated by the exhaust, going from negative-423 degrees F to 1,340 degrees F and changing from a liquid to a hot gas. The hydrogen gas was fed through a turbine, which spun the compressor fans and liquid hydrogen pump via a reduction gear. Some of the hydrogen was sprayed out the burners and ignited. The rest was sent to an afterburner.

The first runs of the 304 engine began on September 11, 1957. In all, twenty-five and a half hours of operation with liquid hydrogen were completed during the next year. Despite failures with the turbines, heat exchanger, and bearing, the development was seen as progressing satisfactorily.[5]

The CL-400 would never get to test its wings, however. By October 1957, the Sun Tan project had effectively ended. The problem was the plane's short range. The end came when Johnson was visited by Assistant Air Force Secretary James H. Douglas Jr. and Lt. Gen. Clarence A. Irvine. They asked how much "stretch" was in the CL-400. Johnson told them only 3 percent. The plane was a flying thermos bottle. The only space was the cockpit, and fuel could not be carried in the hot wing structure. Douglas and Irvine asked Pratt and Whitney how much improvement could be made in the 304's fuel efficiency. The answer was only 5 or 6 percent over five years.[6]

To increase the CL-400's range, its size would have to be increased considerably. The Skunk Works looked at planes as long as a football field. This made the plane even less practical, and Johnson urged that Sun Tan be canceled. The air force was also short of money for several higher-priority projects, and there were doubts Eisenhower would approve overflights. With this, the project ended. The prototype CL-400s were canceled in October 1957, although the engine tests continued through 1958. The formal cancellation was made in February 1959. In all, between $100 and $250 million had been spent. Not until 1973 was the Sun Tan project revealed.[7]

Sun Tan was only one thread in a number of post–U-2 ideas. After the failure of the Dirty Bird U-2s, Johnson studied a large flying-wing design. The span of the swept-back wings was larger than that of the U-2. It was powered by two jet engines fed from a nose intake. Fins were located near the wing tips. In overall shape, it resembled the World War II Go 229 German fighter. The design was capable of very high altitudes, but still at relatively low speeds.

GUSTO

It was not until the fall of 1957 that the emerging high-speed reconnaissance aircraft program began to coalesce. Bissell arranged for a study of how a plane's speed, altitude, and radar cross section affected its probability of being shot down. The study found that supersonic speeds greatly reduced the chances of radar detection. The aircraft would need a top speed of Mach 3, to fly at altitudes over 80,000 feet, and to incorporate radar-absorbing material.[8]

To achieve such speeds was a nearly impossible task. At this time, there had been only one manned Mach 3 flight. On September 27, 1956, the X-2 rocket-powered research aircraft reached Mach 3.196, equivalent to 2,094 mph. The plane went out of control, killing the pilot, Capt. Milburn Apt.[9] Even this had been a brief, rocket-powered sprint. The reconnaissance aircraft would need to maintain these speeds for a prolonged time, while being subjected to more severe airframe heating than on Sun Tan.

To put in perspective what was required, the plane would have a *sustained* speed 60 percent higher than the maximum *dash* speed of any jet then operational. It would have to fly 70 percent higher and have 500 percent better range. Speeds above Mach 2 were unknown territory. The only large, high-speed aircraft was the B-58, and its flight control system was overly complicated, once being described as "designed standing up in a hammock." Nothing then in existence could be used to build such an airplane.[10]

If these speeds could be reached, however, it would vastly complicate the problem facing Soviet air defenses. A U-2 flying directly toward an SA-2 SAM site would be detected about ten minutes before reaching it and would be in range for about five minutes. A Mach 3 aircraft would have a warning time of less than two minutes. Only *twenty* seconds would elapse from the time the aircraft entered the site's range, until it was too close to be fired on. The SA-2 would then have to chase the plane as it flew away from the site. With the missile's top speed of Mach 3.5, it would be a dead heat.[11]

Speed would greatly reduce the reaction time of air defenses. Use of radar-absorbing material would further reduce the range at which the plane could be detected.

An airplane with these capabilities would be very expensive—far more than the U-2 had been. A clear assessment of the plane's feasibility was needed. (Sun Tan had, by this time, proven to be a "wide-body dog.") Bissell put together a panel to provide this assessment. The chairman was Dr. Land, and the panel included two aerodynamic experts and a physicist. The assistant secretaries of the air force and navy for research and development also attended some of the six meetings.[12]

The navy, Convair, and Lockheed were made aware of the general requirements and submitted designs. (As yet, no money or contracts had been issued.) The navy submitted a design for a ramjet-powered aircraft with rubber inflatable wings. It would be carried to high altitude by a huge balloon. The aircraft would then be boosted by a rocket to a speed at which the ramjets could start. The navy proposal proved to be totally impractical. It was determined that the balloon would have to be a mile in diameter and the aircraft's wing area one-seventh of an acre.

Convair proposed a ramjet-powered Mach 4 aircraft that would be launched from a B-58. This proposal, although far more practical than the navy concept, also had shortcomings. The B-58 could not reach supersonic speed with the aircraft attached. Moreover, it was thought the aircraft's ramjet would suffer "blowouts" during maneuvers. The total flight time for the Marquardt ramjet was less than seven hours, but Convair engineers continued to refine the design.

Lockheed and Johnson were studying a wide range of concepts for what was initially called the "U-3" project. Many were based on the Sun Tan airframe, but using kerosene fuel. Different size aircraft were looked at, with both two and four engines. Johnson also looked at exotic concepts. These included towing the U-3 to altitude behind a U-2; using a booster stage; carrying the U-3 to altitude under a balloon; aircraft with jet, rocket, and ramjet engines; designs that used coal slurries or boron fuel; vertically launched aircraft; and a design with inflatable wings and tail. In the end, Johnson rejected them all.[13]

The failure of Sun Tan seems to have had an effect on Johnson's view of the high-speed reconnaissance aircraft. Since the exotic technology of liquid hydrogen had proven impractical, he understood that this new aircraft would have to be based on solid technology.

Johnson began a series of design studies on April 21, 1958. The first was designated "A-1." The U-2 had been called the Angel by Skunk Works engineers. These new designs would fly far faster and higher, so, accordingly, the *A* stood for "Archangel."

In late November 1958, the Land Panel decided that it was possible to build the aircraft. Their report concluded: "The successor reconnaissance aircraft would have to achieve a substantial increase in altitude and speed; be of reduced radar detectability; suffer no loss in range to that of the U-2; and be of minimum size and weight."[14] They further recommended that President Eisenhower approve funding for additional studies and tests. Both Eisenhower and his scientific adviser, Dr. James Killian, had already been briefed on the project. Eisenhower approved the recommendation, and funding was provided to Lockheed and Convair to prepare definitive studies. The effort was code-named "Gusto."

By the spring of 1959, Johnson and his Skunk Works engineers had worked their way up to the A-10, but success seemed elusive. President Eisenhower was intent on a plane with a zero radar cross section. He did not want the Soviets to even know it was there. Kelly Johnson told the CIA that there was no way to accomplish this.

Work continued on reducing the radar cross section. One idea involved adding wedge-shaped chines made of radar absorbing material to the A-10's cylindrical fuselage. Tests of a small model were successful, and by May 1959 the chines had been incorporated into the A-11 design. This showed a reduction of a full 90 percent in radar cross section. Although not invisible, success was now within reach. In July, a final revised design of the A-11 was prepared. It made full use of the chines, as well as elements from the previous designs, and was the sum of fifteen months of work.

After a day and a half of work, the final drawing was completed. The long sheet of paper was presented to Johnson. Ben Rich, one of the engineers who worked on it told him, "Kelly, everything is now exactly where it should be—the engines, the inlets, the twin tails. This is probably as close to the best we can come up with." Johnson took the design and made repeated trips to CIA headquarters.[15]

On July 20, 1959, President Eisenhower was again briefed on Gusto. At the meeting were Allen Dulles and Bissell from the CIA, Defense Secretary Neil McElroy, scientific advisers Dr. Killian and Dr. George Kistiakowsky, Gens. Thomas D. White and C. P. Cabell, and National Security Adviser Gordon Gray. The meeting lasted nearly an hour. Eisenhower gave approval for development to begin.[16]

The Convair and Lockheed designs were submitted to a joint DOD-USAF-CIA selection board on August 20, 1959. The Convair design, called "Kingfisher," was a large delta-wing aircraft 79.5 feet long, with a wingspan of 56 feet and weighing 101,700 pounds. It was to be powered by two J65 jet engines and two Marquardt RJ59 ramjets. The jets would be used for takeoff and climb. Once up to speed, the ramjets would ignite and accelerate the plane to Mach 3.2. During flight, the Kingfisher would climb from an initial altitude of 85,000 up to 94,000 feet. The Convair aircraft had a range of 4,000 nautical miles.

The final Lockheed design, the A-11, was a single-seat aircraft. It had a long fuselage with a delta wing at the rear. The two J58 engines were midway out on the wings. The A-11 was 102 feet long and had a wingspan of 57 feet—a much larger aircraft than the Kingfisher. Yet its weight was 110,000 pounds, only marginally heavier. Its top speed was also Mach 3.2, and it had a range of 4,120 nautical miles. The A-11 had a better altitude capability—at the start of the cruise it would be at 84,500 feet, and this would increase to 97,600 feet. Both aircraft were to be ready in twenty-two months.[17]

The Lockheed A-11 was selected on September 3. The Gusto code name was replaced by "Oxcart." Given the plane's extreme speed, the code name seemed to be "inspired perversity," as the official history put it. There was a subtle symbolism, however. Lockheed aircraft had long carried "star" or astronomical names—Orion, Vega, Sirius, Altair, Electra, Constellation, Starfire, Starfighter, and JetStar. In Europe, the constellation of the Big Dipper is often called a wagon—or an oxcart.

OXCART

Once Lockheed was selected, the CIA gave approval for a four-month series of aerodynamic and structural tests, engineering design, and construction of a full-scale A-11 mock-up.[18] The mock-up was needed to test the aircraft's radar cross section. Due to the complexity of the problem, it was not possible to use subscale models. It was completed in November 1959, then was packed in a huge box and moved by road from Burbank to Groom Lake. The mock-up was then reassembled and mounted on a pylon. For the next eighteen months the mock-up was scanned by radar, while adjustments and modifications were made. This early work was successful, and the CIA gave approval on January 30, 1960, for production of twelve aircraft.[19]

Extreme security measures, tighter even than for the U-2, were used to hide the program. Because knowledge of Lockheed's involvement would create speculation, money to subcontractors was paid through "front" companies. Once the parts were completed, they would be shipped to warehouses, also rented to front companies. The parts would then be sent to Burbank. Few, if any, of the subcontractors knew what the parts were for. Ironically, some drawings were deliberately not classified; the assumption was that if they were stamped "Secret," people would take an interest.

Just over three months after the Oxcart program started, Powers's U-2 was shot down. It was clear to Eisenhower that the United States would never again be able to make overflights of the Soviet Union. This also brought into question the future of Oxcart. The president seemed undecided, saying at one point that he was not sure if it would be best to end development, or if so much had been invested that the United States should capitalize on the effort by carrying it through. In the latter case, the program should be continued, although at a low priority, for use by the air force rather than the CIA. He asked CIA director Dulles to meet with Defense Secretary Thomas S. Gates and Maurice Stans, director of the Bureau of the Budget, to make a recommendation.[20]

A new challenger appeared in the late summer. On August 19, the recovery capsule from the *Discoverer 14* reconnaissance satellite was caught in midair by a C-119 aircraft. This ended eighteen months of launch failures, tumbling satellites, and lost capsules. More important, the capsule carried a twenty-pound roll of film, covering 1 million square miles of the Soviet Union. This one mission provided more coverage than the twenty-four U-2 overflights together had accomplished.[21] More Discoverer satellites were launched and, within a year, they showed there was no missile gap: taken together, Soviet ICBM, submarine-launched missiles, and bomber forces were a fraction of the U.S. total. Satellites could cover the whole of the Soviet Union, without the political risks of aircraft overflights.

Ultimately, Oxcart was seen as needed and was continued under CIA control. Satellites would be restricted to coverage of the Soviet Union for the foreseeable future. It would also be many years before a satellite camera had the resolution of the U-2's B camera. If the USSR was off limits for the U-2, it could still provide coverage of Communist China, Cuba, Vietnam, or the Mideast. In a few years, however, these areas could no longer be overflown with impunity. The Chinese already had SA-2 SAMs, and other countries would have them by the early and mid-1960s. The Oxcart would soon be needed to conduct overflights of even Third-World countries.

INITIAL DEVELOPMENT

Once the future of Oxcart was resolved, the initial development work continued. Temperature affected every aspect of the Oxcart's design. Even though the plane would be flying at the edge of space, friction would raise the skin temperature to over 500 degrees F. The *coolest* part of the engine, the inlet, reached 800 degrees F. The afterburner section would reach 3,200 degrees F.[22] The plane would have to be built of stainless steel or titanium. Stainless steel honeycomb was being used in the Mach 3 XB-70, then under development, but Johnson rejected this when he saw the production problems it entailed. The honeycomb had to be produced in a clean room, under sterile conditions. The Skunk Works motto was "KISS" (Keep It Simple, Stupid). Stainless steel was too complicated and was likely to cause problems.

Johnson decided to use heat-treated B-120 titanium alloy. This was still a major step into the unknown. Although it had been used in aircraft before, nobody had ever tried to build an entire airframe out of the material. Even drilling a hole was a problem, due to titanium's extreme hardness. Drills would be worn out after only seventeen holes. A special West German drill was found that could drill 150 holes before needing resharpening.

Before beginning production, Johnson decided to build a sample of the wing structure and nose section. When the wing structure was put in the "hot box," to simulate the high temperatures, it literally wrinkled. The solution was to put corrugations in the wing skin. At high temperatures, the corrugations only deepened slightly. Johnson was jokingly accused of building a Mach 3 Ford Trimotor (which also had a corrugated skin). The nose segment was used to study requirements for cooling the pilot, camera, and systems.[23]

A continuing problem during development was the poor quality of the titanium. A full 80 percent was rejected; the material was so brittle that it would shatter like glass if dropped. This problem continued into 1961, until a group from CIA headquarters went to the Titanium Metals Corporation and briefed company officials about Oxcart. The supply soon became satisfactory.[24] Lockheed also established an extensive quality-control program. There were times, Johnson later recalled, "when I thought we were doing nothing but making test samples."[25]

Sometimes the problems with titanium bordered on the bizarre. During heat tests, bolt heads would simply fall off after one or two runs. It was found that cadmium plating had flaked off the tools used to tighten the bolts. This was enough to "poison" the titanium, causing a spiderweb network of cracks to form. All cadmium-plated tools had to be thrown in a big vat that was boiling "like a witch's brew" to strip off the plating. It was also found that welds of wing panels done during the summer soon failed, while those made during the winter lasted indefinitely. Again, it was a chemical reaction. The parts were washed before welding, and in the summer, Burbank city water had chlorine added to reduce algae. Even an ordinary pencil was dangerous. A shop worker took a pencil and wrote some numbers on a piece of titanium; a week later, it was discovered the graphite had etched the metal.[26]

Not simply the airframe, but every part would have to withstand temperatures higher than ever before endured by an aircraft. Johnson said later, "Everything on the aircraft, from rivets and fluids up through materials and power plants, had to be invented from scratch." All electrical connections were gold-plated, as gold retained its electrical conductivity better at high temperatures than copper or silver. The control cables were made of Elgiloy, a steel, chromium, and nickel alloy normally used in watch springs.[27] A hydraulic fluid was developed to withstand temperatures of 650 degrees F (150 degrees hotter than normal).[28]

Fuel was a difficult problem. During subsonic cruise, such as during refueling, temperatures would drop to negative-90 degrees F. At Mach 3, the fuel would be heated to 285 degrees F. It would then be pumped through the afterburner exit flaps, acting like hydraulic fluid to control their position. This would raise its temperature to 600 degrees. The fuel would then be pumped into the J58 engine. Conventional fuel would boil and explode at such temperatures. The fuel developed was JP-7, also called LF-2A. It had a low vapor pressure; if a match was thrown into a pool of JP-7, the match would go out.[29]

The internal stress caused by such heat affected the quartz glass window for the camera. The heat had to be even throughout the window, or there would be optical distortion. This one problem took three years and $2 million to solve. The quartz window was fused to its metal frame using high-frequency sound waves.

The effect of these many problems was to delay the program and raise its cost.

THE J58 PROPULSION SYSTEM

Development of the J58 engines and their nacelles proved the most difficult problem. The J58 program was begun in late 1956 to power a navy attack plane with a dash speed of Mach 3. This speed would be maintained for only a few seconds. By late 1959, however, navy interest was fading, and it was decided to cancel the engine. The CIA requested the work be continued and the engine be modified for a continuous speed of Mach 3.2. A contract was issued for three ground test and three flight test engines.[30]

With the many design changes needed to accommodate the extreme heat, virtually nothing remained of the original navy J58 engine when development was finished. To give one example, a standard ground test stand could not simulate the heat and altitude conditions required. Pratt and Whitney built a new test stand in which a J75 engine's exhaust was run through and around the J58. Speeds over Mach 3.6 and altitudes of 100,000 feet could be simulated.[31]

For all its power, the J58 engine alone was not enough to drive the A-11 to Mach 3 by brute force. The nacelles were the key that opened the way to those speeds. They were not simply a place to put the engines, but an integral part of the propulsion system. Up to 1,600 mph, air would come in through the intake and a ring of centerbody bleed vents to feed the engine. As the A-11 approached Mach 3, the flow cycle would change. Air was now vented *out* the centerbody bleed vents. The effects were amazing—at Mach 3, a full 56 percent of the total thrust came from the intake. Another 27 percent came from the afterburner, while only 17 percent came from the J58 engine itself. In effect, the J58 was a flow inducer and the nacelles pushed the airplane.[32]

It was a remarkable achievement, but years of development and flight testing would be needed before the system was reliable.

AREA 51

Once development began in earnest, the question became where to test the A-11. Despite the success of the U-2 flight tests and the A-11 mock-up radar tests, Groom Lake was not initially considered. It was a "Wild West" outpost, with primitive facilities for only 150 people. The A-11 test program would require more than ten times that number. Groom Lake's five-thousand-foot asphalt runway was both too short and unable to support the weight of the Oxcart. The fuel supply, hangar space, and shop space were all inadequate.

Instead, ten air force bases scheduled for closure were examined. (This indicates the scale of operations envisioned.) The site had to be away from any cities and military or civilian airways to prevent sightings. It also had to have good weather, the necessary housing and fuel supplies, and an eighty-five-hundred-foot runway. None of the air force bases met the security requirements, although, for a time, Edwards Air Force Base was considered.

In the end, Groom Lake was the only possibility. Plans were drawn up for the necessary facilities. As cover, the site was described as a radar test range. The remote location was explained as necessary to reduce interference from outside sources. Construction began in September 1960, several months after the CIA U-2 operation closed down. The first construction workers were housed in surplus trailers. A new water well was drilled, but the site still lacked anything but the basics.

The first major construction work was the 8,500-foot runway. This was built between September 7 and November 15, 1960, and required some 25,000 yards of concrete. This was followed by construction of the fuel storage tanks. A-11 test operations would need 500,000 gallons of JP-7 per month. By early 1962, a tank farm with a storage capability of 1,320,000 gallons was completed. Three surplus navy hangars were obtained, moved to Groom Lake, then reassembled at the north end of the facility. The navy also provided over 100 surplus housing buildings. Additional warehouse and shop space was added. Repairs to the existing buildings from the U-2 days were also made. To provide access, 18 miles of highway leading into the site were resurfaced. This work was done on a two-shift basis and continued into mid-1964.

The CIA ran into a legal problem with the construction work. Nevada law required that the names of all contractor personnel who stayed in the state for more than forty-eight hours be reported to state authorities. Listing the personnel and the companies working on the project would reveal the existence of Oxcart. The CIA general counsel discovered a loop-hole—government employees were exempt. Accordingly, all contractor personnel at Groom Lake received appointments as "government consultants." If any questions were raised, it could truthfully be said that only government employees worked at the site.

By August 1961, a year after work began, the basic facilities had been completed to support the initial flight tests. Although work would continue for another three years, Groom Lake had been transformed from a ramshackle collection of hangars and trailers in the desert into a permanent, state-of-the-art flight test center.

At this same time, the radar test program on the A-11 mock-up had been under way. By the time the work was completed in mid-1961, it was found that most of the radar return came from the vertical stabilizers, the engine inlet, and the forward sides of the nacelles. The edges of the chines and wings, as well as the vertical stabilizers, were made of a radar-absorbing laminated plastic. Of course, this plastic also had to withstand the 500-plus-degree F heat. This was the first time plastic had been used as a structural material. Because of the design changes from the radar tests, the aircraft was renamed the "A-12."[33]

Groom Lake was also used for low-speed tests of the A-12's ejector seat. It would have to work from standing still on the runway up to a speed of over Mach 3 at 100,000 feet. Johnson was never convinced that a capsule ejection system, such as that on the B-58 or XB-70, was needed. The pilot would be wearing a pressure suit, which would provide protection from wind blast and heat. Instead, a modified F-104 seat would be used. The system was tested by towing a fuselage mock-up across the lake bed behind a car.[34] Later, in-flight ejection tests were done using a two-seat F-104.

Groom Lake had also, by this time, received a new official name. The nuclear test site was divided into several numbered areas. To blend in, Groom Lake became "Area 51." (Its unofficial name remained the Ranch through the 1960s.)

DELAY

The first A-12 was originally scheduled to be ready in May 1961. Due to problems with wing assembly and J58 engine development, this date was pushed back to August 30, then December 1. Bissell was very upset by the delays: "I trust this is the last of such disappointments short of a severe earthquake in Burbank," he commented.

It was not to be—on September 11, 1961, Pratt and Whitney notified Lockheed of continuing problems with the J58's weight, performance, and delivery schedule. The completion date had slipped to December 22, 1961, with the first flight set for February 27, 1962. Because the J58 would not be ready, it was decided to temporarily install J75 engines (used in the F-105, F-106, and U-2C.) This would allow flight tests up to a speed of Mach 1.6 and 50,000 feet. With this, the A-12 program began to pick up momentum. But there would be more problems.

As flight testing neared, activities at Groom Lake also increased. In late 1961, Col. Robert J. Holbury was named Area 51 commander. A CIA officer was his deputy. Support aircraft began arriving in the spring of 1962. This consisted of an F-104 chase plane, eight F-101s for training, two T-33s for proficiency flights, a helicopter for search and rescue, a C-130 for cargo, and a Cessna 180 and U-3A for liaison use.

At Burbank, the first A-12, Article 121, was undergoing final checkout and tests. Once this was finished, the aircraft's wings were removed and the fuselage was loaded into a boxlike trailer, which hid its shape. Article 121 left Burbank at 3:00 A.M. on February 26, 1962. The route from Burbank to Groom Lake had already been surveyed, and it was found that an object 105 feet long and 35 feet wide could be moved with only a few road signs having to be removed, trees trimmed, and roadsides leveled to provide clearance. By sunrise, the convoy was out on the desert and away from prying eyes. After arrival, work began on reassembling Article 121 and installing the J75 engines. There was a final delay—the sealing compound had failed to stick to the fuel tank's interior. It was necessary to strip the tanks and reline them.

THE ARCHANGEL TAKES FLIGHT

Finally, the A-12 was ready to test its wings. Lockheed test pilot Louis W. Schalk was selected to make the first flight. In preparation, he made several flights in a modified F-100. With the center of gravity aft, it matched the A-12's expected handling characteristics. The first tests in Article 121 were engine runs and low- and medium-speed taxi runs. The prototype A-12 was unpainted and unmarked, with no national insignia, no "U.S. Air Force," no civilian N-number, not even an article number.

All was ready by April 24, 1962, for a high-speed taxi test. Schalk would momentarily lift the plane off the runway. For this test, the A-12's stability augmentation system (SAS) was left disconnected. Because of its design, the A-12 was inherently unstable under some flight conditions, and the SAS was necessary to keep the plane under control. The SAS was triple redundant in yaw and pitch and double redundant in roll.[35]

All went well with the taxi test until the A-12 lifted off. As it did, the plane wallowed into the air, the wings rocking from side to side, and the nose high.[36] Schalk recalled later, "I really didn't think that I was going to be able to sit the aircraft back down on the ground safely." Finally, he was able to regain control and cut the throttles. By this point, the A-12 had flown past the end of the runway. As it touched down on the lake bed, the wheels kicked up a huge cloud of dust, hiding the aircraft. The Groom Lake tower asked what was happening. Schalk radioed an answer, but the antenna was on the plane's underside and he could not be heard. Once the A-12 slowed, Schalk turned and the aircraft emerged from the dust cloud. Everyone breathed a sigh of relief. There was no damage from the near mishap. Schalk judged the A-12 was ready for flight, but added that the SAS should be turned on.[37]

The first A-12 flight was made on April 26. The plane remained aloft for some forty minutes, with the landing gear left down to avoid any retraction problems. Schalk switched off each of the SAS dampers, one by one. The plane remained stable, and he turned them back on and landed.

The official first flight was made four days later, on April 30. As the plane's landing gear retracted and it accelerated, several fuselage and wing fillet panels began falling off. There were no handling problems, and the plane reached 30,000 feet, a speed of 340 knots, and remained aloft for 59 minutes. The loss of the skin panels was solved by filling the cavities with steel wool. The repairs were completed and, on May 4, the A-12 reached Mach 1.1.[38]

After nearly a year's delay, the A-12 had embarked on its flight into the unknown. The new CIA director, John McCone, sent a telegram of congratulations to Johnson.[39] With the first flights completed, the test program now began expanding. Schalk made the first thirteen flights. In late 1962, three more Lockheed test pilots joined the program—William C. Park, Robert Gilliland, and James D. Eastham. The early flights tested aircraft systems, the inertial navigation system, and midair refueling. This was done using KC-135Q tankers of the 903d Air Refuelling Squadron. A maximum altitude of 60,000 feet was also reached by the end of 1962.

Because the J58 engines were not yet installed, little could be done in the way of high-speed flight testing. The CIA pressed Lockheed to make a Mach 2 flight, arguing that if the J75-powered F-106 could reach Mach 2, the A-12 should be able to do the same. Finally, Park put an A-12 into a dive and reached Mach 2.16. The flight proved little. Since the inlet-nacelle design was mismatched with the J75 engines, a "duct shutter" resulted— a vibration caused by airflow within the inlet as the plane neared Mach 2.[40]

As flight testing continued, more A-12s were being delivered to Groom Lake. By August 1962, Article 122 and Article 123 had arrived. Article 124, the A-12T two-seat trainer was moved to Groom Lake in November, and Article 125 arrived on December 17, 1962.

The Oxcart program received a boost during the summer of 1962 when CIA U-2s discovered the deployment of SA-2 SAMs in Cuba. CIA Director McCone asked if the A-12 could take over the Cuban overflights. The A-12 was still at too early a point in the flight-test program to consider such a mission. Following the Cuban Missile Crisis in October, bringing the A-12 to operational status became one of the highest national priorities.

Despite the added A-12s now available, the test program was still handicapped by the delay of the J58 engines. CIA Director McCone decided this was unacceptable. He wrote to the president of United Aircraft on December 3, 1962: "I have been advised that J58 engine deliveries have been delayed again due to engine control production problems. . . . By the end of the year it appears we will have barely enough J58 engines to support the flight test program adequately . . . Furthermore, due to various engine difficulties we have not yet reached design speed and altitude. Engine thrust and fuel consumption deficiencies at present prevent sustained flight at design conditions which is so necessary to complete developments."[41]

The first J58 finally was delivered to Groom Lake and installed in Article 121. The first problem was getting it started. The small-scale, wind-tunnel model did not adequately predict the internal airflow. As an interim measure, an inlet access panel was removed during ground tests. Holes were later drilled in the nacelles to cure the problem. Article 121 made its initial flights with one J58 and one J75. On January 15, 1963, the first A-12 flight with two J58s was made. By the end of January, ten J58 engines had been delivered and were being installed in the A-12s.

RECRUITMENT

Recruitment of the CIA pilots had begun even before the first A-12 flight. The Oxcart pilots would need remarkable skill, due both to the performance characteristics of the A-12 and the demands of flying secret intelligence missions. Air Force Brig. Gen. Don Flickinger was picked to establish the requirements. He received advice from both Johnson and CIA Headquarters. The initial criteria included experience in high-performance aircraft, emotional stability, and good self-motivation. The pilots also had to be between twenty-five and forty years of age. The small size of the A-12's cockpit meant that the pilots had to be under six feet tall and weigh less than 175 pounds.

Air force files were screened for possible candidates. The initial list was further reduced by psychological assessments, medical exams, and refinement of the criteria. The final evaluation resulted in sixteen potential pilots, who were then subjected to intensive security and medical checks by the CIA. Those still remaining were approached to work "on a highly classified project involving a very advanced aircraft."[42] In November 1961, five pilots agreed: William L. Skliar, Kenneth S. Collins, Walter L. Ray, Dennis B. Sullivan, and Alonzo J. Walter. They were a mixed group—Skliar was an Air Force Test Pilot School graduate (Class 56D) and was assigned to the Armament Development Center at Eglin Air Force Base.[43] The others had operational backgrounds. Like the CIA U-2 pilots, they were sheep dipped, leaving the air force to become civilians. Their time with the CIA would be counted toward their rank and retirement. The pay and insurance arrangements were similar to those of CIA U-2 pilots.[44]

The CIA A-12 pilots arrived at Groom Lake in February 1963. Like their U-2 counterparts in the 1950s, the men found Area 51 "desert, windy and hot, windy and cold, isolated, basic." They made several flights in the A-12T trainer (also called the "Titanium Goose"), then began making training and test flights in the single-seat A-12s. Each pilot had a personal call sign—"Dutch" followed by a two-digit number. The unit was designated the 1129th Special Activities Squadron, nicknamed "the Roadrunners."[45]

With the deliveries of the J58 engines, and the arrival of the CIA pilots, the program began a three-shift schedule. This required a large number of engineers, who were also recruited in a clandestine manner. One Lockheed engineer was asked if he wanted to work on a "special job." He would be flown to a site, work there all week, then be flown back to Burbank on Friday. In some cases, the engineers were not told what they would be doing until they actually saw the A-12 for the first time.

SIGHTINGS

The A-12 was a large, loud, and distinctive-looking aircraft. Keeping it a secret would be a problem. During the early test flights, the CIA tried to limit the number of people who saw the aircraft. All those at Groom Lake not connected with the Oxcart program were herded into the mess hall before the plane took off. This was soon dropped as it disrupted activities and was impractical with the large number of flights.

As the flights could range across the southwest United States, sightings away from Groom Lake were also a problem. As the A-12 climbed and accelerated, its sonic boom was heard by "the inhabitants of a small village some 30 miles from the test site." A change in the flight path removed this problem.

Although the airspace above Groom Lake was closed, it was near busy Nellis Air Force Base. So, inevitably, there were sightings. In one case, an air force pilot was flying to a gunnery range in the northwest area of the base. He saw an A-12 climbing through his altitude off in the distance. He could see the shape and realized it was some type of experimental aircraft. In another incident, several pilots in a formation saw an A-12. After they landed, a general told all of them that they were to say nothing. Some Nellis pilots saw the A-12 several times. It was common knowledge that something "weird" was going on out in the desert.[46]

To the southwest of Groom Lake was Edwards Air Force Base. NASA test pilots flew numerous X-15 training and support flights to tracking sites and dry lake beds across California and Nevada. The only areas they avoided were the nuclear test site and the Ranch.[47] At least one NASA test pilot saw an A-12. He radioed the Edwards tower and asked what it was. He was curtly told to halt transmissions. After landing, he was told what he had seen was vital to U.S. security. He also signed a secrecy agreement.[48]

Sightings were even made from the ground. At 5:30 A.M., an air force captain was checking the main runway at Edwards for any debris before flight operations began for the day. Suddenly, an A-12 made a low pass and then climbed away. Its shape was so unusual that he first thought it was two planes in close formation. The captain called the tower and asked, "What was that airplane?" The tower radioed back, "What airplane?"

The major source of A-12 sightings was airline pilots. It is believed that twenty to thirty airline sightings were made. One American Airlines pilot saw an A-12 twice. During one sighting, a pilot saw an A-12 and two chase planes; he radioed, "I see a goose and two goslings."[49] Word of these sightings spread among the aerospace community. *Aviation Week and Space Technology* picked up the rumors. The question became how long the secret could be kept.

The security problem became greater on May 24, 1963. Kenneth Collins was flying a subsonic training-test flight in Article 123. As he descended into clouds, the pitot-static tube became plugged with ice, which caused the instruments to display an incorrect airspeed. The A-12 stalled and pitched up. Collins was unable to control the plane, and he ejected. He landed safely, while Article 123 crashed fourteen miles south of Wendover, Utah.[50]

The Nellis Air Force Base base commander was called. "One of your F-105's has just crashed," he was told. He responded, "But that's impossible. They're all here, out on the field." He was curtly told, "Don't argue. If anyone asks about a plane crash, you just report that one of your 105's crashed on a routine training flight north of Nellis."[51]

The F-105 cover story was issued to the press. It took two days to recover the debris. Persons at the scene were requested to sign security agreements. All A-12 aircraft were grounded for a week following the crash. The grounding order was raised once the cause was traced to icing.

The secrecy held despite the crash. The A-12 was ready to begin its quest for Mach 3.

TOWARD THE UNKNOWN

The years 1963 and 1964 were spent bringing the A-12 to Mach 3-plus speeds and operational status. At times, Lockheed and the CIA despaired of ever succeeding. The problem was with the nacelle system and inlet spike. As the A-12 flew faster, the spike moved back; this regulated the airflow into the engine.[52] The flight-test program had to develop the "inlet schedule," which would be programmed into the pneumatic system. This controlled the spike's position, according to the plane's speed.[53]

The A-12's test speed was increased at one-tenth Mach increments. The plane would take off from Groom Lake, then fly north to Wendover, Utah, and onward to the Canadian border. It would then make a 180-degree turn (with a diameter of 128 nautical miles) and head back to Nevada at 65,000 to 72,000 feet. The flight path was called "Copper Bravo." As the A-12 flew back, it would accelerate to the test speed, then decelerate and land. If trouble appeared, at least the plane would be flying toward home at thirty-five miles per minute, rather than away.[54]

The test flights soon showed that the pneumatic system could not compensate for atmospheric changes. The result was an "unstart." The out-of-position spike disrupted the airflow to the engine, which stopped producing thrust and began overheating. The loss of all thrust on one side caused the A-12 to violently yaw toward the dead engine. This literally bounced the pilot's helmet against the canopy. The pilot had to manually open the bypass doors to break the unstart. As with the U-2's early flame-out problems, the unstart would have to be solved before the A-12 could fly in hostile airspace.

Lockheed engineers tried everything they could think of to cure the problem. The inlet geometry and schedules were changed. The manual trim of the fuel flow, spike position, and bypass door position were speeded up. Yet, nothing helped. The two inlets on each A-12 never seemed to match. This resulted in multiple unstarts on each flight. Even a special task force could not find a solution.

Finally, Johnson decided to scrap the existing pneumatic system and replace it with an electronic unit. Even this had problems. During a ground test, the pilot used the radio; this caused a false signal in the electronics and the spike retracted. Once the electronic interference problem was solved, the system proved far more effective, although at the price of much greater maintenance time. The electronic system was retrofitted to the existing aircraft and all new A-12s from Article 129 onward.

The unstart was only the most spectacular of the A-12's problems. The J58 engines' main shaft had to be redesigned to compensate for the high temperatures. The engine mounting points were also changed. The frictional heating raised the cockpit temperature to 130 degrees. On one flight, the control stick became so hot Park had to change hands to keep from burning himself. Changes in the air-conditioning system reduced the cockpit temperature to a "warm but livable" level.[55]

Another change caused by the heat was in the A-12's finish. The prototype had flown in a bare-metal finish without any markings. By late 1963, the edges of the chines, the spikes, and the cockpit area were painted in a heat-resistant black paint. This reduced the internal heating of the airframe. The aircraft also received a full set of national markings. Finding paint that could withstand exposure to high temperatures and fuel was, like everything about the plane, difficult.

A continuing problem was foreign-object damage. Nuts, bolts, clamps, and other debris were sometimes left in the nacelles during construction. When the engines were run up, the debris would be sucked in and damage the engines.[56] In one case, an inspector's flashlight caused $250,000 damage. The engines would also suck rocks, asphalt pieces, and other debris off taxiways and runways.

Changes in procedures were made, such as cleaning the nacelles with 50-horsepower vacuums, then rolling them and listening for anything rattling around.[57] Taxiways and other areas were swept to remove any debris. After landing, covers were put on the inlets and locked with a "great big padlock." They would be unlocked only after the pilot was strapped in the cockpit for the next flight.[58]

Finally, after fifteen months of painful flight testing, the A-12 was ready to attempt Mach 3. The flight was made on July 20, 1963, by Lockheed test pilot Louis Schalk. Additional Mach 3 flights were made during the summer and fall.[59] In November, the design speed of Mach 3.2 was reached. It had taken sixty-six speed-buildup flights to go from Mach 2 to Mach 3.2.[60]

In these buildup flights, the peak speed was held only momentarily. The next step was sustained Mach 3 flight. This was much more demanding than a brief dash, as the heat would soak into the plane's structure. Lockheed test pilot James Eastham made the first sustained Mach 3 flight on February 3, 1964. The plan envisioned a peak speed of Mach 3.16, which would be held for ten minutes. Eastham began the speed run at 78,000 feet. By the end of the run, the A-12 had climbed to 83,000 feet. Eastham cut the throttles and landed at Groom Lake, the end of what seemed to be a completely successful flight. Lockheed's senior flight test engineer, Glen Fulkerson said, "Turn it around, we'll fly it tomorrow." But it would be another eight weeks before the A-12 would fly again.

During the postflight inspection, it was found that the plane had been "burned to a crisp." There had been an error in the air-speed system: rather than Mach 3.16, the plane had actually approached Mach 3.3. The heating had been far higher than predicted. The wiring had been damaged by 800-degree F temperatures. Nearly all the hydraulic fluid had been lost from the four flight-control systems—only one-half gallon remained out of the original seven and a half gallons. Eastham recalled years later, "About fifteen more seconds at speed and I think I would have been out walking."

The engineers did not know where the hydraulic fluid had gone. There were no leaks in the ground tests. Finally they used heat lamps to raise the temperature to 600 degrees F. As the joints expanded, the 3,300 psi hydraulic fluid literally flowed out. The plane was surrounded by smoke from the vaporizing fluid. Once the hydraulic system cooled, the leaks closed up. The test pilots insisted a hydraulic fluid quantity gauge be added before another Mach 3 flight was made. Article 121 was fitted with the gauges (the only A-12 so equipped). After several maintenance flights, they were ready to try again.

The A-12 took off at first light with Eastham at the controls for a thirty-minute flight. During the Mach 3 run, no leaks appeared and the communications checks were successful. Eastham cut the throttles and descended toward a landing. As the A-12 turned onto the downwind leg, the left hydraulic system failed. Eastham thought, "Oh boy, here we go again." Despite the failure, he landed successfully. A postflight inspection found that the brake manufacturer had put an aluminum plug in the hydraulic system. The high temperature and pressure had blown it out. At the next Monday morning technical meeting, Johnson asked, "How the hell did a piece of aluminum get in this plane?"[61]

A-12 DERIVATIVES

By the end of 1963, nine A-12s were at Groom Lake. They had also been joined by a derivative, the YF-12A interceptor. As the Oxcart program got under way, Johnson realized the basic A-12 airframe had a tremendous growth potential. The first derivative was to be an air-defense interceptor. The aircraft would use its Mach 3 speed to fly out to incoming Soviet bombers, which could then be destroyed well before they neared their targets.

This plane was the ultimate expression of a trend in fighter development under way since the early 1950s. The traditional fighter, with an emphasis on maneuverability, had been replaced by all-missile-armed interceptors. They were not meant to attack other fighters, but rather large, nonmaneuvering bombers. It was widely accepted that any future war would be a nuclear "high noon" with the Soviet Union. There would be no "limited wars" such as Korea. Accordingly, fighter aircraft were oriented toward either nuclear delivery or interception. Ultimately, it was thought, manned aircraft would be replaced by guided missiles.

On March 16 and 17, 1960, Johnson had met in Washington, D.C., with Air Force Brig. Gen. Howell M. Estes Jr. and Courtlandt Perkins, Air Force Secretary for Research and Development, to discuss the interceptor. They were impressed and told Johnson to meet with Gen. Marvin Demler at Wright-Patterson Air Force Base. On October 31, 1960, Lockheed received a contract to build three of the Improved Manned Interceptors. The program was kept separate from the A-12. The planes were built in the same building that held the A-12 assembly line but were screened off to prevent anyone working on one project from knowing about the other.[62]

The A-12's nose and forward chines were cut back and replaced by a bulky-looking radome and two infrared scanners. The aircraft carried a Hughes pulse-doppler AN/ASG-18 radar and fire-control system; this could detect and track aircraft flying at low altitude and long range. The aircraft carried AIM-47 Falcon missiles with either a high-explosive or a low-yield nuclear warhead. The missile had a range of 120 miles and was carried internally. The fuselage underside was redesigned to accommodate the three missile bays. A second seat for a radar intercept officer (RIO) was added to the Q-bay behind the pilot's cockpit. The addition of the radome reduced high-speed stability, but this was corrected with the addition of two fins under the nacelles and a large retractable fin under the rear fuselage.[63]

In early January 1961, several months after the go-ahead was given for the interceptor, Johnson proposed two more A-12 derivatives—the B-12 strategic bomber and the R-12 reconnaissance aircraft. The B-12 design envisioned the nuclear weapons being carried in fuselage bays. The B-12 could carry four short-range attack missiles (SRAMs), six strike missiles, or twelve guided bombs. Targets would be picked up by radar in an A-12-like pointed nose. A second crewman would target the weapons. The B-12's mission would be "cleanup" after the initial ICBM strike. Targets in Eastern Europe and the western Soviet Union could be hit, to include command posts, MiG 25 air bases, missile sites, submarine pens, and SAM sites. For the R-12 reconnaissance version, the weapons would be replaced by cameras, radar, and electronic intelligence (ELINT) equipment. To control the reconnaissance gear, a second crewman was added. The R-12's wartime mission would be poststrike reconnaissance. Internal stowage of the weapons and reconnaissance equipment would keep weight down and preserve the low radar cross section. A pod, like that on the B-58, was not considered.

A mock-up of the "Reconnaissance Strike-71" (RS-71) was inspected by the air force on June 4, 1962. The concept of a strike A-12 ran into problems from two directions. Some in the air force saw it as a threat to the XB-70 program (also called the "RS-70"). More important, Defense Secretary Robert S. McNamara and his "whiz kids" saw no need for manned bombers. In the coming years, the entire B-47 force, as well as early model B-52s, would be retired, and the XB-70 program reduced to an aeronautical research program.

Accordingly, only the reconnaissance version of the RS-71 remained. (It kept the "strike" part of the name, however.) Externally, the plane resembled the A-12, with the nose slightly less pointed and the tail cone extended to hold more fuel. The A-12 was designed for clandestine overflights, with the minimal payload of a single camera. The RS-71, in contrast, carried a larger payload. The nose was removable and could carry either a high-resolution radar or a panoramic camera. Additional cameras or reconnaissance equipment could be carried in the chines. This gave it a much greater capability, not just for poststrike reconnaissance and overflights, but also for peacetime flights along the Soviet border.[64]

On December 27–28, 1962, a contract was issued to Lockheed to build six test RS-71s. By this time, the A-12 had made its first flights and was about to begin its journey toward Mach 3. The first YF-12A, as the interceptor was now known, was also nearing completion.

The first YF-12A (Article 1001) was moved to Groom Lake in July 1963. Final assembly was completed, and it made its first flight on August 7. The pilot was Eastham, who had written the manual and had made flight tests of the radar and missile systems in a modified B-58. During the flight, the YF-12A went supersonic. The second YF-12A (Article 1002) was flown on November 26, 1963, by Schalk. Because the A-12 had already proved the design's Mach 3 performance, the initial YF-12 flights were tests of the radar systems.[65]

SURFACING THE OXCART

Despite all that had happened—the crash of Article 123, the achievement of Mach 3, the sightings and rumors, the *Aviation Week* report, and the hundreds of people involved in building the A-12—the secrecy held. Because the A-12 was the only aircraft capable of Mach 3 cruise flight, its technology would be very valuable for the emerging U.S. supersonic-transport program. But the data could not be used as long as the plane remained secret.

And 1964 was also a presidential election year.

Lyndon B. Johnson had been briefed on the A-12 project a week after becoming president. He directed a plan be developed for an announcement in the spring of 1964.[66] The expected Republican candidate was Barry Goldwater, a right-wing senator who had long accused the Democrats of being soft on defense and communism. "Surfacing" the A-12 was an obvious tool to counter such charges. (Although started under Eisenhower, a Republican, the plane had first flown under John F. Kennedy, a Democrat.)

By the end of February 1964, the time was judged right. At a February 29 meeting of the National Security Council, the members were briefed on the A-12 by McCone and McNamara. They were then asked for approval for a public announcement.[67] Later that day, President Johnson read a statement to the press.

> The United States has successfully developed an advanced experimental jet aircraft, the A-11, which has been tested in sustained flight at more than 2,000 miles per hour and at altitudes in excess of 70,000 feet. The performance of the A-11 far exceeds that of any other aircraft in the world today. The development of this aircraft has been made possible by major advances in aircraft technology of great significance for both military and commercial applications. Several A-11 aircraft are now being flight tested at Edwards Air Force Base in California. The existence of this program is being disclosed today to permit the orderly exploitation of this advanced technology in our military and commercial program.

The debut generated considerable press attention. The stories claimed the United States had a dozen "A-11s" flying (true) and that they had already made overflights (false). President Johnson's use of "A-11" was deliberate. This was the original designation of the Oxcart, before the antiradar modifications were made. Should A-12 become public, it would appear it was a follow-on to the A-11, rather than the original airplane.[68]

The personnel at Groom Lake knew an announcement was near but did not know the exact timing. Accordingly, they were taken by surprise. No A-12 or YF-12A had ever operated from Edwards Air Force Base, so the two YF-12s were hurriedly flown over by Schalk and Park. They taxied up and made a 180-degree turn in front of their new hangar. As they did, the hot exhaust was blown into the hangar and triggered the fire extinguisher valves. Water came flooding down.[69]

The third YF-12A (Article 1003) made its first flight on March 13, 1964, and was soon transferred to Edwards. The planes embarked on tests of its missile system. It proved successful in launching missiles at Mach 3 and intercepting the target aircraft. But this was already a dead issue—McNamara had no more interest in air defenses than he did in manned bombers. His "cost-effectiveness" studies concluded that the Soviet bomber force was a minimal threat. Over the next several years, McNamara withheld funding to build ninety-three improved F-12B interceptors, even though funding had been approved by Congress. The existing F-101, F-102, and F-106 air-defense squadrons, radar sites, and SAM sites were closed over the next several years. The Soviets, on the other hand, were starting an aggressive bomber-development program. This resulted in the Tu-22 Backfire and the Tu-160 Blackjack—exactly the aircraft the F-12B was designed to intercept.

Four months after the A-11 announcement, there was another. According to legend, Johnson asked an aide what the RS-71 was for. The aide responded, "strategic reconnaissance." Thus, when he announced the existence of a new reconnaissance aircraft, on July 24, 1964, President Johnson called it the "SR-71." As a result of switching the letters, twenty-five thousand drawings had to be changed.[70] The prototype SR-71 (Article 2001) was delivered to Palmdale, California, on October 29, 1964. It made its first flight on December 22, 1964, with Robert Gilliland at the controls. Unlike the secret flights from Groom Lake, the SR-71's takeoff could be seen from the surrounding area.[71]

Later that same day, another member of the A-12 family also made its first, secret, flight.

President Johnson's announcements created an unusual security situation. Both the YF-12A and the SR-71 were White (i.e., the fact they existed was not a secret). In contrast, the A-12 was still Black. Its existence would remain a secret until 1981. To maintain the secret, all those involved were told of the coming A-11 announcement and warned to keep the A-12 separate.

One aspect of this effort was the A-12's paint finish. From 1963 and into 1965, they had a half bare metal, half black finish. "U.S. Air Force," "USAF," and the serial numbers were in black. The SR-71 made its first flight in an all-black finish with white lettering. The A-12s were soon painted in an identical scheme. This improved temperature control, as the black paint radiated heat better than the bare metal. It also meant that it was nearly impossible to tell the difference between an A-12 and an SR-71 at any distance. (The two aircraft had nearly identical shapes.)

On July 9, 1964, Park was involved with the second A-12 crash. He had completed a high-Mach check flight in a new aircraft, Article 133, and was on approach to the runway. At an altitude of 500 feet, the plane began a roll to the right. Park could not control it. When the plane reached a bank angle of 45 degrees and was only 200 feet above the ground, he ejected. Park separated from the seat and his parachute opened. As he swung down to the vertical, his feet touched the ground. The official history called it "one of the narrower escapes in the perilous history of test piloting." The A-12 hit the ground and exploded. The crash was traced to the right outboard roll and pitch control, which had frozen up. No word of the crash leaked out.[72]

WAITING TO LEAVE THE NEST

In early 1964, with a limited Mach 3 capability at hand, CIA headquarters began thinking about Cuban overflight missions. For several months, Fidel Castro had been threatening to attack the U-2s making overflights of Cuba. Secretary of State Dean Rusk suggested to President Johnson that a diplomatic note be sent to the Cubans, warning against any attempt to shoot down the planes. He added that, as a further deterrent, Castro should be given the word through Black channels that the United States had taken note of the statements, interpreted them as a threat, and that "we would like nothing better, and we are prepared to react immediately to such an eventuality."[73]

In reality, there was little besides diplomatic warnings and veiled threats to protect the U-2s. Adding electronic countermeasures (ECM) equipment to jam the SA-2 radars was studied, but it was determined this would not keep them safe from attack.[74] The U.S. government began looking at alternative means of making the overflights. The A-12 was finally selected.

The initial plan for contingency A-12 overflights of Cuba was code-named "Skylark." Park's crash delayed the plan for a time, but on August 5, acting CIA Director, Gen. Marshall S. Carter, ordered that Skylark have an emergency operational readiness by November 5, 1964. To meet the deadline, the five CIA pilots would have to be qualified to Mach 2.8 speed and 80,000 feet altitude. The camera system would also have to be proven. One major problem was the lack of ECM equipment for the A-12s. Only one complete set of ECM gear would be ready by November. An intra-agency group was organized to study the risk. They decided the first few Cuban overflights could be made safely without full ECM equipment. Later overflights, however, would require full defensive setup. The ECM delivery schedule could meet this requirement.

After completing training missions that simulated the Cuban overflights, the CIA A-12 unit was judged ready to undertake Skylark. It would take two-weeks' notice before an overflight could be made, and it would be done with fewer pilots and aircraft than had been planned. The next step was to convert this emergency capability into a sustained program. Training flights were conducted to determine range and fuel consumption of the A-12, to finish qualification of the pilots, and to prepare a number of Skylark mission profiles. By the end of 1964, five pilots and five aircraft stood ready to undertake sustained Cuban overflights. There were now eleven A-12s at Groom Lake—four test aircraft and seven for the CIA detachment.[75]

As Skylark was becoming operational, the Lockheed test pilots were bringing the A-12 up to its full design capability. On January 27, 1965, one of the test aircraft made the first high-speed, long-range flight. The mission took one hour and forty minutes and reached speeds above Mach 3.1, at between 75,600 and 80,000 feet.

With Skylark, the continued A-12 test flights, and the start of SR-71 test work at the site, Groom Lake was at a peak of activity during 1965.[76] Construction was finished, and the population reached 1,835 (equivalent to a small town). Lockheed-operated Constellation airliners made daily flights between Burbank and Groom Lake. There were also twice-daily C-47 flights to Las Vegas.[77]

The Groom Lake facility had grown considerably since the U-2 days. The original U-2 facility at the edge of the lake bed was much expanded, with four new, larger hangars. Just south of this was the housing area, with neat rows of buildings. Conditions at Groom Lake were more livable—a movie theater replaced the projector on a mess hall table, and a baseball diamond was built. Nonetheless, the site was still isolated, hot, and barren. At the south end of the facility was the A-12–Lockheed area. It included individual hangars, each of which housed an A-12. The hangars provided protection against both the sun and blowing dust, as well as hiding the aircraft from the prying cameras of Soviet reconnaissance satellites. The main run way ran up to the edge of the lake bed. A long asphalt overrun strip extended out across the lake bed.[78]

The year 1965 also saw recruitment and training of the second group of CIA A-12 pilots. There were only three members, all from operational backgrounds—Mel Vojvodich, Ronald J. Layton, and Jack C. Weeks.[79] As had the XP-59A pilots of two decades before, they lived a unique existence. They were flying the fastest airplane in the world, but not even their wives knew what they were doing. Like the Bell pilots, they used symbols to define their secret brotherhood. These took the form of flight suit patches. After seeing the A-12, Weeks dubbed it "Cygnus," after the constellation of the swan. Patches showing the constellation and "Cygnus" were made. Another patch showed a cartoon roadrunner (the unit's nickname) and the words "Road Runners" and "Beep Beep." An emblem showed a swan-shaped dragster and the words "1129th SAS The Road. Runnin'est."[80]

While the CIA pilots awaited orders to overfly Cuba, a new target appeared for the A-12. On March 18, 1965, CIA Director McCone warned McNamara that reconnaissance operations over Communist China were facing increased threats. Since 1962, four U-2s flown by Nationalist Chinese pilots had been shot down over the mainland. The A-12 was the clear alternative. It was decided to start construction of the facilities needed for the A-12 on Okinawa. This stopped short of deploying the A-12, however; a decision to overfly China could be made only by the president.

Four days later, the A-12 operational plan, code-named "Black Shield," was ready. Initially, three A-12s would be deployed for a sixty-day period, twice a year. The aircraft would fly from Kadena Air Base on Okinawa;

later, a permanent detachment would be established at Kadena. Funding was released, with the A-12 support facilities to be ready in the fall of 1965.[81]

Reconnaissance operations over North Vietnam also were being threatened. On April 5, 1965, an SA-2 SAM site was photographed in North Vietnam. Soon, more sites were spotted near Hanoi. General William Westmoreland, commander of U.S. forces in South Vietnam, wanted to attack the sites but was ordered not to by McNamara. His rationale was this: the SA-2s had been sent by the Soviets to appease the North Vietnamese; if the United States did not attack the SAM sites, it would send a "signal" to the North Vietnamese, who would then not use the SAMs. (Johnson and McNamara saw the Rolling Thunder bombing campaign as a means of "signaling" North Vietnam to negotiate.)

Despite his belief that North Vietnam would not use the SAMs, McNamara asked the under secretary of the air force on June 3, 1965, about using the A-12 to replace the U-2 missions. He was told the A-12 could begin operations over North Vietnam as soon as the final qualification flights had been made. On July 24, 1965, the North Vietnamese, ignoring the U.S. signal, used the SA-2s to shoot down an air force F-4C. More U.S. aircraft were lost, and it was clear the days of the U-2 over North Vietnam were numbered.[82]

With deployment of the A-12 seemingly at hand, the CIA unit began the final steps toward operational status. The first three "H cameras" were delivered in April 1965. Originally designed for the U-2, the H camera had a sixty-inch focal-length lens. It used Kodak 3414 film, which had a low ASA 8 rating and a frame size of four and a half inches square. The resolution from 80,000 feet was two inches. The other camera that could be carried in the A-12's Q-bay was the KA-102A. This had a forty-eight-inch focal length and carried a seven-hundred-foot-long roll of film that could provide 1,675 frames. Both the H camera and the KA-102A used a motion compensator to prevent the A-12's high speed from blurring the photos.[83]

The pilots also began the final qualification flights. These were to prove aircraft and system reliability at speeds of Mach 3.05, altitudes of 76,000 feet, at a range of 2,300 nautical miles, with three aerial refuelings. These longer flights revealed new problems. The most important were with the electrical wiring. It was exposed to prolonged temperatures of more than 800 degrees F, as well as flexing of the structure, vibration, and shock. The wiring could not withstand the conditions, which caused malfunctions in the inlet controls, communications equipment, ECM systems, and cockpit instrumentation. There were also continued problems with the fuel tank sealing.

The problems, severe enough to threaten the Black Shield schedule, were traced to poor Lockheed maintenance. On August 3, 1965, CIA Deputy for

Technology John Paragosky met with Kelly Johnson. They had a "frank discussion" on what was needed to fix the shortcomings. Johnson decided he would have to personally supervise activities at Groom Lake on a full-time basis; the following day, he began working at the site. The official history said Johnson's "firm and effective management" put Black Shield back on schedule.

Four A-12s were selected to make the deployment. During the final qualification flights, the A-12 reached a speed of Mach 3.29 and an altitude of 90,000 feet. The maximum duration above Mach 3.2 was one hour and fourteen minutes. The total flight duration was six hours and twenty minutes.

On November 22, 1965, Kelly Johnson wrote: "Over-all, my considered opinion is that the aircraft can be successfully deployed for the Black Shield mission with what I would consider to be at least as low a degree of risk as in the early U-2 deployment days. Actually, considering our performance level of more than four times the U-2 speed and three miles more operating altitude, it is probably much less risky than our first U-2 deployment. I think the time has come when the bird should leave its nest."

The decision for the A-12 to "leave its nest" rested with the 303 Committee, the board that oversaw intelligence operations. On December 2, the 303 Committee received a formal request that the A-12 be deployed to Kadena. The committee refused but ordered that a quick-reaction capability be established. This would allow the A-12s to deploy within twenty-one days of an order, any time after January 1, 1966.

The year ended on a sour note. On December 28, 1965, Vojvodich took off in Article 126 to make a check flight after major maintenance. Seven seconds after he left the ground, the plane went out of control. Vojvodich had no chance to deal with the problem and ejected at an altitude of 150 feet. He narrowly missed the fireball as Article 126 exploded, but he survived unharmed.[84]

The accident investigation board found that a flight-line electrician had reversed the connections of the yaw and pitch gyros, which reversed the controls. CIA Director McCone ordered the Office of Security to investigate the possibility that it had been sabotage. No evidence was found, but they discovered the gyro manufacturer had earlier warned such an accident was possible. No action (such as color-coding the connections) had been taken on the warning. As with Park's crash the year before, no word leaked out about the accident.[85]

Throughout 1966, there were frequent requests to the 303 Committee to allow the A-12 to be deployed. The CIA, the Joint Chiefs of Staff, and the President's Foreign Intelligence Advisory Board all favored the move, while the State and Defense Departments opposed it. The A-12's supporters argued that there was an urgent need for intelligence data on any possible Chinese moves to enter the Vietnam War. Those opposed to deployment felt the need was not sufficient to justify the risks to the aircraft, and the political risks of basing it on Okinawa. Japan had powerful left-wing groups who were protesting U.S. involvement in Vietnam. On August 12, 1966, the disagreement was brought to President Johnson, who refused to approve deployment.

As the 303 Committee debated, the Black Shield plan was further refined. The new plan cut the original twenty-one-day deployment time nearly in half. The first loads of personnel and equipment would leave Groom Lake for Kadena on the day deployment was approved. On the fifth day, the first A-12 would takeoff on the five-hour-and-thirty-four-minute, 6,673-mile flight. The second A-12 would follow on the seventh day, and the third on the ninth day. Two A-12s would be ready for an emergency overflight eleven days after approval was given. A normal mission could be flown after fifteen days. A Skylark mission over Cuba could be flown seven days after the go-ahead.

The A-12 also showed what it could do. On the morning of December 21, 1966, Park took off from Groom Lake. He flew north to Yellowstone National Park; turned east to Bismark, North Dakota, and Duluth, Minnesota; then flew south to Atlanta, Georgia, and on to Tampa, Florida. He turned west, flying across the country to Portland, Oregon, then south to Nevada. He again turned east, flying to Denver, Colorado; St. Louis, Missouri; and Knoxville, Tennessee. He turned west, passing Memphis before finally landing back at Groom Lake. The flight covered 10,198 miles, involved four flights across the United States, several in-flight refuelings, and still had taken only six hours.[86]

But following this success, the Oxcart program had its first fatal accident. On January 5, 1967, Walter Ray was flying a training mission in Article 125. As he descended, a fuel gauge malfunctioned, and the plane ran out of fuel about seventy miles from Groom Lake.[87] Ray ejected, but the seat separation device failed when his parachute pack became wedged against the head rest. He died when the seat hit the ground.[88]

The air force made an announcement that an SR-71 on a routine test flight out of Edwards Air Force Base was missing and presumed down in Nevada. The pilot was described as a civilian test pilot, and newspapers assumed he was with Lockheed.[89] The wreckage was found on January 6, and Ray's body was recovered the next day. The A-12s were grounded pending an investigation of the fuel gauge and ejector seat failures.

The third and final group of CIA A-12 pilots began training at Groom Lake in the spring of 1967. They were David P. Young, Francis J. Murray, and Russell J. Scott. Scott was an Air Force Test Pilot School graduate

(Class 62C and ARPS IV) while the others came from operational backgrounds.[90]

BLACK SHIELD

In May of 1967, the roadblock to the Black Shield deployment finally ended. Fears began to grow that surface-to-surface missiles might be introduced into North Vietnam. Aggravating matters were concerns that conventional reconnaissance aircraft lacked the capability to detect such weapons. President Johnson requested a study of the matter. When told that the A-12's camera was far superior to those on the U-2, and that the plane was less vulnerable, State and Defense representatives who had opposed deployment began to reconsider. CIA Director Richard Helms submitted another proposal to the 303 Committee for A-12 deployment. He also raised the issue at President Johnson's "Tuesday lunch" on May 16. Johnson finally agreed to the deployment. The formal approval was made later that day. Black Shield was under way.[91]

The airlift to Kadena began the next day. On May 22, the first A-12, Article 131, was flown by Vojvodich from Groom Lake to Kadena in six hours and six minutes. Layton piloted Article 127 to Kadena on May 24, while Article 129 with Weeks as pilot, left on May 26. Following a precautionary landing at Wake Island, it continued on the following day. By May 29, 1967, the A-12 Oxcarts were ready to make their first overflight. After ten years of work, it was time.

Project Headquarters in Washington, D.C., had been monitoring the weather over North Vietnam. At the May 30 mission alert briefing, the weather was judged favorable, and the A-12 unit was ordered to make an overflight the next day. The alert message also contained the specific route it was to take. At Kadena, the message set events in motion. Vojvodich was selected as the primary A-12 pilot with Layton as the backup pilot. The two planes, a primary and backup A-12, were inspected, the systems were checked, and the camera was loaded with film. Like the CIA U-2s, these planes carried no national markings, only a black paint finish and a small five-digit serial number on the tail fins.

Twelve hours before the planned takeoff time (H minus twelve), a second review of the weather was made. The forecast continued favorable, and the two pilots were given a detailed route briefing during the early evening. On the morning of May 31, the pilots received a final preflight briefing—the condition of the two aircraft was covered, last-minute weather and intelligence reviewed, and any changes in the flight plan gone over. At H minus two hours, a final "go-no-go" review of weather was made by headquarters. This covered not only North Vietnam, but the refueling areas and the takeoff and landing sites. The only problem was at Kadena—it was raining heavily. Ironically, after all its testing, the A-12 had never flown in the rain. The target area weather was clear, however, and the decision was made to carry out the flight. A "go" message was sent to Kadena.

With the final authorization, Vojvodich underwent a medical examination, got into his pressure suit, and was taken out to the primary aircraft, Article 131. If any problem appeared in the preflight checkout, the backup plane could be ready to make the overflight one hour later. Finally, with rain still falling, the A-12 taxied out, ignited its afterburner, and took off into the threatening skies.

The first Black Shield mission made two passes. The first went over Haiphong and Hanoi and left North Vietnam's airspace near Dien Bien Phu. Vojvodich refueled over Thailand, then made a second pass over the Demilitarized Zone. The route covered 70 of the 190 known SAM sites, as well as 9 other primary targets. The photos were judged "satisfactory." The runs had been made at a speed of Mach 3.1 and an altitude of 80,000 feet. No radar signals were detected; the mission had gone unnoticed by the North Vietnamese and Chinese. The total flight time was three hours and forty minutes. Vojvodich needed three instrument approaches amid driving rain before landing back at Kadena.

Between May 31 and July 15, a total of fifteen Black Shield missions were alerted. Of these, seven were flown. Four of the overflights detected radar-tracking signals, but none of the A-12s were fired on. By mid-July it was clear there were no surface-to-surface missiles in North Vietnam. The early overflights showed how good the A-12 was, and the hesitation to use it ended.

Between August 16 and the end of the year, twenty-six missions were alerted and fifteen were flown. A typical Black Shield mission would involve an aerial refueling south of Okinawa soon after takeoff, one or two photo passes, and a second refueling over Thailand before the return to Kadena. Due to the plane's huge turning radius, some mission profiles required the A-12 to enter Chinese airspace. On a single-pass mission, the A-12 would spend only twelve and a half minutes over North Vietnam. If two passes were made, the A-12 would spend twenty-one and a half minutes in hostile airspace. Once back at Kadena, the exposed film would be unloaded and placed aboard a special plane for shipment to the processing facility. For the first overflights, this was the Eastman Kodak plant in Rochester, New York. By late summer, an air force processing center had been set up in Japan. The data would be in the hands of U.S. commanders within twenty-four hours of an overflight.

Despite the speed and altitude of the A-12, the risks of overflights were clear. On September 17, a SAM site tracked an A-12 with its acquisition radar. The Fan Song guidance radar was unable to gain a lock on the plane, however. On October 28, a North Vietnamese SAM site fired a single SA-2 at an A-12 flown by Sullivan. The plane's camera photographed the smoke from the site, then the missile's contrail. The ECM equipment worked well and the SAM missed.[92]

Sullivan's next overflight, on October 30, 1967, resulted in a "hit." On his first pass, Sullivan noted the plane was being tracked, with two SAM sites preparing to fire. On the second pass, as he flew toward Hanoi from the east, the North Vietnamese were ready—at least six SA-2s were fired at the A-12. This was the first of many concerted efforts to bring down an A-12/SR-71. Sullivan saw contrails and the detonation of three missiles.[93] The bursts appeared, then seemed to collapse instantly as the A-12 sped away.[94] Unlike a tactical fighter, the A-12 could not evade a missile by maneuvering. The pilot had to continue on his course and trust the ECM equipment would protect him.[95]

When Sullivan landed back at Kadena, a postflight inspection discovered a piece of metal had hit the lower right-wing fillet area and become lodged against the wing tank support structure. The fragment was not a warhead pellet, but a very small piece of the brass fuze from one of the missiles. This was the only hit scored on an A-12 or SR-71 in over one thousand overflights. Sullivan kept the fragment as a souvenir of the mission.

The new year brought new crises, and the A-12 was in the midst of them. On January 23, 1968, the U.S.S. *Pueblo* was captured by the North Koreans. There were fears that this was the prelude to full-scale military action, and an A-12 overflight of North Korea was authorized. On January 25, the first attempt was made by Weeks, but a problem caused an abort shortly after takeoff. The next day, January 26, Murray took off. The mission was to locate the ship and then determine if an invasion of South Korea was about to occur. Murray made his first pass down the east coast of Korea: "As I approached Wonsan I could see the *Pueblo* through my view sight. The harbor was all iced up except at the very entrance and there she was, sitting off to the right of the main entrance."

Murray made a total of four passes over North Korea, from the DMZ to the Yalu River, covering the entire country. The A-12 was tracked by the Chinese, but no missiles were fired. When the photos were analyzed, they showed no evidence that a ground attack was imminent.[96]

A second overflight of North Korea was requested, but the State Department was reluctant, fearing political problems should the A-12 be shot down. Secretary of State Dean Rusk was briefed on the mission: the plane would spend only seven minutes over North Korea, and even if a problem occurred, it was highly unlikely the A-12 would land inside China or North Korea. Even so, Rusk suggested changes in the flight path before giving his approval. Rusk thus became the A-12's highest-ranking flight planner. (This was not unique; President Eisenhower had made changes in U-2 overflights.) The mission was flown by Layton on May 8, 1968.

Between January 1 and March 31, 1968, four North Vietnamese overflights and one North Korean overflight were made (out of fifteen alerted). Between April 1 and June 9, 1968, two North Korean overflights were alerted; only the May 8 mission was flown. May 8 also was the last A-12 overflight. In all, the A-12 made twenty-six overflights of North Vietnam and two of North Korea. No overflights of China were made (although several flights did enter Chinese airspace during turns). No Skylark missions over Cuba were flown, as the U-2 proved adequate. Now, the A-12 Oxcart was to pass from the scene.[97]

THE END OF OXCART

Starting in November 1965, even as the A-12 was declared operational, doubts were expressed about the cost of operating the two separate groups of A-12s and SR-71s. After a year or more of debate, it was decided on January 10, 1967, to phase out the A-12 program. The first four A-12s were to be put in storage in July 1967, another two by December, and the final two by the end of January 1968. At the same time, the SR-71s would be phased into operation. This, it should be noted, was before the A-12 had undertaken a single overflight.

Once the overflights began, the A-12 demonstrated an exceptional technical capability. As the scheduled phaseout date neared, concerns were expressed by high officials. Walt Rostow, the president's special assistant, members of the president's Foreign Intelligence Advisory Board, the president's Scientific Advisory Committee, and several congressmen all expressed doubts about the phaseout.

In the meantime, SR-71s began arriving at Kadena, starting in early March 1968. The first SR-71 overflight of North Vietnam was made on March 21, 1968. By gradual stages, they took over the Black Shield mission, until the A-12 became the backup to the SR-71. After the final North Korean overflight on May 8, the unit was told to prepare to return home.

Eight days after the final A-12 overflight, Defense Secretary Clark Clifford reaffirmed the phaseout decision. On May 21, President Johnson agreed that the A-12s would be put into storage. The aircraft at Groom Lake would be placed in storage at Palmdale by June 7. The A-12s at Kadena would be restricted to flight safety and pilot proficiency missions; June 8 was selected as the date they would return to the United States.[98]

Virtually on the eve of the return, the A-12 program suffered its final loss. On June 4, 1968, Jack Weeks took off in Article 129 to make a check flight. An engine had been changed, and it had to be tested before the redeployment. Weeks was last heard from when the plane was 520 miles east of Manila. Then all contact was lost. No debris was found, nor was a cause ever determined. An air force press release identified the plane as an SR-71.

A few days later, the final two A-12s returned to Groom Lake. The final A-12 flight was made on June 21, 1968, when Article 131 took off from the Ranch. Frank Murray landed it thirty-five minutes later at Palmdale. The first A-12 to make an overflight, and the last to fly, had its fuel and oil drained. It was then placed in storage. The Oxcart story had ended.[99]

On June 26, 1968, an awards ceremony was held at Groom Lake. Vice Admiral Rufus L. Taylor, deputy director of Central Intelligence, presented the CIA's Intelligence Star for courageous action to Kenneth S. Collins, Ronald J. Layton, Francis J. Murray, Dennis B. Sullivan, Mel Vojvodich, and, posthumously, to Jack C. Weeks, for their roles in the Black Shield missions. Weeks's widow accepted his award. Colonel Hugh C. Slater, commander of the Kadena detachment, and his deputy, Col. Maynard N. Amundson, received the Air Force Legion of Merit. The 1129th Special Activities Squadron and its support units received the U.S. Air Force Outstanding Unit Award. The wives of the pilots were also present and learned for the first time just what their husbands had been doing for the past several years.[100]

Although the Oxcart was gone, its descendant, the SR-71, would continue to fly intelligence missions for the next twenty-two years. It covered trouble spots such as North Vietnam, North Korea, the Mideast, Libya, kept watch on Eastern European borders, and tracked Soviet submarines. Finally, in 1990, the SR-71 was retired. Like the A-12, the reason was cost. The surviving A-12s and SR-71s were originally to be scrapped, but the air force relented, and they were sent to museums.

The A-12 was the most exotic Dark Eagle ever built. No other Black airplane has posed so great an aerodynamic and engineering challenge. The A-12 was the final expression of a trend that had been under way since World War I—aircraft trying to evade air defenses by going higher and faster. This was true of the B-17s and B-29s of World War II. As the Cold War began, B-47s, B-52s, and British V-bombers flew at 50,000 feet and near supersonic speeds. The B-58 raised this to supersonic speeds, while the U-2 could reach altitudes of 70,000-plus feet.

It was the Soviet SA-2 SAM that brought this era to a close. Bombers would now have to attack at low altitudes—a few hundred feet above the ground. The Mach 3 XB-70 was canceled, while the B-58's service life was cut short.

The A-12 was the last of its line. The Oxcart was so much faster, flew so much higher, and had a reduced radar return. The combination of these factors resulted in an airplane that was unstoppable. For so critical a mission, it was possible to justify so expensive and specialized an aircraft.

But there were other trends in Black reconnaissance aircraft . . .

CHAPTER 4

Alone, Unarmed, Unafraid, and Unmanned
The Model 147 Lightning Bug

Probe him and learn where his strength is abundant and where deficient.

Sun Tzu
ca. 400 B.C.

Black reconnaissance aircraft, such as the U-2 or A-12, faced two basic problems. The first was technical—to build an aircraft with altitude and speed performance superior to enemy air defenses. The second proved more difficult—to convince the president that the risks of a plane being lost and the pilot being captured were acceptable. Reconnaissance pilots said they flew their missions, "Alone, Unarmed, Unafraid." During the 1960s and early 1970s, a series of Dark Eagles added "Unmanned" to this motto. These drones were used to cover targets that were too heavily defended, or too politically sensitive, to risk a manned reconnaissance aircraft.

BEGINNINGS, 1959–1962

As with the U-2, the effort began small. In September 1959, Col. Harold L. Wood, chief of the Reconnaissance Division at Air Force Headquarters, and his deputy, Lt. Col. Lloyd M. Ryan, met with Raymond A. Ballweg Jr., vice-president of Hycon Manufacturing Company, which made the U-2s' cameras. The conversation came around to the risk of manned reconnaissance. Ballweg responded, "Hell, Lloyd, why don't you have us install a camera in a jet target drone? No reason it can't be programmed to do the recon job for you and bring back pictures." Colonels Wood and Ryan asked, "What drone?" Ballweg explained that Ryan Aeronautical Company built the Q-2C Firebee jet-powered target drone, which might be usable.

Several weeks later, Lieutenant Colonel Ryan made a call to Ryan Aeronautical to suggest a meeting. At first there seemed little interest in a photo reconnaissance Firebee; finally, an agreement was reached for Ryan Aeronautical and Hycon to do a joint study. As with other Black airplane projects, a small group would conduct the effort. On January 21, 1960, Robert R. Schwanhausser was named to head the reconnaissance drone group. He was told to take six or eight people and get started. Schwanhausser was reluctant, saying, "I don't see much future in the reconnaissance drone stuff."

Setting to work, he estimated the Firebee's range could be extended to allow it to make overflights from the Barents Sea, across the Soviet Union, to a recovery in Turkey. With longer wings, the drone could reach altitudes above that of the regular Firebee. The drone would also have a reduced radar return, making it virtually undetectable. Launch would be from either a C-130 transport plane or a ground launcher.[1]

In mid-April 1960, the Reconnaissance Panel of the air staff was briefed on the project. Two weeks later, Powers's U-2 was lost. On July 1, 1960, an RB-47 on an ELINT flight over international waters was shot down—only two of the six-man crew survived, and they were captured.

On July 8, the air force issued a $200,000 contract. Ryan Aeronautical made reflective measurements of one-fourth and one-eighth scale models of the Firebee. These showed the radar return could be reduced by putting a wire screen over the intake, painting the nose with nonconductive paint, and placing radar-absorbing blankets on the sides of the drone.

This was confirmed by the flight-test program, which was conducted between September 16 and October 12, 1960, at Holloman Air Force Base in New Mexico. The tests showed the radar return of a drone at 50,000 feet could be reduced without causing aerodynamic problems. A cover story was also created should one of the modified drones come down outside the Holloman test range: the drone was a "Q-2D," a "ground-controlled target" used to test SAM missiles at altitudes of 60,000 feet. This would conceal its true reconnaissance role.

The test data was to be used for Ryan's proposed Model 136 reconnaissance drone. It used long, straight wings, a horizontal stabilizer with inward-tilted rudders at their tips, and a jet engine mounted on top of the airframe to reduce the radar and infrared signatures. Both the test flights and the Model 136 were code-named "Red Wagon." (A Boeing design was called "Blue Scooter.")

As with Bell and Lockheed, Ryan set up its own Black production facility for the reconnaissance drone program, in a warehouse on Frontier Street in San Diego, California. The people needed for the effort were recruited without being told what they would be doing. Only after arriving at the warehouse did they learn the project dealt with drone reconnaissance.

Behind the scenes, there was considerable debate over the future of reconnaissance programs. The end of Red Wagon came on election day 1960.

President John F. Kennedy and the Democrats would have their own ideas about reconnaissance, so any new projects were put on hold.

Despite this, interest remained in drone reconnaissance. Ryan Aeronautical proposed a new system called "Lucy Lee" (also "L Squared"). It was to undertake photo and ELINT missions outside Soviet airspace. Lucy Lee would use a modified Firebee rear fuselage, long straight wings, and a new forward fuselage with an intake above the nose. It would fly at altitudes between 65,000 and 72,000 feet, and the radar return of Lucy Lee would be reduced. Ryan Aeronautical also proposed that $500,000 be used to modify a standard Firebee drone to a reconnaissance configuration.

By mid-summer 1961, it seemed Lucy Lee would succeed. Then, despite support at nearly every level, the project was canceled in January 1962. Work at the warehouse ground to a halt; it was down to "one light bulb, one engineer, one secretary, and a guard." Finally, Ryan Aeronautical management issued orders to close down the warehouse.[2]

A half hour later, the air force called.

147A FIRE FLY

The air force had accepted the Ryan Aeronautical proposal for a modified Firebee reconnaissance drone. Rather than an expensive, all-new drone, and the unknown this involved, the proven Firebee would be used. The money came from a program called "Big Safari," which had been established in the early 1950s as a means to modify existing aircraft for reconnaissance missions in a very short time. On February 2, 1962, a $1.1 million contract was issued to Ryan to modify four Firebee target drones as "special purpose aircraft." Code-named "Fire Fly," the Model 147A drones were to have a 1,200-mile range, a cruising altitude over 55,000 feet, and a photo resolution of two feet. They were to be ready by May 15.

The first 147A drone was to be a standard Firebee with a simple guidance system—a timer-programmer and an MA-1 gyro compass. (A telephone stepper switch was used which cost $17.) The other three 147As were "stretched" with a thirty-five-inch plug added to the fuselage. This carried an additional sixty-eight gallons of fuel. The nose was also modified to carry a camera. Again to speed things up, the optics from a U-2 were borrowed, and mounted in a homebuilt frame. Due to security reasons, it was not called a camera but rather a "scorer." The test program would use two of the drones—147A-1 would test the navigation systems, while the stretched 147A-2 would check out the camera and other modifications. Once the system was proven, the other two 147A drones would be placed on alert for deployment in a crisis.

The first flights of 147A-1 were made from Holloman Air Force Base, New Mexico, in April. In its first off-range flight, the drone flew from

Holloman to the Wendover Air Force Range, Utah, and then back to the White Sands Missile Range, New Mexico, without any commands from the B-57 chase plane or the ground station. In all, three flights of 147A-1 were made to prove out the navigation system. This was followed by four test flights of the 147A-2 drone in late April and early May. When the film from the scorer was developed, it showed very good resolution. The Fire Fly had an adequate range, an altitude and resolution better than required, and the ability to fly the desired track.

With this, the 147A reconnaissance drones were considered operational. Two drones and their DC-130 launch aircraft were placed on seventy-two-hour alert at Holloman. This very limited capability was operated by the Strategic Air Command (SAC).[3]

In the summer of 1962, it was decided to run a simulated deployment to test the drones under operational conditions. The operational test and evaluation would involve two reconnaissance flights over the Atlantic Missile Range at Cape Canaveral and three "live-fire" tests at MacDill Air Force Base. During the two reconnaissance flights, fighters swarmed aloft in simulated interceptions; they ended up chasing each other. The radar blankets around the drone were effective, and ground radar never picked it up.

The live-fire tests at MacDill also showed the 147A's low radar return, small size, high altitude, and subsonic speed combined to make it highly survivable. The drone's main problem was the contrail—a giant banner that gave away the drone's location. (Neither of the two successful shoot downs would have been possible without the contrail as guide.) A "no-con" (no contrail) program was quickly started to find means to suppress it. It was eventually decided that a chemical agent would be added to the exhaust.[4]

As the no-con program began at Tyndall Air Force Base, events were unfolding that would bring the world to the brink of nuclear war, and Fire Fly to within moments of making an operational mission.

THE MISSILES OF OCTOBER

For several years, CIA U-2s had been directed against Communist Cuba. The missions of August 29 and September 5, 1962, revealed a major change—the Soviets had introduced SA-2 SAMs. Eleven sites, which covered most of the island, were found. The risks of Cuban overflights had increased.[5]

The dangers were underlined four days later. Nationalist Chinese U-2 pilots had been conducting overflights of mainland China since late 1960, making as many as three overflights per month. On September 9, Radio Peking announced: "A U.S.-made U-2 high-altitude reconnaissance plane of the Chiang Kai-shek gang was shot down this morning by an Air Force unit of the Chinese People's Liberation Army when it intruded over east China." The pilot, Col. Chen Huai Sheng, was severely injured when his U-2 was hit by an SA-2. He was captured and taken to a hospital but died that night.[6]

Secretary of State Dean Rusk and presidential adviser McGeorge Bundy were worried about the political effects of a U-2 being shot down over Cuba. Rusk seemed obsessed with the idea that continued U-2 overflights would increase tensions to the point of war.[7] At a September 20 meeting, the air force proposed that the Fire Fly drones be used over Cuba. No interest was expressed in their use. At that time, there were only two drones and they were still in the test phase.

By early October, there had been no U-2 coverage of the interior of Cuba for a month. Finally, a single U-2 overflight of western Cuba was authorized. Unlike previous U-2 missions, it would be flown by an air force pilot. Two SAC U-2 pilots, Majors Richard S. Heyser and Rudolf Anderson, were checked out in the CIA's U-2F version.

At 8:30 P.M. PST on October 13, Heyser took off from North Base at Edwards and headed toward Cuba. He started his run over Cuba at 7:31 A.M. EST, October 14. The overflight was made at 72,500 feet, with a flight path that went south to north across the island. At 7:43 A.M., he left Cuban airspace and turned toward McCoy Air Force Base. After landing, the film was removed and flown to Washington, D.C., for analysis at the National Photographic Interpretation Center (NPIC).[8]

The following day, interpreters noticed six long canvas-covered objects in the San Cristobal area. They were about seventy feet long—too large to be SA-2s. They were identified as SS-4 Sandal medium range ballistic missiles (MRBMs). Three MRBM sites under construction were discovered—at San Cristobal, Los Palacios, and San Diego de los Banos. The Cuban Missile Crisis had begun.[9]

On the morning of October 16, President Kennedy learned about the missiles. At the same time, three U-2As of the air force's 4080th Strategic Reconnaissance Wing were alerted. Over the next week, some twenty U-2 missions were flown. These spotted two more SS-4 MRBM sites at Sagua la Grande. Two SS-5 Skean intermediate range ballistic missile (IRBM) sites were found at Guanajay and a third SS-5 site was discovered at Remedios. On October 22, Kennedy announced the existence of the missiles and the imposition of a blockade. He also warned that any nuclear attack from Cuba would be met with a full retaliatory response by the United States.[10]

In the following days, as U-2s flew on high and air force RF-101s and navy RF-8s went in at 200 to 500 feet, work continued on the missile sites.

On October 27, all twenty-four SS-4 pads were considered operational. In anticipation of a U.S. invasion, the nuclear warheads for a Soviet Frog short-range missile battery, IL 28 light bombers, and the SSC-2B Samlet coastal defense missiles were readied. They would have caused tens of thousands of casualties among U.S. troops hitting the beaches. At Key West Airport, four Pershing missiles stood ready. Upon a presidential order, their nuclear warheads would be launched to destroy Havana. Around the world, some 1,200 U.S. bombers and nearly 400 missiles were prepared to hit their targets in the Soviet Union.[11] The Soviet's own smaller nuclear force, 180 bombers, some 20 ICBMs, 100 submarine-launched missiles, and the 24 Cuban sites were on alert. For the first time, Soviet ICBMs were fueled and made ready to fire. Armageddon loomed.[12]

At 8:10 A.M. on October 27, Major Anderson took off from McCoy Air Force Base. As he flew over the Banes naval base, a salvo of SA-2s was fired. One of the missiles exploded above and behind his U-2.[13] One or more fragments penetrated the cockpit and hit Anderson at shoulder level. The cockpit depressurized and his damaged suit failed to inflate. Anderson lost consciousness within seconds and died.[14]

When word reached Washington that Anderson's plane was overdue, most feared it was a direct Soviet escalation. The pressure to take military action—either striking the SAM sites, or an invasion—was growing. The situation seemed ready to explode with little or no warning. October 27 later became known as "Black Saturday." That evening Robert Kennedy met with Soviet Ambassador Anatoly Dobrynin and delivered an ultimatum. If the Soviets were unwilling to remove the missiles, the United States would attack within a day or two. This was coupled with a deal—the United States would promise not to invade Cuba if the missiles were removed. And, once the crisis was over, the United States would remove the Jupiter IRBMs based in Turkey.[15]

The death of Anderson had a similar impact in Moscow. There had been strict orders not to fire on the U-2s. The attack had been ordered, without authorization, by Gen. Igor D. Statsenko, a senior Soviet commander in Cuba.[16] Soviet Premier Nikita Khrushchev realized that if SAMs could be launched without authorization, so could other missiles. Robert Kennedy's ultimatum also made it clear he had very little time.[17] Then he received a report that President Kennedy would address the nation "at 5 o'clock." The Soviets believed it would be an announcement of an air strike or invasion. To forestall this, a message was hurriedly drafted and rushed to a radio station. At 9:04 A.M. EST, October 28, Radio Moscow broadcast a statement announcing the Soviets would remove the missiles. The crisis had ended.[18]

FIRE FLY IS ALERTED

U-2 overflights had been halted following Anderson's death, and there were no guarantees they would not be fired on again when operations resumed, or even if they could resume.[19] The uncertainty caused Undersecretary of the Air Force Dr. Joseph Charyk to reverse the earlier decisions not to use the Fire Fly drones over Cuba. The unit was alerted to prepare for a two-drone mission on short notice.

The Fire Fly's planned mission was different from that envisioned by the test flights. The 147A would fly at a medium altitude, 30,000 feet, rather than 50,000-plus feet. The drones' autopilot was reprogrammed, and the cameras were modified slightly to accommodate this lower altitude. The scorers were then serviced, loaded with film, and installed in the drones. Everything was ready for the Fire Fly's first overflight. The DC-130 was on the flight line, with all four engines running, awaiting clearance to head down the taxiway to the end of the runway. They were moments away from starting when the mission was aborted on orders of Gen. Curtis E. LeMay, the air force chief of staff. LeMay supported the drone effort but wanted to save the capability for something bigger. U-2 overflights resumed on November 5. Although radar continued to track the planes, no SAMs were fired.[20]

LIGHTNING BUG

Shortly after the aborted launch, the air force issued a contract for a family of operational drones to undertake different types of missions. The first was the 147B, a specialized high-altitude drone. The wingspan was extended from the 13 feet of the 147A to 27 feet, which raised the altitude ceiling to 62,500 feet. Two test vehicles and seven production 147B drones were to be built.

It would take several months for the 147Bs to be ready. To provide an immediate reconnaissance capability, the air force ordered seven 147Cs, production versions of the 147As. The wingspan was increased from 13 to 15 feet, and the contrail suppression system was added. Three of the 147Cs were then modified to produce the 147D. This drone was designed to undertake a mission that would be impossible for a manned aircraft. The air force needed data on the proximity fuze of the SA-2; to get the data, the drone would have to be hit by the SAM. The three 147Ds were delivered on December 16, 1962. Six weeks after the Cuban Missile Crisis ended, the United States had a limited unmanned reconnaissance capability based on the 147C and D drones.

Because the Fire Fly code name had been compromised, a new one was needed. In keeping with the insect trend of earlier names, the new drones were called "Lightning Bugs."

On July 1, 1963, the 4028th SRS(W)—Strategic Reconnaissance Squadron (Weather)—was declared operational and was placed on seventy-two-hour alert. The unit was initially equipped with two 147Cs and two 147Ds, pending arrival of the first of the production 147Bs. It was located at Davis-Monthan Air Force Base, outside Tucson, Arizona, in an old World War I hangar. Like the A-12 unit, the 4028th had its own patch. It showed a cartoon of a fire fly (with a lightbulb in its tail). From its antennas came two lightning bolts—hence a Lightning Bug.

The unit's activities were considered highly secret—the deputy commander for maintenance at the base was not told of their operations. He only learned of it by accident. When he asked the wing commander, he was told that he would be given no information until he received a "need-to-know" clearance.[21]

On December 20, 1963, the secretary of the air force approved a follow-on contract for fourteen more 147B drones. In January 1964, three 147Es were delivered. These were B models fitted with the equipment from the D version. To support the expanding production, Ryan moved its Black operation from the Frontier Street warehouse to a secure factory in Kearny Mesa, an industrial park a few miles north of downtown San Diego.

In early 1964, Castro began making threats over U-2 overflights.[22] On May 2, 1964, President Johnson ordered a review of alternatives. The drones quickly emerged as the preferred method. A memorandum of May 5 noted:

> The examination of alternative means of overflights . . . has led to a sharp rise in support for handling this matter by drones. It appears that we have drones which *might* do this job with a level of efficiency which would enable them to continue even if Castro tried to bring them down, because new drones could be supplied faster than he could bring them down. This at least is the position of the Defense Department civilians. The Joint Chiefs have not yet expressed a final view. Probably the result of today's discussion should be a direction to accelerate preparation for the use of drones, and production of additional drones in case we decide to shift to them.[23]

The following day, the *New York Herald Tribune* newspaper published a story on possible use of the drones. This was the first public suggestion that the United States had developed an unmanned reconnaissance capability. Under the headline, "U.S. Studies Drones For Use Over Cuba," it read:

> Washington—A missile or pilotless plane to replace manned U-2s for surveillance flights over Cuba is being given serious consideration here, it was learned yesterday.

> The use of a drone craft, some administration officials believe, would reduce the chances of a brink-of-war confrontation between East and West if the Castro regime decides to shoot down a U.S. reconnaissance vehicle in Cuban air space.
>
> If an unmanned spy craft were brought down by Cuban antiaircraft missiles, it is felt, the incident would not be likely to require the same drastic countermeasures as the capture or death of a U.S. pilot.
>
> There is still considerable controversy both within the administration and the Pentagon as to whether pilotless spy flights would produce the quality of photographs that high-altitude U-2s and low-level F-104 [*sic*] and F-8U Crusader jets are getting.
>
> There is no technical barrier to sending pilotless craft over Cuba and taking photographs, military sources here said.[24]

Obviously, the discussions about the drones had leaked, and leaked very quickly. In any event, the A-12 was selected to back up the U-2s for the Skylark missions.

Three months later, the drones were at war.

LIGHTNING BUG GOES TO WAR

On the afternoon of August 2, 1964, the destroyer U.S.S. *Maddox* was sailing in international waters off the North Vietnamese coast when it was attacked by three North Vietnamese PT boats, which were sunk. The Tonkin Gulf Incident set events in motion and the Vietnam War began in earnest.[25]

President Johnson and his advisors feared Chinese intervention in Vietnam. The drones were seen as a way to watch for any buildup. At 4:00 P.M. on August 4, the 147B drones were ordered to Kadena Air Base on Okinawa, in preparation for overflights of Communist China. Specific targets were southeast China, near the border with North Vietnam and Laos, Hainan Island, and the coastal areas. The two launch DC-130s and four drones, 147B-8, B-9, B-10, and B-11, were made ready. The preparations were interrupted for several days by a typhoon alert.

On the morning of August 20, 1964, the DC-130 launch aircraft rolled down the runway and took off. It carried two drones—147B-9, the primary drone, and 147B-8, the backup. As the launch aircraft approached the Chinese coast, the crew checked out B-9. Everything was in readiness; the DC-130 began its run to the launch point, the release was pressed . . . and nothing happened. B-9 would not come off the shackle, even with the emergency release.

The launch crew regrouped, checked out B-8, and made a successful launch. The drone climbed to its programmed altitude and set off for its overflight of southeast China. The DC-130, with B-9 still on the launch rack, headed back to Kadena. Thirteen minutes after B-8 was launched, B-9 just fell off the rack. Only the dye marker showed the impact point.

In the meantime, B-8 continued across China. At an altitude of about 62,000 feet, the sky above was a deep blue black. The drone's black paint finish hid it from visual sightings, while the radar blankets concealed it from electronic detection. The drone's navigation was later described as "not spectacular," but it did cover a number of primary targets and returned with "significant information." Once the photo runs were completed, B-8 turned east, toward Taiwan. When the recovery team picked it up on radar, it was only a few miles to the right of the desired track. The radar transmitted the recovery signal, and the drone descended under a 100-foot parachute. The recovery zone was a half mile wide and two miles long. B-8 landed in a rice paddy, but the parachute release did not operate. The wind dragged the drone until it flipped over, causing major damage. The drone was picked up by a helicopter and later returned to Kadena.

Unlike the U-2, Lightning Bug overflights were made every few days. The second mission was flown by B-11 on August 29. Everything seemed to work satisfactorily until the recovery. A short had caused the programmer to stop operating, and the drone would not accept the recovery command. B-11 kept flying, past the recovery zone and out to sea, until it ran out of fuel.

The third mission, of B-10 on September 3, had better luck. The only mishap was an engine flameout during the recovery sequence. It landed successfully with only minor damage. When processed, the photos were good.

On September 9, a pair of missions was attempted. Both ended in failure. B-13 flew its mission, but as it descended through 30,000 feet toward Taiwan, the engine flamed out. The parachute was deployed, but the drone was lost at sea. The day's second mission never got started. B-6 was launched from the DC-130 and began its climb. Soon after, it crashed into the jungles of Laos.[26]

Thus, of the first five missions, only two had been successful. It was clear technical problems had to be resolved. Still, the Nationalist Chinese were very enthusiastic about the drones, due to the continuing U-2 losses to SA-2s over the mainland. A second U-2 had been shot down on November 1, 1963, and its pilot, Maj. Yei Chang Yi, was captured. A third U-2 loss occurred on July 7, 1964. Lieutenant Colonel Terry Lee died with his plane. The fourth Nationalist Chinese pilot lost was Maj. Jack Chang, who took off from Taiwan in the early evening of January 10, 1965. An SA-2 ended the mission forty-five minutes after he crossed the coast. Chang bailed out but landed so hard he broke both his legs. Medical attention saved his legs, but he and Yei would not be released until October 1983. A Nationalist Chinese U-2 pilot was required to fly ten overflights. Very few survived a tour.[27]

Although none of the drones had yet been lost over China, they were being jumped by Chinese MiGs. On the September 25, 1964, flight, B-14 was followed by two pairs of MiGs about 10,000 feet below the drone. On September 29, a single MiG came within 5,000 feet of B-10. It was clear the Chinese were making an intense effort to down the drones. As yet, they were having no more luck than U.S. pilots had during the test missions.

In anticipation of the loss of a drone, thought was given to equipping the drones with a destruct system and removing the manufacturers' name plates. Finally, it was decided to do nothing. If a drone was lost, and the Communist Chinese announced it, the United States would say only "no comment." There would be no cover story or acknowledgment that it was a secret project.[28]

After two months of operation, it was clear the recovery zone on Taiwan was not working out. Winds dragged the drones after landing, causing major damage. Launch operations were moved to Bien Hoa Air Base, outside Saigon, South Vietnam. The landing zone was near Da Nang, on the coast. The shift also marked a change in targeting. The early missions were directed against Communist China. Now North Vietnam was also to be covered.

The first drone flight from Bien Hoa was made on October 11, 1964, but B-14 was lost during recovery when it descended through a rainstorm and the parachute tore off. Despite the loss, the next several missions were successful. Launches were often made over Laos; in some cases, the DC-130 would nearly reach the Chinese border. Two more drone missions were flown over North Vietnam on October 22 and 27. No further drone missions were flown until November 7. It, too, was successful, bringing the total to five in a row.[29]

FIRST LOSSES

On November 15, 1964, 147B-19 was launched for a mission over China. It climbed to its programmed altitude, then crossed the border. During this "penetration phase" of the flight, MiGs jumped B-19. It was later reported that between sixteen and twenty MiGs went after the drone, making between thirty to fifty passes before it was shot down.[30]

The Chinese announced the incident the next day. The statement by the Hsinhua press agency said, "A pilotless high-altitude reconnaissance military plane of U.S. imperialism, intruding into China's territorial airspace over the area of central south China on November 15, was shot down by the air force of the Chinese People's Liberation Army." A separate report said that Marshal Lin Piao, minister of national defense, had commended the air unit responsible. He added, "This major victory was scored" because the unit "firmly carried out orders, maintained its combat readiness, seriously studied the enemy situation, did their best to master tactics and technique, overcame difficulties, and displayed a spirit of heroism in fighting." Lin continued that he hoped the unit would "be ready to deal blows at any invading enemy aircraft and win greater victories."[31]

In Washington, according to a *New York Times* article, "officials professed to be baffled by the Chinese description of a 'pilotless' plane. The United States has some aerial-photographic drones capable of short-range reconnaissance over a battlefield. But as far as has been revealed there are no pilotless planes capable of long-range flights."[32]

The State Department said it had no information to support the Chinese claim. White House Press Secretary George E. Reedy said, "I know nothing about it. This is the first I've heard of it." The Defense Department said only, "No comment."[33] Defense Secretary Robert McNamara said he would not comment on the report and could add nothing to the statements issued by the State and Defense Departments.[34]

Despite the international publicity, the loss of B-19 seemed to have little impact. The stories were printed and forgotten. At Bien Hoa, operations halted for thirty days following B-19's loss, to allow things to cool off following the news reports. The first 147B mission after the loss of B-19 was made on December 15 and was called the best to date. About six more operational missions were flown before the end of the year. In all, twenty 147B missions were flown during 1964.[35]

The year 1965 began with another burst of publicity—on January 2, 147B-21 became the second drone shot down by the Chinese. Lin Piao called it a "major victory." The Chinese press stated that the drone had been shot down "by the air force" over south central China. The U. S. Defense Department had no comment about the incident.[36] Press interest was also fading—the articles on the November 15 drone shoot down had been on the front page; B-21's loss made page 3 with no follow-up.

On March 31, B-20 was shot down over China. This time the news coverage was vanishingly small—the entire *New York Times* article read: "The Peking radio said an unmanned United States reconnaissance plane was shot down over south China today by a naval anti-aircraft battery."[37]

On April 2, the Chinese put B-21 on display in a military museum in Peking. The aircraft was readily identifiable as a modified Firebee drone. Some thirty thousand Chinese marched past the wreckage. The Hsinhua press agency said the shooting down of three of the drones was "a serious warning to United States marauders who are now extending the flames of their aggressive war in Indochina and conducting constant military provocations against China." The *New York Times* article on the U.S. response to the display read: "The Defense Department has not denied occasional reports that the United States has been sending drones over Communist China for reconnaissance purposes. Its standing policy has been to refuse comment on such reports."[38]

The following day, April 3, the Chinese announced the fourth drone had been shot down. B-23, nicknamed "Crazy Legs," was lost over central south China. Hsinhua said it was "the second espionage plane of the same type to be shot down within three days." Again, the press took little notice.[39] The fifth 147B was lost over China on April 18, 1965.[40]

On April 20, the Chinese put three of the captured drones on display at the Chinese People's Revolutionary Museum in Peking. A photo of the drones was published on the front pages of American newspapers the next day. The *San Diego Union* noted, "The new photo and others released in the past have left little doubt the planes in Chinese hands are Firebees." Ryan responded by neither confirming nor denying it.[41]

On August 21, a sixth 147B drone was shot down over Hainan Island by an air unit of the Chinese navy. Again, the U.S. press paid little attention.[42]

The Chinese issued their statements and photos and displays, but few took notice of those small, almost toylike airplanes. Within a few days, events pushed them aside, and the memory faded. The concept of using an unmanned drone for high-risk overflights, and simply not commenting on any losses, had proved valid. The Lightning Bug was *still* a Black airplane.

CHANGING OF THE GUARD

By New Year's of 1965, it was clear the 147B drones were a success. Soon after the end of the holidays, Ryan Aeronautical began work on an improved version of the high-altitude 147B. This was done without a formal contract. Ryan and the air force had developed a good working relationship, and the effort could get under way without waiting for the paperwork.

The new version was the 147G drone. The major change was a more powerful engine for a higher altitude over the target area (at the cost of a shorter range than the 147B). A contrail suppression system was also added to the engine. This would lessen the chance of a visual sighting. The fuselage was stretched to 29 feet, while the wings spanned 27 feet. The formal contract for the 147G was issued in March 1965, and the first was delivered in July 1965.[43]

At this same time, the first changing of the guard was made in the skies of North Vietnam. During a photo run over Haiphong Harbor, an air force U-2 was fired on unsuccessfully by an SA-2. In response, a dual mission was planned—a 147B drone was to fly over an area defended by the SAMs, while a U-2 remained just out of range to observe. As the U-2 pilot watched in awe, an SA-2 rose up and consumed the drone. U-2 overflights of North Vietnam ended. They were shifted to "signals intelligence" missions. This involved flying long hours outside North Vietnamese airspace to pick up radio and radar transmissions.[44] The 147B reconnaissance drones took over high-altitude, photo-reconnaissance missions over North Vietnam.

That fall, there was another changing of the guard. The first 147G mission was flown in late October 1965. The new drone soon replaced the 147B, which made its last flight in December. The first 147G lost over China was shot down on February 7, 1966. (This was the seventh drone brought down by the Chinese.)[45] The Chinese destroyed another 147G drone on March 5, 1966, over the central south region. They said the drone was on "a provocative reconnoitering flight." The *New York Times* carried the story on page 54.[46]

As the U-2s were giving way to the 147Bs, and they, in turn, were replaced by the 147Gs, work also began on a drone using a completely different mission profile. The normal weather pattern over North Vietnam was for clear skies between May and September. During the winter monsoon season, between November and March, there were thick clouds and heavy rain, with ceilings down to 500 feet.[47] Even with clear weather during the summer, smoke and ground haze would often cause the photos from high-altitude drones to be poor.

It was clear the drones would have to go in *under* the clouds. This would also help them evade North Vietnamese air defenses and make them more difficult to track by radar. The SA-2s could not engage a target below 1,500 feet, but the drones would still face 37mm cannons, .50-caliber machine guns, and even rifle-armed peasants.

In October 1965, a contract was issued to produce the 147J. This was a 147G modified for low-level operation, with a new altitude-control system and camera package. Development proved difficult, with the loss of three prototypes and damage to a DC-130 when 147XJ-2 collided with it just after launch.

To test the 147J's control system under combat conditions, two of the old 147C drones were modified. A new "duck head" nose was added to house a larger camera. Both were soon lost; the C was marginal at best for so demanding a mission.[48]

In all, seventy-seven missions were flown over China and North Vietnam during 1965. These were made up of the last of the 147B drones, the first of the 147Gs, the two 147Cs flown to test low-altitude operations, two 147Ds flown as decoys, and the first of the 147E ELINT drones, in a program called "United Effort."[49]

UNITED EFFORT

With SA-2 sites spreading throughout North Vietnam, the need for the fuze data was all the greater. This mission was undertaken by 147E drones. These were 147Bs with their cameras replaced by special ELINT equipment.

The data would be retransmitted to an RB-47, even as the drones themselves were destroyed. Three 147Es were sent to Bien Hoa in October 1965.

The first three 147E missions were not successful, due to ELINT package failures. The 147E drones were withdrawn from operations and underwent environmental chamber tests. The problem was traced to overheating of the ELINT equipment, and the drones returned overseas in early 1966.

Success came on February 13, 1966, with the fourth attempt. The ELINT equipment relayed data on the SA-2's proximity fuze, radar guidance after the fuze activated, and the blast overpressure that destroyed the drone. United States intelligence had been trying for years to get this data. This mission was later described as the most significant ELINT mission since the start of the Cold War. It paid for the whole 147 program. The data was incorporated into the design of new electronic countermeasures equipment (ECM).[50]

The first of this new ECM equipment, the ALQ-51 "Shoehorn," was then tested aboard a drone against SA-2 missiles. Ryan modified a single 147 drone, B-7, to the 147F configuration. The Shoehorn was a large package, and it was difficult to fit it into the drone. The 147F drone was sent overseas and made several flights in July 1966. It was lost on July 22, 1966, but not before ten or eleven SAMs had been fired at it. The 147F was able to prove out the Shoehorn without risk to a pilot.[51]

BORN TO LOSE—THE 147N AND NX DECOY DRONES

North Vietnamese air defenses had continued to expand and were taking a toll on the drones. Of a series of twenty-four missions, sixteen drones were lost. This loss rate was too high and, in early 1966, the air force asked Ryan to build a decoy drone. This led to yet another branch in the 147 family.

The new decoy effort was given priority to bypass normal procedures, even those of Big Safari. Over a ten-day period, ten Firebee target drones were modified with traveling wave tubes to make them look like larger aircraft. They were designated 147N drones. As the 147Ns were never meant to survive, they had only a ninety-minute fuel supply and no recovery parachute.

The first 147N mission was launched on March 3, 1966. The decoy 147N and a 147G were released from the DC-130 almost simultaneously. They flew a parallel course until they approached the target area. The two drones then diverged, giving the North Vietnamese two possible targets. The 147N, with a larger radar return, was flying at a vulnerable altitude. As expected, they went after the 147N, while the 147G returned to Da Nang.

Although all were lost, the eight 147Ns were credited with five MiG "kills." In the first case, a 147N headed out over the Tonkin Gulf with a MiG hot on its tail. The MiG pilot ran out of fuel and had to eject at sea.

Other kills were "friendly fire"—a SAM was launched at a drone but destroyed a MiG. Still another MiG was shot down by its wingman.[52]

Their success led to a follow-on decoy. Despite expectations, several of the 147Ns had survived to reach Da Nang. Without a parachute, however, they could not be recovered. In August 1966, another order was placed for ten decoys. Like the 147Ns, these were to be Firebee target drones equipped to make them appear as bigger targets. Unlike the earlier decoys, the 147NX also carried an inexpensive, low-resolution camera. From medium altitude, it could take photos with a six-foot resolution. The 147NX could be used to spot trucks and provide general indications of activity. The first 147NX missions were flown in November. They would function as a confusion factor for the high-altitude 147Gs, rather than cover specific targets. If the 147NX made it back, the photos were a bonus.[53]

LOW OVER THE RICE PADDIES—THE 147J

In late March 1966, the 147J began low-level operations over North Vietnam. It used a barometric low-altitude control system (BLACS) to remain a preset height above the ground. A dual camera system was added—one camera looked front and rear, the second looked left and right. The most visible difference with the 147J was the paint finish. All the earlier, high-altitude drones had been painted black; because the 147Js would fly at low altitude, they were painted gray on the upper surfaces and white underneath. The 147Js soon were showing greater survivability over the target. They flew below the effective altitude of the SA-2 SAM. J-14 came back with photos of a SAM being fired at it. The SAM missed, and J-14 made it home. (The last of the 147Ns were used as decoys for the 147Js.) J-4, the prototype drone, was sent to Bien Hoa and flew five successful missions over three months.

The 147Js also showed an improvement in recovery. There had always been some damage due to ground impact, so a midair recovery system was used for the 147J (as well as later drones). A helicopter grabbed a small parachute with cables rigged between two poles. Once it was secure, the main parachute was released, and the helicopter's winch pulled up the drone.[54]

The 147Js, along with the high-altitude 147Gs, were used in support of attacks on North Vietnam's supply of petroleum, oil, and lubricants (POL). By the end of July 1966, 70 percent of North Vietnam's known bulk storage of POL had been destroyed. The effort was frustrated, however. Drone photos showed the North Vietnamese were storing oil drums along the streets of villages, which they knew the United States would not bomb.[55]

In all, 105 missions were launched in 1966 over North Vietnam and China. This consisted of 147Gs, which made up the bulk of the flights, the

147J drones, the 147E and 147F ELINT missions, and the 147N and NX decoys.[56]

ONE BY NIGHT, TWO BY DAY—THE 147NRE, NP, AND NQ

By early 1967, the 147J missions had shown the value of low-level coverage. There were concerns the supply of 147Js could run out before a specially designed low-altitude drone could be ready. As a short-term solution, additional 147Gs were converted to the J configuration. Ryan was also requested to build an interim low-altitude, day photo reconnaissance drone, based on a new version of the original 147A. This was the 147NP. It had a stretched 28-foot fuselage, a 15-foot Firebee wing, and the original low-powered engine.

Before the 147NP was ready, a new requirement emerged. Much of the supplies the North Vietnamese were sending south were moved at night. Four of the 147NP drones were diverted from the production line and modified as night reconnaissance drones. These were fitted with a two-camera package sequenced with a flashing white light mounted in the drone's belly. The planned altitude and ground speed had to be programmed into the strobe. For an altitude of 1,200 feet, the light would flash at the rate of once per second. The drone itself could not be seen, but the flash was very visible. If the drone was in clouds, the whole sky would be lit up. The new system was designated the 147NRE (night reconnaissance-electronic).

Two of the drones were sent to Point Mugu for testing. The results were so encouraging that the air force decided to send all four NRE drones to Vietnam. Supplies were very short—there was literally only one box of the special film used by the camera.

NRE-1 was flown on May 25, 1967. The launch was successful, but the drone did not return. Despite the failure, the NRE had beaten the NP into action. The first 147NP day reconnaissance drone was flown a week later, in early June. Unlike the black-painted NREs, the NPs had a camouflage finish.

NRE-2 was flown in the early morning hours of June 5, 1967. It survived its trip north and headed to the recovery zone. As the recovery sequence began, the main parachute separated, and the drone fell toward the jungle north of Da Nang. The small midair recovery parachute caught the tops of the trees and the drone landed intact.

Ed Christian, a Ryan camera specialist, volunteered to go after it. Armed only with an M16, a pistol, an axe, and a safe-conduct pass, Christian was lowered into the jungle. A Viet Cong patrol was also after the drone. Christian chopped open the fiberglass cover and removed the film from the two cameras. These were sent up a cable to the waiting CH-3 helicopter. He then destroyed the cameras with the M16 and tried to punch holes in the fuel tank so he could set the drone on fire, if necessary. By this time the helicopter was low on fuel and it headed off, leaving Christian in the jungle. Two marine gunships soon arrived and started strafing the Viet Cong.

A second helicopter arrived to try to recover the drone. Christian attached the cable, but as the drone was lifted, the fuel poured out, and he was sprayed with it. Another five minutes passed before the first helicopter returned and lifted him out of the jungle, even as the Viet Cong neared.

The photos from NRE-2's mission showed the drone had covered the target, but subsequent flights indicated the 147NRE's navigation system lacked sufficient accuracy; the field of view of the strobe was so small the drone would have to fly directly over the target. There was, however, a great deal of bonus intelligence picked up by flying the 147NREs almost at random. There was also harassment value due to the brilliant strobe light. It was decided to build a specialized drone for night reconnaissance.

In all, seven 147NRE missions were flown between May and September 1967, while the 147NP drones flew nineteen missions between June and September. As it turned out, the supply of 147Js proved adequate.

The 147NP was followed by another low-altitude drone, the 147NQ. It was equipped with a higher-resolution camera than the NX. The main difference between the 147NQs and earlier drones was the control system. Rather than being controlled by a flight programmer, it was hand flown by a crewman aboard the DC-130. Its primary target was shipping in Haiphong Harbor. Missions were flown nearly every day between May and December 1968, when the last one was lost.[57]

THE GREAT WHITE HOPE—THE 147H HIGH-ALTITUDE DRONE

While the low-altitude 147Js, NREs, and NPs were making an increasing share of the drone missions over North Vietnam, the high-altitude 147Gs continued operations. By March 1967, the third-generation, high-altitude 147H drone was ready to begin operations. It used the same engine as the 147G, but with a highly modified airframe. The wings were stretched from the 27 feet of the 147B–G to 32 feet, and fitted with internal fuel tanks to increase its range. The longer wings, plus a lighter airframe, meant the 147H could reach altitudes of over 65,000 feet. The 147H was also equipped with a new Hycon camera that could photograph an area 780 nautical miles long and 22 miles wide, with a better resolution than the earlier drones.

With the growth of both Chinese and North Vietnamese air defenses, the 147Gs were suffering an increased loss rate over the 147Bs. The radar-absorbing blankets and the 147N and NX decoys were not enough. Accordingly, the 147H was also fitted with several different types of countermeasures. These included "Rivet Bouncer," which jammed the SA-2's guidance radar; a coating in the intake to reduce its radar reflection; systems that would trigger evasive maneuvers if the 147H was illuminated by either MiG or SAM tracking radars; and an improved contrail suppression system.

The 147H was one of the most difficult of the drones to develop, and it took nearly two years before it was ready. With its higher altitude, longer range, and countermeasures equipment, the 147H was the "great white hope" of the drone program.[58]

The first 147H mission was flown in March 1967. The 147Hs and Gs continued to operate side by side until the final 147G flight in August. In some cases, a DC-130 would carry one G and one H under its wings.[59]

The start of 147H operations in the spring of 1967 coincided with an increase in the number and intensity of U.S. airstrikes on North Vietnam. The first targets hit were power plants in the Hanoi-Haiphong area. By mid-June, 85 percent of North Vietnam's electrical capacity had been destroyed. In late July, attacks were approved on more targets within Hanoi and Haiphong. Starting on September 4, navy planes began cutting rail, road, and canal links to isolate Haiphong. In the end, the 1967 bombing effort proved futile. The North Vietnamese put out "peace feelers," and President Johnson ordered a bombing halt of targets in central Hanoi. The pattern reverted to that of 1965–66—a greatly reduced scale with frequent interruptions.

On January 31, 1968, the Viet Cong launched attacks in cities throughout South Vietnam—the Tet Offensive had begun. At home, protest rallies grew and became both more violent and more pro–North Vietnam. On March 31, President Johnson halted all bombing north of the nineteenth parallel. Peace talks opened in Paris on May 13, and on November 1, Johnson ordered a halt to all bombing of North Vietnam.[60]

The 147Hs were also continuing overflights of Communist China, now engulfed by the madness of Mao Tse-tung's "Great Proletariat Cultural Revolution."[61] On April 30, 1967, the shooting down of a drone over south China was announced.[62] Another announcement followed on June 12.[63] A total of fourteen drones had been shot down by the end of 1967. A fifth Nationalist Chinese U-2 was also shot down on September 9, 1967. The pilot, Capt. Tom Hwang Lung Pei, was killed when an SA-2 hit his plane.[64] In early 1968, it was decided to end the Nationalist Chinese U-2 overflights, due to the risk. From now on, the 147H drones would carry the burden of watching China.

The new role had a cost—on January 20, 1968, the Chinese shot down 147H-25. The *Peking Review*'s announcement reflected the political madness sweeping China: "The Air Force of the heroic Chinese People's

Liberation Army, which is boundlessly loyal to Chairman Mao Tse-tung's thought and Chairman Mao's proletarian revolutionary line, shot down a U.S. imperialist pilotless high-altitude military reconnaissance plane when it intruded into China's air space over southwest China for reconnaissance and provocation." In March 1968, a "bad streak" of three drone losses over China brought to eighteen the number of drones lost on China overflights since November 1964. Despite this, the 147H's loss rate was below that of the 147B and G drones.[65]

THE DEFINITIVE DRONE—THE 147S BUFFALO HUNTER

December 1967 saw the debut of the first of a family of low-altitude drones, a series that would form the backbone of the final years of 147 operations. The 147J was not the ideal low-altitude drone. Its long, flexible wings, originally built for the high-altitude 147G, were somewhat unstable at low altitude and prevented the 147J from making sharp turns. The 147NPs and NQs were a quick, short-term effort.

There was a need for a low-cost drone specifically designed for the low-altitude mission. Ryan was told that if they could produce a cheap drone to replace the 147J, the production run would amount to several hundred, compared to only thirty to forty of the other models.

In building the new drone, Ryan went back to the basic Firebee design. The Firebee's 13-foot wing was used; this was much cheaper than the long wing of the G and J models. The fuselage was 29 feet long and carried a redesigned camera system. Rather than two cameras, as on the 147J, the new drone had a single camera. This provided an 80 percent increase in coverage. The contract for the new 147S drone was issued in December 1966. Unit cost was about $160,000—40 percent less than that of the G and J drones.[66]

It would take a year to get the 147S into operation. Five of the last 147Gs were modified to test the new "poly-profile" low-altitude control system. In the first test of the system, the drone flew into the water. There was a lag in the system, with corrections coming too late. It took six weeks to develop a fix for the problem. It was known as the "Polly Get Well Kit."

When the 147S test flights were completed, there were a number of proposals for modifications which would give it additional capabilities. These included antiflutter kits, different yaw rate gyros, multiple altitude settings, radar altimeters, and a digital programmer. It was decided to group the changes in production blocks. All the drones in a specific block would have identical configurations.

The first such block was the 147SA. A total of forty were produced. The first operational mission was made in December 1967. The drone's camera could produce photos with a one-foot resolution along a sixty-mile-long strip of North Vietnam. In some cases, objects as small as six inches could be identified. On one mission, a stack of truck tires in a storage yard was photographed. The trademark could be read.

Most of the 147SA missions were directed at the main bridges around Hanoi. These were under repair, and it was necessary to have regular coverage. Other targets were supply lines and SAM sites. In some cases, winds or navigation problems would cause the drone to go off the track, but the targets of opportunity picked up would often be more valuable.

The early 147SA missions were very successful, and, in March 1968, a second block of forty drones was ordered. The 147SB carried the multiple altitude control system (MACS). The 147SB could be programmed to fly at three different altitude settings between 1,000 and 20,000 feet. It could also vary between the three settings throughout the mission, making it much less predictable. The drone was also equipped with new yaw gyros that allowed tighter, more precise turns.

The first 147SB missions were flown in March 1968, overlapping with the 147SAs. One early mission was flown by a pair of drones over Haiphong Harbor. The Soviets claimed one of their freighters had been torpedoed by the United States as it entered the harbor. The drone unit was ordered to photograph the ship to see if it had been damaged. The drones' flight paths were to cross over the ship. The two drones flew as programmed and returned with photos looking directly into the cargo hold. They showed no damage at all. The North Vietnamese, however, caused some damage when they opened fire at the low-flying drones. Shooting at a nearly flat trajectory, the shells hit the ground throughout the harbor area.

The 147S drones were the source of a number of "war stories." None matched the adventure of 147SB-12 on October 6, 1968. After launch, the MACS had a problem and rather than flying at an altitude of 1,500 feet, the drone flew at 150 feet above the ground. Its programmed flight path took it under a line of high-tension power lines. The photo showed the tower looming above the drone, while on the ground, people were looking upward at the low-flying plane. The unit commander posted the photo on the bulletin board with a note saying, "The FAA frowns on this bullshit!"[67]

Such achievements were not without cost. With the partial bombing halt on March 31, the drones became a prime target. Flying at 1,000 to 1,200 feet, they were taking heavy losses. On April 21, SA-17 was lost over Haiphong.[68] On June 8, a drone was reported shot down over Hanoi.[69]

A change in profile was necessary—the drones were set to fly at 500 feet and 500 knots. This put them below the minimum altitude of heavy antiaircraft guns, while the high speed made them difficult to hit with light antiaircraft guns or small arms.[70] Still, by late 1968, the North Vietnamese had shot down a total of about forty of the drones.[71]

The North Vietnamese also sought to end the drone flights through political means. When the Paris Peace Talks opened, the North Vietnamese demanded an end to all reconnaissance flights. This was described as the first order of business: the halt must be "without delay" and "definite and unconditional," and continuation of the talks was dependant on U.S. acceptance of the demand.[72] This effort also proved ineffective. United States reconnaissance flights continued to monitor North Vietnamese activities.

These reconnaissance photos showed that, within two weeks of the bombing halt, the North Vietnamese had repaired all the bombed-out bridges between the seventeenth and nineteenth parallels. Roads had also been made passable, and troop and truck traffic had quadrupled to some four hundred trucks per day.[73]

The next version of the 147S family made its debut in November 1968. This was the 147SRE night reconnaissance drone. They were equipped with a near infrared strobe. In flight, this was visible as a small red light; it was hard to see unless someone looked directly at it, making the drone much harder to track than the white-light strobe on the NRE. The film was geared to near infrared, and the camera had a filter to cut down the effects of haze. A doppler navigation system also provided better accuracy than the NRE.

The first flight, by 147SRE-1, was made on November 7, 1968. A total of five flights were made in November and six more in December. Missions were flown in the predawn hours, when activities were just starting. When compared to photos taken later in the day, this would give an indication of activities.

The December 19 flight of SRE-2 was an adventure. As it flew over a SAM site, it was fired on. The camera photographed the SAM overtaking the drone, then exploding behind it as the SAM hit the ground. A second SAM was launched, which passed so close the photo was burned out by the exhaust flame. The flight was intended to cover the Haiphong docks and seaplane base, then turn west to cover an airfield near Hanoi. The doppler system was not set correctly, however, and the turns were coming late. This caused the drone to fly to an area northwest of Haiphong, where it missed colliding with a ridgeline seven times. The photos showed it only ten or twenty feet above the trees. The final 147SRE flight was made in October 1969.

Although the SREs were successful, less use was made of the system than was possible. The photo interpreters were not trained in analyzing the near infrared images. Many times targets were missed.[74]

A total of 340 drone missions were launched in 1968. Of these, 205 were 147S drones, while only 67 were 147H high-altitude flights. Clearly, there was a shift in operations. The original concept of high-altitude, covert reconnaissance, similar to that of the CIA U-2 overflights, had been replaced by the much simpler low-altitude mission.

In January 1969, the 147SC was introduced. The SC drones had a cross-correlation doppler radar and a digital programmer to improve navigation accuracy. Of the total of 437 launches made in 1969, 307 were SC drones, known as "Buffalo Hunters." They provided photos with a three-to-five-inch resolution and were used to provide technical intelligence. In contrast to the huge numbers of SC drones, there were only twenty-one high-altitude 147H missions during 1969. These were conducted between January and June; it would be seven months before another was flown.[75]

The drone program did suffer losses, but throughout the war the North Vietnamese grossly inflated the American losses. On April 19, 1969, they reported shooting down a drone, which they claimed was the 3,278th U.S. aircraft downed over the north.[76] In fact this total was more than twice the true number. The actual drone loss rate in 1969 was 24 percent. Even if hit, the drones often survived. During H-58's seven flights during 1969, it was damaged twice. SC-75, dubbed "Myassis Dragon," was hit seven times by shell fragments during its eighth mission. After recovery, SC-75 was "awarded" a Purple Heart. (SC-75 was finally "killed in action" on its tenth mission.)[77]

BELFRY EXPRESS

Up to this point, the drones had been solely an air force operation. The navy had access to the information, but thought it was not timely enough. Now the navy wanted to test the idea of drones being launched from ships. This would give the task force commander the ability to cover targets immediately. A contract was issued to modify several SC drones for surface launch. These were the 147SKs; they used the SC's 29-foot fuselage, but with 15-foot wings (two feet longer than the SC's).

Test launches were done from Point Mugu and from the U.S.S. *Bennington* before deployment off Vietnam. The drone got its initial boost from a rocket, which would burn out and separate. An E-2A aircraft would guide the flight to the initial point, where the drone's own system would take over and fly the programmed mission. After it was completed, a midair recovery would be made. The program was code-named "Belfry Express."

For the operational missions, three 147SK drones were loaded aboard the U.S.S. *Ranger*. The first flight was made on November 23, 1969, to cover North Vietnam's Highway 1, which ran parallel to the coast a few miles inland. SK-5 was launched successfully, but the carrier was two miles out of position. This meant the ground track was shifted, and the drone photographed an area two miles seaward of Highway 1.

A second mission was flown on November 27, which followed the planned route. The third Belfry Express mission was made on November 30. This time, a midair recovery was made by an air force helicopter. After landing on the *Ranger*'s deck, the pilot climbed out with a large American flag and announced, "I claim this island for the United States Air Force!"

By February 10, 1970, fifteen Belfry Express missions had been flown. SK-5 was launched on mission sixteen and ran into problems. The tracking beacon could not be picked up, and the drone was lost. When the drone ran out of fuel, a radio signal to deploy the parachute was transmitted. A helicopter was sent to its estimated position, but nothing was found.

Several days later, the Chinese announced they had "shot down" SK-5 over Hainan Island. This was the 20th drone to be lost over China. A newspaper report said, "A broadcast from Canton describing the downing of the plane said that the craft 'cunningly' changed altitude as it veered over Hainan but 'could never escape the eyes of our radar operators.' Chinese Navy men were said to have 'shot down' this U.S. pirate plane at once 'while cherishing infinite loyalty to our great leader Chairman Mao and harboring bitter hatred for the U.S. aggressors.'"

A total of fifteen more Belfry Express missions were flown after SK-5's capture. The three missions of April 18, 22, and 27, 1970, were particularly effective, providing photos of SAM and antiaircraft gun sites at Vinh and Than Hoa, as well as railroads, bridges, pipelines, truck parks, storage yards, and anchorage areas. The only disappointment was the loss of SK-3 on April 24. The mission was successfully flown, but the drag and main parachutes failed to deploy.

The final flight, by SK-10, was made on May 10, 1970. After a near perfect mission, the main parachute failed to open, and the drone was destroyed. The problem was later traced to salt water contamination of the parachute actuation circuit cable. On this note, Belfry Express, and the navy experience with drone reconnaissance, ended.[78]

LAST OF THE BREED—THE 147T

Although the 147S family made up the bulk of flights in 1969, the year also saw introduction of the final high-altitude drone—the 147T. It had been ordered in early 1967 as successor to the 147H. It used the basic 147H airframe and camera but had a new engine that increased the maximum altitude to 75,000 feet. The 147T also carried the Rivet Bouncer SA-2 jammer and the radar-absorbing inlet coating.

The first 147T missions were flown in April and May 1969, followed by a second series in October and November 1969. One of the new drones, T-17, was shot down over China on October 28. Further 147T missions were flown during February–May 1970, then again in September 1970. With this, the program ended after only twenty-eight missions over two years. The same pattern was true for the 147Gs. There were only nineteen 147G flights in 1970, and a mere nine in 1971. In both years, the 147Gs were flown between March and June. This brought the era of high-altitude drone photo reconnaissance to a close. The 147T would find success in another mission.

On April 18, 1969, a navy EC-121 ELINT aircraft was shot down by North Korean MiGs over international waters. All thirty-one crewmen were killed. Lieutenant Colonel Andy Corra, head of unmanned reconnaissance systems, learned of the incident as he left his hotel for a meeting at Ryan Aeronautical to review the 147T program. When he arrived at the plant, he suggested using the 147T as an ELINT drone. It would carry receivers that would pick up radar and radio transmissions. The data would then be relayed from the drone to a ground station. Operators on the ground would control its operations.

The first briefing was ready a week after the EC-121 was lost. Four 147T drones were modified into a TE configuration. The first test flight was made on November 25, 1969. They were then sent to Osan, South Korea, for operational testing. It was a schedule that many in the National Security Agency had said was physically impossible to meet.

The first 147TE mission was flown on February 15, 1970, beginning a two-month operational test program. The drone was equipped with ten receivers for radio traffic. A ball-shaped radome on the drone's tail relayed the transmissions. Each receiver was individually controlled from a ground station. The 147TE could fly a preprogrammed mission or be controlled by the DC-130 launch aircraft. In some cases, it would have to fly a very tight "race track" pattern to remain within the signal beam. Following the initial 147TE missions, a contract was issued for fifteen production 147TE drones, in a program code-named "Combat Dawn." The first flight of the production TEs was made on October 10, 1970.

Unlike the other drones, the 147TEs remained at least fifty nautical miles offshore. The ELINT drones flew two types of missions—over the Yellow Sea between North Korea and China, and along the Demilitarized Zone between North and South Korea. Several times, MiGs were sent out after the drones. When MiGs were detected, the drone was maneuvered to avoid the fighters. None were lost to enemy action. Late in the TE program, external tanks were added to the drone, which increased the flight time from five hours to nearly eight. The 147TEs provided about ten thousand hours of intelligence data per year. A total of 268 147TE missions were flown up until the replacement of the drone in June 1973.

The replacement was an improved version, the 147TF. They were fitted with the external tanks and ELINT equipment that could pick up either radio or radar transmissions. The 147TF was introduced in February 1973 and would make 216 flights over the next two years, until the end of the drone program.[79]

SON TAY

The most significant of the 276 147SC missions flown in 1970, out of a total of 365 drone flights, was a "package" of seven. Their target was a small, isolated, walled compound twenty-three miles west of Hanoi. It stood on the bank of the Song Con River, outside the provincial capital of Son Tay. It was a POW camp, and the United States was planning to raid it. The camp was identified in May 1970. Comparison of old and new reconnaissance photos showed a guard tower and new wall had been added. It was also noticed that some uniforms had been spread out on the ground to spell out "SAR"—search and rescue. In one corner of the compound the letter *K* had been stomped in the ground—the code letter for "come get us." The POWs were calling for a rescue mission.[80]

The 147SCs were to provide "prisoner verification" and "positive identification of the enemy order of battle"—whether the POWs were at the camp, and the defenses in the area. Tragically, the drones went through another bad streak of losses. At least two were shot down, and another four had mechanical failure. The final drone, on July 12, suffered an even more frustrating failure. Two of the POWs, air force Lt. Col. Elmo C. Baker and Capt. Larry E. Carrigan, saw it coming and started waving. The drone was supposed to show "the height, color, eyes, and facial expressions" of every man in the compound, but the programmer was slightly off; the drone banked a moment too early and the photos showed only the horizon.

Because too many drones flying over so isolated a site would tip off the North Vietnamese, it was decided to switch to SR-71s. The data they brought back was ambiguous—the camp was not as active as before, but there did seem to be someone there. The raid was given a go-ahead.[81]

It started at 2:18 A.M. on November 21, 1970, when an HH-53 helicopter flew over the camp and blasted the guard towers and a guard barracks with minigun fire. Moments later, an HH-3 crash-landed inside the compound with the assault team. They quickly secured the camp and killed the remaining guards. Two more HH-53s with troops landed outside the camp; a third HH-53 mistook another set of buildings four hundred yards away for the camp and landed there. After a short but fierce firefight, the helicopter picked up the troops and flew them to the camp. The raid took only twenty-seven minutes, but no American POWs were found. Unknown to U.S. intelligence, the POWs had been moved out of Son Tay.[82]

COMING OUT OF THE BLACK

In July 1971, the world's political landscape profoundly changed—President Richard M. Nixon announced he would be making a trip to Communist China. Soon after, it was reported that the United States was suspending overflights of China.

As the political situation eased, so did the shroud of secrecy that had enveloped the drones. Since the first loss of a 147 drone in November 1964, the U.S. government had held to a strict "no comment" policy. During 1970–71, this began to change. The November 9, 1970, issue of *Aviation Week and Space Technology* carried an article on the drones, based on off-the-record information. By the spring of 1971, the air force allowed release of photos of the 147 drones and a very general statement that the Air Force had developed drones that could be used for reconnaissance. What still could not be discussed were code names, technical details, or any references to operational missions. These were still Black.

The year 1971 saw an increase in drone activity, with a total of 406 missions. Unlike previous years, which had seen a host of different 147 versions, only three types of drones were flown. There were 277 SC missions, 120 TE flights, and the final nine 147Hs.[83]

In the war itself, there was little movement. The peace talks, both the public ones in Paris and secret discussions conducted by presidential adviser Henry Kissinger, were deadlocked. The number of U.S. troops declined, even as the antiwar movement grew. The North Vietnamese had used the respite to rebuild and prepare for a ground invasion of the South. As 1971 ended, it was becoming clear that would not be far off.

LINEBACKER I

The drone operation was the first to reflect the impending North Vietnamese invasion. In late December 1971, the sortie rate of SC drones increased to 1.2 per day, twice what it had been. On March 20, 1972, the Easter Offensive was launched. On May 8, Haiphong Harbor was mined, cutting off the North's main source of supply. The Linebacker I bombing campaign began, hitting such targets as bridges, barracks, barges, and rail lines.

With the North Vietnamese invasion, the drones were launched at an average rate of nearly two per day. Some days saw as many as five launches, with nine drone missions flown over a three-day period. This was far higher than during Rolling Thunder. The 147SC drones covered areas that were denied to manned reconnaissance aircraft as too dangerous. This included not only Hanoi and Haiphong, but all of North Vietnam and even occupied areas of South Vietnam.[84]

As the bombing continued during the summer and fall, a new version of the 147SC was introduced. The SC/TV was first flown in June 1972. As the name suggests, this was a standard SC fitted with a television camera. The images were transmitted to a controller aboard the DC-130. He would then guide the drone over the target. This made it possible to cover exactly that part of a target needed. On one SC/TV mission, eight out of nine targets were covered and three bonus targets were also spotted, this despite visibility of down to two miles.[85]

SPECIAL MISSIONS—THE LITTERBUG AND COMPASS COOKIE

The following month the last of the 147N family began combat operations. The 147NC had an unusual history. It originally was built for dropping radar-jamming chaff and had been operated by the tactical air command for several years. It had not, however, been sent to Vietnam. The chaff was carried in two external pods. It was realized the pods could also carry propaganda leaflets over North Vietnam.

Between July and December 1972, twenty-eight missions were flown by the 147NC drones. The biggest problem faced by the drones was predicting the wind over the target at the time of the drop. The leaflets would drift on the wind after release. Several missions were ineffective because the winds carried the leaflets away from the target. The project's official name was "Litterbug." The working troops called the drones "bullshit bombers."

In September 1972, a final series of four 147H missions was flown. The cameras were replaced with ELINT equipment. As with the United Effort missions of 1965–66, the purpose of this special project, called "Compass Cookie," was to gain radar and fuze data on the SA-2. A number of new versions had been introduced since 1966, and the mission would provide an update. The September 28 mission was fired on by three SA-2s but transmitted the data before being destroyed.[86]

LINEBACKER II

By the fall, a peace agreement seemed complete. Nixon ordered a bombing halt above the twentieth parallel on October 24, while Kissinger declared, "Peace is at hand." The final details proved elusive, however. The North Vietnamese reopened several issues and finally broke off talks on December 13. Five days later, B-52s began hitting targets in Hanoi and

Haiphong in the most intense air campaign in history. Linebacker II became known as the Eleven-Day War. The B-52s took heavy losses, but devastated airfields, factories, railyards, warehouses, and SAM sites.

Linebacker II saw the drone's heaviest use of the entire war. Between December 20, 1972, and January 19, 1973, over 100 missions were flown. Two-thirds of these were bomb damage assessment (BDA) missions in support of the B-52 strikes. The photos showed the targets were turned into cratered moonscapes. The drones also photographed POW camps in the Hanoi area. Throughout the war, POWs had seen or heard the drones many times. During the 1968–72 bombing halt, they were one of the few things sustaining the prisoners' morale.

The final B-52 strikes were flown on December 29, 1972. Following a New Year's halt, bombing was restricted to below the twentieth parallel. On January 15, 1973, agreement was reached and all bombing of North Vietnam stopped. It was announced that reconnaissance flights would continue over North Vietnam. They would be conducted by SR-71s and low-altitude drones. One Pentagon official said, "The use of pilotless drones is no change and is one method we have used whenever bombing missions over North Vietnam have been halted." It was as close to an official acknowledgment as had been made. The final flight before the cease-fire went into effect was a 147SC/TV, dubbed "The Last Picture Show." A total of 570 drone missions had been launched in 1972. Of this, 466 were SC drones (52 were lost), while the 147TEs amounted to a mere 69 flights.[87]

AFTERMATH

The cease-fire agreement was signed on January 27, 1973. The drones were placed on a "hold/standby" status. This lasted only five days. Operations resumed with a pair of flights on February 5. The following day, the first of a series of SC/TV missions was launched. Policing the cease-fire was nearly as demanding as Linebacker had been: 444 drone missions were launched in 1973.

The year saw the debut of the final two members of the 147S family, the 147SD and SDL. The SD was designed with an improved navigation system with an accuracy of 1.1 miles per 100 miles. (The SC's accuracy was 3 percent.) The SD also had an improved radar altimeter, a new cooling system to cope with low-altitude flight in hot tropical weather, and external tanks to extend the range. The first 147SD mission was flown in June.

The other was the SDL. This was a 147SD equipped with a Loran radionavigation system, which provided even greater accuracy. The first two missions had actually been flown in August 1972, but both were lost. The cause was traced to interference from the navigation system. Normally, the drone would bank up to a maximum angle. Once at this point, the guidance system would not accept any further bank commands. It was found that the Loran was generating an override signal to the roll control; the bank angle increased and the drone went out of control.

Survivability of these last-generation drones was also phenomenal. By November of 1973 there had been 100 launches without a loss. The 147SC drones were designed for an average lifetime of two and a half missions each. They would far surpass this. The record holder was "Tom Cat," with 68 missions, each covering an average of twelve targets. The runners-up were "Budweiser" (63 missions), "Ryan's Daughter" (52 missions), and "Baby Buck" (46 missions). A 147SC/TV flew 42 missions, an SD made 39, while an SDL made 36 missions.

The final eighteen months of drone operations—between 1974 and early June 1975—saw a total of 518 flights. From the start, it was discovered that the North Vietnamese were violating the peace agreement. Troops, tanks, and SAMs poured into the South. The United States, its spirit broken by the war and increasingly obsessed with the Watergate scandal, was both unwilling and unable to do more than issue feeble protests.

By early 1975, the North Vietnamese began their final offensive. The South Vietnamese army was driven back, while the U.S. Congress cut off all aid. By late April, Saigon was surrounded. The United States began an evacuation, and South Vietnamese aircraft and helicopters began to flee. The final 147S-series mission was flown on April 30, 1975—the day Saigon fell.

With the fall of Saigon, the 147SC and SD drones were put into storage. The 147TF drones continued a little longer. The final flight was made on June 2, 1975. Then they, too, were stored. Although some in Congress objected to the loss of so valuable a capability, the decision stood.

The story of the Model 147 drones was an amazing chapter in the history of U.S. Black aircraft. Using the existing Firebee drone gave it flexibility, while new versions were developed on a short-time scale at low cost. The result was a reconnaissance capability that was unmatched by manned aircraft. In all, 3,435 drone missions were flown against Communist China, North Vietnam, and North Korea. Of these, 1,651 were by 147SC drones. A total of about 1,000 147SC drones were built, in nineteen different versions.

A total of 578 drones were lost—251 were confirmed kills, the vast majority to North Vietnamese air defenses. Another 80 were possible losses to enemy action, 53 were lost in the recovery sequence, 30 in retrieval, and the remainder in other ways.[88] The drones survived the heaviest air defenses built up to that time. One drone had an SA-2 explode within twenty to thirty feet of it and still made it home. One of the high-altitude drones evaded eight MiG intercepts, three air-to-air missile firings, and nine SA-2 launches.[89]

Their accomplishments were many and varied. They had provided the first photographs of North Vietnamese SA-2 construction, MiG 21s, and helicopters; arming and fuze data on the SA-2; and the only low-altitude BDA coverage of Linebacker. The total number of photos they took, over areas too physically or politically dangerous for manned aircraft, is estimated to be 145 million.

The photos these Dark Eagles brought back cast a long shadow. For more than a decade after the defeat in Vietnam, the images of collapse and failure raised doubts about the ability and even competence of the U.S. military. In the years to follow, it was depicted as unable to win and equipped with weapons that did not work.

Until another Dark Eagle, and a night of thunder.

CHAPTER 5

Orphaned Eagle
The Model 154 Firefly

. . . of the four seasons, none lasts forever; of the days, some are long and some short, and the Moon waxes and wanes.

Sun Tzu
ca. 400 B.C.

Despite the failures of the Model 136 and Lucy Lee proposals to gain approval, Ryan Aeronautical remained interested in an advanced drone. There was only so much growth in the basic Firebee airframe. Also, a greater altitude and range, as well as further reductions in radar cross section, would need a completely new design.

The target area for the advanced drone was Communist China. The nuclear test site at Lop Nor, as well as the reactors and reprocessing plants, were beyond the reach of the 147 drones. Even the U-2s were hard-pressed to cover these targets. The losses suffered by Nationalist Chinese U-2 pilots made a long-range drone program that much more attractive.

HATCHING THE FIREFLY

With most of Ryan's efforts directed toward the 147 program, work on the advanced drone remained at a low level. Understanding how shape affected radar return was one major area of study. Unlike the 147 drones, which relied on radar-absorbing blankets, the new design would use shape to make it hard to detect. From time to time, the advanced drone was proposed to the air force, but a place could not be found for it in their plans or funding.

The CIA expressed an interest to Ryan about developing a separate drone program, and a formal proposal was put together. It was given the designation Model 150 "Red Book." Ryan felt uncomfortable about going "behind the back" of the air force with the proposal, and told the CIA that if they did not respond within thirty days, Ryan would feel free to deal with the air force. Within a week the CIA rejected the Model 150 proposal, suggesting Ryan talk to the air force about the project. The CIA had its own, very different, drone project.

Ryan renamed the project "Blue Book," which sounded better and was less suggestive of a project aimed at Communist China. The Model 150 was also revised to the Model 151 through the Model 154 designs. After several years of work and proposals, Ryan felt the time was right for a major effort. This included a formal briefing at SAC Headquarters by the company's founder, T. Claude Ryan. By this time, the 147G drones were conducting overflights of China, and the 147H was beginning development, so the presentation was successful. Because this was a whole new aircraft, rather than a simple conversion of an existing target drone, a design competition was started.

The competition pitted North American Aviation against Ryan. North American had set up a separate division to undertake drone work. Additionally, Northrop tried to enter the emerging competition, but the air force refused its proposal. The range and altitude requirements for the drone were similar to what Ryan had proposed, and the company felt confident its design would be selected. North American put up a strong challenge, and for a time it seemed likely to win. Ryan emerged victorious, however, and won the development contract in June 1966.[1]

THE MODEL 154 FIREFLY

The new drone was called the Model 154 Firefly. The fuselage resembled that of the Model 136, with the engine over the fuselage and inward-canted fins. The sloped, flat sides were designed with reduced radar return in mind. The fins were tilted inward to both reduce radar return and shield the exhaust. Much of the airframe was made of plastic, which also absorbed radar signals. Infrared suppression was provided by placing the engine above the fuselage and mixing the hot exhaust with cool intake air. Active ECM equipment would provide further protection.

The total length was 34.2 feet, while the swept-back wings spanned 47.68 feet. The 154 had a maximum altitude between 72,000 and 78,000 feet. As with the 147 drones, it would be launched by a DC-130 and recovered in midair by a helicopter. It was equipped with a KA-80A camera able to provide coverage along a 1,720-mile strip. A highly precise doppler-inertial system handled the navigation. Due to the sensitive onboard equipment, the 154 was also to be fitted with a destruct system on operational missions. In every aspect, it pushed the state of the art in drone technology.

It was not surprising that the 154 program was soon behind schedule. The major problem was the guidance system, which was supposed to be accurate to 5 miles per 1,000 miles flown. The system had five different operating modes. Even if it suffered a complete failure, the drone could still automatically fly to a recovery zone. It was soon clear that everyone had been overly optimistic. The project was also overmanaged. Unlike the streamlined Big Safari management, as many as two hundred people attended the monthly progress meetings.[2]

The first 154s were delivered in early 1968. A total of twenty-eight Model 154 drones was produced. This consisted of one static test vehicle (STV), two captive test vehicles (CTV), five flight test vehicles (FTV), and twenty production vehicles, numbered P-1 through P-20.

The initial tests included both captive flights aboard the DC-130 launch aircraft and tests of separation characteristics and recovery parachute operations. The first powered flight was made on September 10, 1968, at Holloman Air Force Base. The early free flights were restricted to the White Sands Missile Range and tested aerodynamics, performance, and stability. Starting in 1969, testing picked up. A total of forty-two free flights were made, including long-range flights from White Sands to Utah and back again. Between April and July, four Model 154 drones were lost due to control problems and recovery accidents.[3]

To this point, the Model 154 Firefly, like the 147 drones, was a Black project. That changed on August 4, 1969.

EXPOSED

154P-4 was on a long-range test flight when a warning light came on at the Holloman control center. A control surface actuator had failed, and the drone was seconds from going out of control. Ground control triggered the parachute recovery to save the drone, but the troubles of 154P-4 were only starting. It was descending toward the Los Alamos complex, during the lunch hour. Thousands of people saw it coming down under a 100-foot-diameter parachute. Suddenly, there were bright flashes as doors on its underside were blown off, and the bags used to cushion the landing impact inflated.

The drone missed a three-story building and landed on a road just inside the complex. A noontime jogger was starting his fourth lap when 154P-4 came down in front of him. The tip of the right wing slid under the guardrail at the edge of the road.

The 154P-4 was undamaged, but there was still worse to come. Although the landing site itself was secure, only a few feet away was the fence marking the boundary with public land. Just across a narrow canyon was a residential area, and word quickly spread of what had happened. A few Los

Alamos employees realized the strange airplane was probably classified, and hurriedly covered it with tarps. But before 154P-4 could be hidden, the press arrived and were able to photograph it from the perimeter fence. By the time an air force–Ryan recovery crew arrived an hour after landing, the fence was lined with reporters and television crews. Others had climbed trees for a better look. The Model 154 Firefly had made a very public debut.

The *Albuquerque Journal* carried the headline "Secret Something Falls to Earth." The article said that "the emergency descent by parachute of a super secret unmanned aircraft . . . ripped some security wraps off 'Firefly.'" Two photos of 154P-4 were published in the Los *Alamos Monitor,* despite air force requests that they be withheld.[4] The *New York Times* carried a small, two-sentence report on the incident on page 24.[5]

The air force released a cover story that the Firefly was simply a "relatively high altitude test of an Air Force target drone" (the story originally developed for the "Q-2D" test flights in 1960). No one was fooled; it was clear the Firefly was a secret project. As the *Albuquerque Journal* said: "If Firefly is simply a high altitude target drone for testing missile systems, the reason behind the strict Firefly or drone aircraft security lid remains a mystery."[6]

The failure was traced to use of low-temperature solder. When it got hot, the solder softened, the wire pulled free, and the secret was out.

FINAL TEST FLIGHTS

Following the accident, the 154 was grounded for several weeks while an investigation was conducted. When flights resumed, they were restricted to the White Sands area. A flight by P-5 in September was successful, and the range restrictions were lifted. During subsequent 154 flights, a CH-3 helicopter was placed on alert at Holloman Air Force Base. Should the 154 land outside the recovery zone, the crew would fly out and secure the drone before it could be further compromised. Two long-range flights were made by P-4 without problems. Another flight on November 21, 1969, almost resulted in more publicity.

The ill-fortuned P-4 was flying over the Navajo reservation in northern Arizona when there was a circuit failure. The 154 went into an automatic recovery. The DC-130 crew saw a group of people around the drone. The plane buzzed them to warn them off. When the recovery crew arrived, they were pleased to discover the 154 had been secured by the tribal police. One of the people who found it was an ex–air force sergeant who realized the drone was a secret aircraft. He called the tribal police, saw that a perimeter guard was set up, gathered up the parachute, and took charge in a very professional manner.

Three long-range navigation flights were conducted in early 1970, followed by a final series of eight tests at Edwards Air Force Base between August and December 1971. These flights reached altitudes of 81,000 feet. The sixth and seventh flights also involved simulated Soviet Fan Song B and E radars at the navy's China Lake facility. These were the radars used by the SA-2. It was found that the 154 was nearly impossible to detect. The drone's small radar cross section alone was enough to protect it. By December 1971, the problems with the 154 were finally solved. The drone program had achieved a capability rivaling that of the U-2.[7]

But now it had no place to go.

THE END OF THE FIREFLY

The Model 154 had been designed for overflights of China. This required a low radar return to prevent detection, a very high altitude to avoid interception, and a precise navigation system to cover the target. With President Nixon's trip to China, this possibility ended. There was no interest in using the 154s over North Vietnam because of the success of the 147SC low-altitude drones. There were suggestions that the 154 be used over Cuba during the spring of 1972. Following the outbreak of the Yom Kippur War in October 1973, use of the 154 was again proposed. Once more, it was turned down. The drones were placed in storage, then scrapped.[8]

Although the 154 had its share of problems, these did not cause its downfall. The Firefly had been overtaken by events. Starting in 1969, the high-altitude drone mission had started to fade. The 147T program was cut short, while the number of 147H missions in 1970–71 was reduced to a mere handful. Nixon's halting of Chinese overflights ended a mission that was already coming to a close. Without its primary mission of Chinese overflights, however, the Model 154 Firefly was left an orphan. It was not needed for flights over North Vietnam, while the other possible targets were covered by SR-71 overflights. Both the A-12 and the 147 drones had to wait for an opportunity to show what they could do. When given that chance, these Dark Eagles excelled. The Model 154 was never given that chance.

These circumstances also affected another Black drone . . .

Chapter 6

The Last Blackbird The D-21 Tagboard

Rid plans of doubts and uncertainties.

Sun Tzu
ca. 400 B.C.

The prototype SR-71 made its initial test flight on December 22, 1964. Its takeoff and landing at Palmdale was a public, very "White" debut. On hand were a number of Lockheed dignitaries. Once the SR-71 landed, they boarded a transport and took off. They did not, however, fly west, toward Burbank. Instead, the plane headed east, to Groom Lake.

Another of the Blackbird family was also to make its first, very Black, test flight that day.

EARLY EFFORTS

The next step in reconnaissance drone development was obvious—use the technology of the A-12 to build a very high-speed, high-altitude drone. With performance superior to that of the 147 drones, it would be much more likely to survive than the modified Firebees. The drone could also have a longer range than the Model 147. This meant they could be used to cover targets otherwise out of range.

Following the loss of Powers's U-2 over the Soviet Union, there were several discussions about using the A-12 itself as a drone. Although Kelly Johnson had come to support the idea of drone reconnaissance, he opposed an A-12 drone, contending that the aircraft was too large and complex for such a conversion. Another possibility was to use the A-12 as a launch aircraft for an unmanned QF-104 reconnaissance drone. Several times the possibility was examined, but the CIA expressed no interest.

Although the CIA turned down the idea, Johnson found an ally with Brig. Gen. Leo Geary, director of air force special projects. General Geary arranged $500,000 from the Black projects contingency fund to begin a drone study in October 1962.

To speed up development of the drone, which was initially called the "Q-12," Lockheed planned to use a Marquardt ramjet engine from the Bomarc SAM. On October 24, in the midst of the Cuban Missile Crisis, Kelly Johnson, Ben Rich, and Rus Daniell met with Marquardt representatives. From these discussions, it was clear that the Bomarc ramjet would have to be modified for use on the Q-12.[1]

A section of the Skunk Works shop in Burbank was walled off for use by the Q-12 team. The effort was kept isolated from the A-12, YF-12A, and SR-71 development. Just as the A-12 was more secret than the U-2, the Q-12 was more closely held. It required special passes to enter the walled-off section, which was dubbed the "Berlin Wall West."[2]

Unlike the Model 147 drones, the Q-12 would not be recovered intact. This was done for cost reasons. Trying to put in a recovery system would make the vehicle bigger. It would require one or more large parachutes to lower it to a soft landing. A system that would allow a runway landing would add complexity, take up space, and increase both the cost and weight. One major problem was simply slowing the drone down to a speed at which any recovery system could work.[3]

Instead, Johnson studied the possibility of recovering the Q-12's nose section by parachute. This would include not only the entire camera, but the guidance system. These were more costly than the airframe itself.

By December 7, 1962, the Q-12 mock-up was completed. On December 10, it was sent to a test site for eleven days of radar cross section (RCS) measurements. These showed it had the lowest radar cross section of any Lockheed design—a record that would be held for the next decade. The Q-12 was then returned to Lockheed for test fits of a mock-up of the camera. The drone would use a Hycon camera which used a rotating mirror to provide panoramic photos. The goal was six-inch resolution from 90,000 feet. Wind-tunnel testing of the design was also beginning.

At the same time, engine tests of the Marquardt RJ43-MA-3 ramjet were conducted in a wind tunnel that simulated the Q-12's flight profile. One concern about using a ramjet was "blowouts." When the drone turned, it was feared that the airflow into the engine would be disrupted. The effect was like blowing out a candle. Both Lockheed and Marquardt were amazed by the results. The ramjet could be shut down for as long as forty-five seconds and still relight. The only ignition source was the hot engine parts.[4]

When Johnson presented the results to the CIA in February 1963, he found them unenthusiastic. With the A-12 still far from Mach 3, the U-2 operations against China, and other secret air operations such as Air America, the CIA was overextended and unable to undertake another risky project. In contrast,

Air Force Secretary Harold Brown was interested in the Q-12 as a possible nuclear-armed cruise missile, as well as a reconnaissance drone.[5]

The air force interest seems to have moved the CIA to take action. On March 20, 1963, the CIA issued a contract to begin full-scale development. It assigned responsibility to Lockheed for the airframe, navigation system, and the ramjet. Funding and operational control was split between the CIA and the air force.

DOUBTS ABOUT THE LAUNCH PROFILE

Once the Q-12 got under way, it was clear a major problem in development was the aerodynamics of the Q-12/A-12 launch profile. The Mach 3 shock waves between the two aircraft could interact, creating high temperatures and stresses on the airframe. The aerodynamics of the separation also had to be determined. The Q-12 would have to pass through the shock wave formed by the A-12's forward fuselage. The Q-12 would have to separate cleanly, as there were only a few feet between its wingtips and the A-12's inward-canted fins. The center of gravity and center of lift of the combination would have to be very carefully controlled.

Wind-tunnel tests were made using metal models of the Q-12 fixed to the back of an A-12. For tests of the separation, a more complex arrangement was needed. The Q-12 model was mounted on a movable arm. This would lift it off the back of the model A-12 and provide data on the aerodynamic effects and stresses on the combination. It was not possible to reproduce actual free flight in the wind tunnel, however.[6]

Problems with the launch profile appeared early and continued into 1964. The wind-tunnel data and calculations indicated getting the drone through the shock wave formed by the launch aircraft's fuselage would prove difficult. The ramjet would be at full power, and there could be air-fuel mixture and engine unstart problems as it passed through the shock wave. The launch would also have to be done in a pushover maneuver. It was not until May 1964 that Johnson began to feel any confidence. Although there were still problems with the launch profile, the drag of the combination, and the equipment, Johnson felt they could "haul the thing" through Mach 1.

Ironically, the actual design and production of the drones had gone much more smoothly. The slight changes in the drone design from the RCS and wind-tunnel testing had changed the size of the payload bay. This forced Hycon to redesign the camera. By August 6, 1963, this had been completed, with no loss of resolution caused by the change. By October 1963, the drone's design had been finalized.

At the same time, the Q-12 underwent a name change. To separate it from the other A–12-based projects, it was renamed the "D-21." (The "12" was reversed to "21.") "Tagboard" was the project's code name. The A-12 launch aircraft were similarly renamed, becoming the "M-21." The *M* stood for "mother," while the *D* was for "daughter."[7]

THE D-21 TAGBOARD

The D-21A resembled an A-12 nacelle, wings, and vertical tail. The D-21A was 42.8 feet long, with a wingspan of 19.02 feet. The airframe was built of titanium. The leading edges of the wings, the control surfaces, and the inlet spike were all made from radar-absorbing plastic. The D-21's fueled weight was about 11,000 pounds. The reconnaissance equipment was carried in a 76-inch-long Q-bay just behind the intake. The Hycon HR-335 camera was mounted on the recoverable "hatch." This also carried the inertial navigation system, the automatic flight controls, and the command and telemetry electronics. These were the high-cost elements of the drone. The D-21 itself would self-destruct.[8]

A Marquardt RJ43-MA-11 ramjet, a modified version of the engine originally designed for the Bomarc SAM, powered the D-21. This could propel it at a speed of Mach 3.35. The normal cruising speed was Mach 3.25 at an altitude of 80,000 to 95,000 feet. This was far above that of the 147 drones. The ramjet burned JP-7 fuel, the same as the A-12. The D-21's range was over 3,000 nautical miles.[9] The ramjet's burn time of two hours represented a major technical accomplishment—the Bomarc's ramjets had operated for only ten minutes.[10]

The D-21 was carried on a pylon, which held it at a nose-up attitude. The pylon had latches to hold the D-21, a compressed air emergency jettisoning system, and a fuel line to transfer fuel to the D-21 from the M-21's own tanks. Two additional launch aircraft, M-21 Articles 134 and 135, were built. Article 134 would be used for captive test flights, while 135 would make the actual launches. As with the other A-12 derivatives, the M-21s were two-seat aircraft. The launch control officer (LCO) sat in the Q-bay.

When the D-21/M-21 reached the launch point, the sequence would begin. The first step would be to blow off the D-21's inlet and exhaust covers. These would fragment and leave the ramjet ready to start. With the D-21/M-21 at the correct speed and altitude, the LCO would start the ramjet and the other systems of the D-21. This was controlled via a panel—green lights would indicate the status of each system. The LCO could watch the D-21 through a periscope on the instrument panel. The M-21 launch aircraft had a camera mounted in what was called a "hot pod" to film the separation. With the D-21's systems activated and running, and the launch aircraft at the correct point, the M-21 would begin a slight pushover, the LCO would push a final button, and the D-21 would come off the pylon.[11]

The first D-21 was completed in the spring of 1964. As with the U-2s and A-12s before them, the D-21s were given Article numbers. The first was Article 501, with seven D-21s planned for completion by the end of the year. The early D-21s were in a natural metal finish with the outer half of the wings in black. This marked the radar-absorbing plastic material.

After four more months of checkouts and static tests, the aircraft was shipped to Groom Lake and reassembled. Lockheed test pilot William Park was selected to make all the captive and launch tests. Everything was judged ready for the first captive flight. A bridge crane on the ceiling of the hangar was waiting to lift the D-21, swing it over, and lower it onto the M-21.

TAGBOARD FLIGHT TESTS

The first D-21/M-21 captive flight was scheduled for December 22, 1964. This was also the date for the SR-71's first flight at Palmdale. The dual first flights would be a big plus for Lockheed. The SR-71 test flight was planned for 8:00 A.M., with the D-21/M-21 takeoff to follow at 12:00 P.M. This would allow time for Kelly Johnson to supervise the SR-71 flight, then fly out to Groom Lake. Minor problems pushed back the SR-71's flight until about noon, however. At Groom Lake, William Park and the rest of the D-21/M-21 team awaited their chance. Finally, at about 2:00 P.M., a Lockheed Jetstar landed, Johnson ran up the ladder, patted Park on the head, and told him to take off.

As the D-21/M-21 flew above the snow-dusted hills, Park checked out its handling. Despite the added drag and weight of the D-21, and low-power engines in the M-21, Park found the combination's flight characteristics were the same as the A-12. For this first flight, the M-21 had only a partial fuel load. As with the SR-71 earlier in the day, the M-21/D-21 went supersonic. Due to the late takeoff, it was getting toward dark when the D-21/M-21 and its F-104 chase planes landed.[12]

Because the combination represented a whole new design, the speed and altitude envelope had to be explored in small steps. Johnson had hoped to make the first D-21 free flight on his birthday, February 27, but it was not to be. A problem with control surface flutter appeared with the D-21. During one captive flight in April 1965, both the D-21's elevons were ripped off by flutter. This required a redesign, which involved adding balance weights and control surface locks. By May, the flutter problem had been corrected, and the combination had reached Mach 2.6.

It was at this time that the LCOs for the test launches were selected. Ray Torick and Keith Beswick would alternate as launch officer on the missions. The launches were to be made over the ocean, west of Point Mugu, on the California coast.

Progress was painfully slow during the summer and fall of 1965; the D-21, its launch mode, the complexity of the D-21's systems, the aerodynamic combination of the two aircraft, and the technical problems of operating a high-speed drone all proved difficult. A new test range was needed, as the D-21/M-21 could not accelerate sufficiently within the range being used for the early tests. This was followed by performance problems; the weight and drag of the D-21 cut into the M-21's speed and range. The D-21/M-21 also showed poor transonic acceleration, particularly on hot days. Several attempts were made to fly to Point Mugu for launch practice, but the plane could not make the range. More powerful engines were fitted to the M-21s, along with a new inlet control system, but the problems persisted.

The old uncertainties with the launch profile also reappeared. The instrumentation system's strain gauges could not measure the launch forces, due to the heat of Mach 3-plus flight. Johnson was unwilling to commit to a launch until he was sure the separation maneuver was understood. The program was effectively stalled. The first batch of D-21 drones had been completed, but Lockheed was yet to launch a single free flight, and Johnson was unwilling to recommend building any more D-21s until it had been proven in flight.

A final problem facing the D-21 program was how to separate the intake and exhaust covers. Fragments could enter the ramjet and strike the M-21. These fears were justified on the first, and only, attempt to separate the covers. The pieces tore up the chines of 503, causing major damage.

It was decided to leave the covers off. Ironically, this also provided the solution to the drag problem. With the covers off, the ramjet could act as a third engine for the M-21 during the acceleration to launch speed. It would be started at Mach 1.24. Just before launch, fuel from the M-21's tanks would be transferred to top off the D-21, replacing that used during the run-up to the launch point.

By late January 1966, more than a year after the first captive flight, everything seemed ready. The launch forces were understood, the ramjet operations and hatch recovery had been proven, and the launch maneuver had been practiced. Although the Minneapolis-Honeywell guidance system was not ready, this would not affect the test flights.[13] As with the other A-12s, Article 135 was painted all black with white markings. The D-21s were also painted all black. (Article 134 would remain in the silver and black finish.)

D-21 LAUNCHES

The first D-21 launch was made on March 5, 1966. William Park was the pilot, while Keith Beswick acted as LCO. On this first mission, D-21 Article 503 had only a partial load of fuel. After takeoff, the combination headed east. Over Texas, the M-21 rendezvoused with a KC-135 tanker, refueled, and started the run to the launch point. This began over the town of Dalhart, Texas, on a direct course toward Point Mugu. The D-21/M-21 accelerated slowly at first as it crossed New Mexico. When the proper speed was reached, the ramjet was started. The combination began to pick up speed.[14]

As the D-21/M-21 neared the launch point, it was flying at a speed of Mach 3.2 and altitude of 72,000 feet. Once the checkout was completed and the tanks of the D-21 were topped off, Park began a slight climb, followed by a pushover into a slight dive, holding 0.9 gs on the M-21's precise g-meter. Once Park was satisfied with the profile, he gave clearance to launch. Beswick then pressed the release button.[15]

Through the periscope, Beswick could see the D-21 rising off the pylon. There was a small puff of vaporized fuel as it separated. Article 503 held steady as it rose away from the M-21. Then, it stopped and seemed to hang perhaps twenty feet above the back of the M-21. For two or three seconds, the D-21 flew in formation with the M-21. Beswick said later at the debriefing that it seemed to have flown in formation for "two hours." The onboard camera had recorded the sight. Finally, it passed out of the view of the periscope. Article 503 flew for 120 nautical miles before it ran out of fuel.[16]

Despite the successful flight, CIA and air force interest remained limited. This placed a burden on Lockheed, as the D-21 program was strapped for money. Johnson had discussions with air force officials and offered, if necessary, to have Lockheed crews launch the early operational missions. He was also looking at a new launch profile—attaching a rocket booster to the D-21 and launching the assembly from a B-52. Despite the successful launch, he was still concerned about the risks of the M-21 profile.

A second D-21 launch was made on April 27. This time Torick was the LCO. Article 506 flew for 1,200 nautical miles, holding its course within a half nautical mile throughout the flight. The peak speed was Mach 3.3, while an altitude of 90,000 feet was reached. The flight ended when a hydraulic pump burned out. Subsequent investigation indicated it had been run unpressurized several times during ground tests, which damaged it.

The two successful launches sparked renewed interest, and an order for a second batch of fifteen D-21s was issued on April 29. In May, Johnson formally proposed the new launch profile. Using a B-52H as the launch aircraft, he told the air force, would improve safety, cut costs, and extend the deployment range over the short-range M-21.

The third D-21 launch was made by Park and Beswick on June 16. Article 505 flew 1,550 nautical miles and made eight programmed turns to photograph the Channel Islands, San Clemente Island, and Santa Catalina.

The flight was perfect until hatch separation. Due to an electrical problem, this did not occur.[17]

The fourth D-21 free flight was set for July 30, 1966. The D-21, Article 504, would carry a full fuel load for the first time. This meant it was heavier than on any of the previous launches. The launch would be made at a slightly higher speed and at exactly 1.00 g. Park, as before, was flying the launch aircraft, Article 135. His launch control officer was Ray Torick. Article 134 was used as the chase plane for the launch. It was flown by Art Peterson, with Keith Beswick in the backseat to film the separation. The chase M-21 flew about three hundred feet to the right and about one hundred feet behind the launch M-21, at a speed of more than Mach 3.[18]

As the two planes flew in formation, Park reached the launch speed of Mach 3.25, began the shallow dive, and Torick started the separation sequence. From analysis of the data, it appears that Article 504 climbed more slowly through the shock wave and suffered an unstart. This caused it to strike the back of Article 135. The M-21 pitched up and the aerodynamic loads tore off the forward fuselage. Park and Torick were subjected to incredible g forces as the fuselage tumbled. The cockpits depressurized and Park and Torick's suits inflated. The two crewmen ejected and landed in the ocean 150 miles offshore. Park was picked up by a helicopter, but Torick, having survived a Mach 3 breakup and ejection, drowned when sea water entered his pressure suit.

Kelly Johnson was devastated by the death of Torick and personally canceled the D-21/M-21 program.[19] He had long feared launch problems and was unwilling to see any more pilots killed. He concluded that the Mach 3 launch of so large an aircraft could not be justified from a safety point of view.[20] A number of D-21s had already been produced, and rather than scrapping the whole effort, Johnson again proposed to the air force that they be launched from a B-52H. This, however, would require major modifications to the D-21. It would take a year to complete the work.

THE D-21B

Using the B-52H as drop plane entailed a complete rebuilding of the D-21s. The process involved removing the outer wing panels, inlet cone, and ramjet from the airframe. Attachment points were added to the top of the fuselage for the pylon and to the bottom for the rocket booster. Once this was completed, the drone would be reassembled and its systems checked out. Due to the major changes made, the drone was redesignated the "D-21B." The program code name remained Tagboard. Conversion work was under way in Burbank in late 1966 and early 1967.

The rocket booster used to propel the D-21B to ramjet ignition speed was longer and heavier than the D-21 itself. The booster was 44.25 feet long, had a diameter of 30.16 inches, and weighed 13,286 pounds. It was cylindrical with several ridges, giving it the appearance of a water pipe. At the pointed nose was the propeller of a ram air turbine, which spun to provide electrical power. The solid rocket motor produced an average thrust of 27,300 pounds and burned for 87 seconds. To stabilize the assembly during the burn, a fin was attached to the bottom of the booster. To provide ground clearance, it folded to the right while attached to the B-52. The total weight of the D-21B and its booster was over 24,000 pounds.

Two B-52Hs were modified to act as launch planes. The major modification was the addition of two large pylons to hold the D-21B. These were much larger than the pylons for the Hound Dog cruise missile normally carried by B-52Hs. They bolted to the existing attachment points and involved no changes to the wings' structure. Inside the two B-52s, two LCO stations were added to the rear of the flight deck; each station was independent, with its own command and telemetry system, as well as a periscope.

The command system allowed the LCO to activate postlaunch functions normally operated by the drone's programming (engine ignition, booster jettison, telemetry, hatch ejection, and destruct). This provided a backup should the programming fail, or if it was necessary to change the timing.

The telemetry system recorded data on the functioning of the flight control, propulsion, fuel, booster, electrical and hydraulic systems, engine and equipment temperatures, as well as the D-21B's Mach number, direction, and location. This information was used to monitor the launch, and for postflight analysis of any problems. The command and the telemetry systems were duplicated for reliability. A stellar navigation system was also added to the B-52. This was used to update the D-21B's own inertial navigation system during the long flight to the drop point.

Finally, an air-conditioning system provided air to the D-21B for temperature control and to drive the auxiliary power unit (APU).[21] Temperature control was critical, as the D-21B would be "cold soaked" by the negative-58 degree F conditions during the long flight to the launch point. After the drop, the D-21 would be suddenly heated by the acceleration to Mach 3-plus cruise. This put severe thermodynamic stresses on the vehicle and its systems.[22]

The first of the B-52s arrived at Palmdale on December 12, 1966, to begin the modifications. The 4200th Test Wing at Beale Air Force Base was assigned to undertake both the test launches and the operational missions.[23]

Before launch, the LCO would lower the booster fin, turn on the telemetry, test the automatic flight control system, and turn on the fuel and the observation camera. The drop would be made at about 38,000 feet. The assembly would fall free for a moment, then the booster would ignite. It would accelerate forward, then pitch up into the climb. At 50,000 feet, it would go into the final climb trajectory. As it passed through 63,000 feet, the destruct system would be activated; the LCO could destroy the vehicle should it go off-course. At 74,000 feet, the ramjet would ignite. Soon after, the APU would take over the electrical load. When the D-21B reached an altitude of about 80,000 feet, two explosive bolts would fire and the booster would drop off. The D-21B's automatic flight control system would then go to a preprogrammed Mach number. During the first ten minutes of the cruise, the LCO could send a destruct command to the D-21B. After this point, the telemetry and command system was turned off.

The D-21B would be controlled during the overflight by an onboard inertial navigation system. This was preprogrammed with the headings, bank angles, and the camera on-and-off points. Should there be a problem, the D-21B would automatically destruct. This would be triggered by any loss of altitude.

When the mission was completed, the D-21 would fly to friendly airspace. Once clear, the command system, recovery beacons, and telemetry would be turned back on. The D-21B could now be controlled by the JC-130 recovery aircraft. Over the recovery zone, the ramjet's fuel supply would be cut off. The D-21B would slow, then go into a dive. As it passed through 60,000 feet, at a speed of Mach 1.67, the latches on the front edge of the hatch would release. It would open like a door and separate from the D-21B's airframe. The recovery parachute would open, and the hatch would be caught in midair by the JC-130. (The procedure was similar to that used to recover reconnaissance satellite capsules.) If the midair recovery was unsuccessful, the hatch would float, and could be picked up by a ship. The D-21B would continue its descent, until it self-destructed at 52,000 feet.[24]

Support for the Tagboard program was starting to grow. On January 18, 1967, Kelly Johnson had a meeting with Deputy Secretary of Defense Cyrus Vance. Vance said he was very much for the project and asked Lockheed to press forward with it. He said that the U.S. government would never again fly a manned aircraft over enemy territory in peacetime.

D-21B/B-52H TEST LAUNCHES

By late summer of 1967, the modification work to both the D-21Bs and the B-52Hs was complete. The test program could now resume. In effect, the development program would have to begin again from scratch. The three successful D-21 flights had provided limited data. They had flown only half the D-21's full range, and the hatch had not yet been successfully recovered. Checks of the navigation system and camera were yet to be made.

The test missions were flown out of Groom Lake, with the actual launches over the Pacific. The first D-21B to be flown was Article 501, the prototype. The first attempt was made on September 28, 1967, and ended in complete failure. As the B-52 was flying toward the launch point, the D-21B fell off the pylon. The B-52H gave a sharp lurch as the drone fell free. The booster fired and was "quite a sight from the ground." The failure was traced to a stripped nut on the forward right attachment point on the pylon. Johnson wrote it was "very embarrassing."[25]

The first actual D-21B/B-52 test launch was made on November 6, 1967. It was also a failure—Article 507 was boosted to altitude, but nosed over into a dive after flying only 134 nautical miles. A second launch on December 2 flew a total distance of 1,430 nautical miles before Article 509 was lost. The third attempt, with Article 508 on January 19, 1968, flew only 280 nautical miles. Several other attempts had to be aborted before launch due to technical problems. Johnson felt another D-21B failure would result in the program being canceled, so he organized a review panel to look at the problems.[26]

The resumptions of D-21 tests took place against a changing reconnaissance background. The A-12 had finally been allowed to deploy, and the SR-71 was soon to replace it. The Model 147 drone program was in full swing, with both the high- and low-altitude drones being flown. Finally, the Nationalist Chinese U-2 program was being ended. The latter was the most important—the 147 drones could not cover targets deep inside mainland China. With a 3,000-nautical-mile range, the D-21Bs could act as a replacement.

At the same time, new developments in reconnaissance satellite technology were nearing operation. Up to this point, the limited number of satellites available restricted coverage to the Soviet Union. A new generation of reconnaissance satellites could soon cover targets anywhere in the world. The satellites' resolution would be comparable to that of aircraft, but without the slightest political risk. Time was running out for the Tagboard.

It was not until April 30, 1968, that the next D-21B launch was made. Article 511 suffered the same fate as the earlier missions, flying only 150 nautical miles.

The next mission, by Article 512 on June 16, was everything Johnson had hoped for. It flew 2,850 nautical miles—the design range—and also reached an altitude of 90,000 feet. During turns, the ramjet blew out, but it reignited each time. This confirmed the wind-tunnel results. At the end of the mission, the hatch was successfully recovered. Although it had carried no camera, Article 512 had demonstrated the complete mission profile. Events, however, would show there was still much work to be done.

The next two missions ended in failure. Article 514 traveled only 80 nautical miles before it failed, followed by an even less successful flight of 78 nautical miles by Article 516. All these failures put Lockheed in a bind—the D-21B was overrunning costs, and Lockheed had to put its own money into the program. Although Johnson still felt the project had a great deal of promise, he knew there were still very difficult technical challenges ahead.

The final flight of the year gave some optimism. On December 15, 1968, Article 515 flew 2,953 nautical miles. This mission carried a camera; when developed, the photos proved to be fair.

The mission of February 11, 1969, was the first attempt to fly a "Captain Hook" mission profile. This involved a launch near Hawaii, a flight path taking the D-21 over Christmas Island or Midway, then back to Hawaii and the recovery zone. It simulated an operational mission. The D-21B was lost after flying 161 nautical miles. Lockheed believed the cause was water in Article 518's guidance system, but this could not be proven.

This was the final disappointment of the test program. Two fully successful flights followed—Article 519 on May 10, 1969, and Article 520 on July 10, 1969. Both missions were in excess of 2,900 nautical miles, and both hatches were recovered. The photos from the first mission were fair, while the second were considered good.

The twin successes demonstrated the D-21B's design performance, and the air force now considered it ready for operational flights. The air force and CIA proposed to President Nixon that the D-21B be used on "hot" missions over Communist China. Kelly Johnson felt the probability was high that approval would be given. If the D-21B was successful, he felt it had a great future. In anticipation of the approval, the remaining D-21Bs were brought up to the configuration of Articles 519 and 520. Johnson also began looking at ways to recover the complete airframe. The initial recovery studies looked promising.[27]

In the early fall, Nixon gave the go-ahead.

SENIOR BOWL—THE TAGBOARD OVER CHINA

With the approval to start operational missions, activities shifted to Beale Air Force Base. The two B-52Hs were moved to the base. It was home of the 465th Bomb Wing, so the two modified aircraft would not be noticed. The D-21B overflight missions were code-named "Senior Bowl."

The first China overflight by a D-21B was made on November 9, 1969. The B-52H, loaded with two D-21Bs, took off from Beale in the predawn hours and flew west. After twelve hours in the air, it was refueled by a tanker. The B-52H reached the launch point after fourteen hours. This was outside the Chinese early-warning radar network. Article 517 was successfully launched and headed toward China. Its target was the nuclear test site at Lop Nor. The guidance system had all the checkpoints programmed into it. Once the target was reached, the D-21 was to repeat the maneuvers in reverse order to return to the recovery area. After the telemetry was shut off, however, Article 517 just disappeared. The Chinese never detected it, but Article 517 never reached the recovery zone.[28]

Following the loss, Johnson had the guidance system reprogrammed. This allowed the D-21B to miss a checkpoint and still be able to go on to the next. Inability to do this was the suspected cause of 517's disappearance.[29]

In the wake of the loss, it was decided to fly another test mission. This was made on February 20, 1970, with Article 521. It flew a Captain Hook mission with a total distance of 2,969 nautical miles. It reached an altitude of over 95,000 feet. It followed the programmed flight path within two or three miles. The hatch was recovered, and the photos were good. Lockheed was told to be ready for a second operational mission in March 1970. As events turned out, political considerations caused the program to be idle for nearly a year. It was not until late 1970 that a second "hot" mission was authorized.

The launch was made on December 16, 1970. Article 523 flew the mission to Lop Nor successfully. Over the recovery zone, the hatch separated, but the parachute failed to open and it was destroyed on impact. Several more months went by before authorization was given for additional flights.

The third operational flight was launched on March 4, 1971. The flight of Article 526 was successful, but the attempt to recover the hatch ran into problems. The midair recovery was unsuccessful, and the parachute was damaged. The hatch splashed down, and a U.S. Navy destroyer headed toward the floating payload. During the recovery, the ship "keelhauled" the hatch and it sank. Another ship spotted the floating Article 526 airframe, but was unable to get a cable around it before the D-21B sank.

The fourth, and what proved to be the last, D-21B overflight was made on March 20, 1971. Article 527 flew 1,900 miles into Chinese airspace, but was lost on the final segment of the route, over a very heavily defended area. Published accounts do not indicate if it was lost to Chinese air defenses or due to a malfunction.[30]

THE END OF TAGBOARD

On July 15, 1971, Kelly Johnson received a wire canceling the D-21B program. The remaining drones were transferred from Beale Air Force Base by a C-5A and placed in dead storage. The tooling used to build the D-21Bs was ordered destroyed. On July 23, Johnson went to Beale to hold a final farewell to mark the disbanding of the 4200th Test Wing. He concluded by saying, "The remarkable part of the program was not that we lost a few birds due to insufficient launches to develop reliability, but rather that we were able to obtain such a high degree of performance with such low cost compared to any other system."[31]

In all, thirty-eight D-21s had been built between 1964 and 1969 (Articles 501 to 538). Of these, twenty-one were flown—four in the M-21 launches, thirteen in B-52H test missions, and four in the overflights. Although two operational D-21Bs were able to reach the recovery zone, no photos were recovered.

Like the A-12 Oxcart, the D-21B Tagboard drones remained a Black airplane, even in retirement. Their existence was not suspected until August 1976, when the first group was placed in storage at the Davis-Monthan Air Force Base Military Storage and Disposition Center. A second group arrived in 1977.[32] The seventeen survivors were Articles 502, 510, 513, 522, 524, 525, and 528–538. They were labeled "GTD-21Bs" (*GT* stood for ground training). Davis-Monthan is an open base, with public tours of the storage area, so the odd-looking drones were soon spotted and photos began appearing in magazines.[33]

The early reports about the D-21Bs underline how elusive the facts about a Black aircraft can be. The early stories speculated that they had been "proof-of-concept" test vehicles for the A-11, that they had been an interim reconnaissance aircraft, used until the SR-71 was operational, and that they had been carried under the belly of the A-11/YF-12A.[34]

It was not until 1982 that a single photo each of the D-21/M-21 and D-21B/B-52H combinations were released. By the mid-1980s, more substantial information was available. The details of the loss of Article 135 were published. The accounts also said that "fewer than five" overflights were made, and that one camera package had been lost during an ocean recovery.[35] The B-52H portion of the Tagboard program remained a blank—it was not clear what year the test launches began nor when the overflights were made. Performance of the D-21 was also not clear. Published accounts gave estimates as high as Mach 4 and 100,000 feet. The published range varied between 1,250 and 10,000 miles.[36]

In 1993, a film entitled *"Kelly's Way"* was produced for the Edwards Air Force Base Flight Test Museum. It included shots of the D-21 being loaded on the M-21, in-flight shots, and film of two successful D-21 launches. There was no footage of the B-52 launches.

It was not until publication of Jay Miller's book, *Lockheed's Skunk Works: The First Fifty Years,* in late 1993 that the details were finally released. Miller's book used Kelly Johnson's own logs and official documents to tell the story of the "Blackest" of the Blackbird family. A year later, Ben Rich's book, *Skunk Works,* gave a personal account of the Tagboard.

That same year, the surviving D-21Bs were released to museums. The Air Force Museum at Wright-Patterson Air Force Base received one, as did the museum at Dover Air Force Base, and the Pima County Air Museum.[37] Another was particularly appropriate. After the M-21 program was canceled, Article 134 was placed in storage. When the A-12s and SR-71s were sent to museums, Article 134 was given to the Seattle Museum of Flight. It was still in the original silver and black finish, but the pylon had been removed. When the D-21Bs were released, museum volunteers built a new pylon. D-21B Article 510 was then mounted atop Article 134. On January 22, 1994, the new display was opened.[38]

NASA also saw the possibilities of the D-21Bs as test aircraft. The Ames-Dryden Flight Research Center was able to get four of them, Articles 513, 525, 529, and 537. No test program had been determined, but the opportunity to acquire such a vehicle was too good to pass up. One obvious program would be testing of scramjet engines, similar to those planned for the X-30. The launch would be made from one of NASA's SR-71s. It was felt that the unstart problem that had caused the fatal crash had been solved. Test instrumentation would replace the camera package. As with the earlier mission profile, the D-21B would be lost at the end of the flight. Any such NASA D-21B flight program would be made a quarter century after the last operational mission. In the meantime, D-21B Article 525 was loaned by NASA to Blackbird Park, where it joined the prototype A-12 and an SR-71 on display.

THE TAGBOARD ASSESSED

The D-21B Tagboard was the ultimate expression of the Black reconnaissance aircraft. The end of the D-21 program brought to a close an era of Black airplane development. The first Black airplanes, the XP-59A and P-80, had been tactical fighters. Starting in the early 1950s, the emphasis shifted to reconnaissance aircraft. The following two decades saw the X-16, U-2, Sun Tan, A-12, 147, 154, and, finally, the D-21.

On the surface, the Tagboard program was cut short by President Nixon's ending of China overflights. As with the Model 154, there was also a deeper reason. The cancellation of the D-21B program was not only the result of changing politics, but also a changing reconnaissance situation.

On June 15, 1971—one month before the D-21B program ended—the first Big Bird reconnaissance satellite was orbited. It was built around a large telescope and had a resolution of six inches from over 100 miles high. Improved Big Birds would operate for as long as 275 days.[39] Johnson had long realized the effect satellites would have. In 1959, when the A-12 project was just getting started, he asked the CIA whether there would be one round of aircraft development or two before the satellites took over.

Both agreed there would be only one.[40] This proved accurate—while the SR-71 served for a quarter of a century, the A-12 and D-21 both had brief operational lives.

The D-21 Tagboard was as challenging as anything undertaken by the Skunk Works. The development problems were not unusual, given the complexity of the D-21's mission profile. The development program was more akin to that of a missile than an aircraft. Like a missile, each D-21 would be lost at the end of the flight—whether it succeeded or failed. All that Lockheed had to go on to determine the cause of any problems was the telemetry. Development of the early missiles was a prolonged process.

GHOST

In February 1986, D-21B Article 517 finally came home. After the guidance system malfunctioned on the first overflight, it had kept going and reached Siberia before self-destructing. The shattered debris rained down from the sky. One of the pieces, a panel from the engine mount, was found by a shepherd, who turned it over to Soviet authorities. Seventeen years after the November morning it was launched, a CIA official walked into Ben Rich's office with the panel. Rich, now head of the Skunk Works after Kelly Johnson's retirement, asked where he had gotten it. The CIA official laughed and said, "Believe it or not, I got it as a Christmas gift from a Soviet KGB agent." The panel, composed of radar-absorbing material, looked as if it had just been made.[41]

As they talked, another Dark Eagle was being built in the same hangar that had seen production of the D-21s. It did not have the thundering speed of the D-21; in fact, this new plane was a subsonic attack aircraft. Unlike the sleek, manta ray shape of the D-21, the new Dark Eagle was angular. It had a form that was a violation of every aerodynamic principle built into airplanes since the Wright Brothers. This strange shape, crafted with the utmost care, had only one virtue.

It was invisible.

Chapter 7

The Dark Eagles of Dreamland
The Have Blue Stealth Aircraft

I conceal my tracks so that none can discern them; I keep silence so that none can hear me.

Sun Tzu
ca. 400 B.C.

A common thread running through the history of the postwar Black airplanes was the quest for a reduced radar cross section. It was hoped that the U-2 would fly so high it would be difficult to pick up on radar. Tests over the United States seemed to justify this hope, but once overflights began, the Soviets had no major difficulties tracking it. Attempts were made to reduce the U-2s' detectability, but these proved ineffective.

Based on this experience, Kelly Johnson realized the A-12 would have to be designed from the start for a reduced radar cross section. The important word was *reduced*—the North Vietnamese and Chinese were able to detect the A-12s. Taken together, the A-12s' speed, height, and reduced radar return made them unstoppable.

With the Ryan drones, both approaches were taken. The Model 147 Lightning Bug drones were modified with radar-absorbing blankets. This made the former target drones difficult to shoot down, as the Chinese and North Vietnamese soon learned. With the advanced Model 154, a reduced radar cross section was built in.

In all these cases, however, the reduced radar cross section was only one of the design considerations. The maximum possible altitude was the driving requirement in the design of the Black reconnaissance airplanes. But by the early 1970s, a reduced radar cross section became the dominant consideration in the design of new aircraft. This became known as "stealth."

A PREHISTORY OF STEALTH

The first attempt to build an "invisible" airplane was made in 1912. Petrocz von Petroczy, an officer with the Austro-Hungarian air service, covered a Taube with clear sheets of a celluloid material called Emaillit. The theory was that a transparent covering would make the plane harder to see and hit with ground fire compared to a fabric-covered plane silhouetted against the sky. The Taube was test flown in May and June 1912. *Flight* magazine reported that the plane was "unable to be located by those present on the ground when flying at an altitude of between 900 and 1,200 feet . . . [At 700 feet] only the framework is dimly visible and this and the outline of the motor and pilot and passenger present so small an area for rifle or gun fire, that . . . accurate aiming at such surfaces becomes nearly impossible."

It was the Germans who soon took the lead. An engineer named Anton Knubel built two monoplanes with clear coverings in 1913–14. The second of the planes had its framework painted a blue gray color to make it even harder to see against the sky. In August 1914, World War I started. In 1915, Knubel built a biplane to test its military applications. Unfortunately, Knubel was killed in a crash of the plane on September 8, 1915.

The idea was seen as having promise, and three Fokker E III fighters were delivered in the summer of 1916, covered with Cellon. Unlike celluloid, it did not burn or shatter. Cellon had found wide use in the automotive industry as a glass substitute. Cellon was soaked in water to expand the sheets. It was then attached to the plane's framework and allowed to dry to a taut finish. The material was called D-Bespannung (*Durchsichtige Bespannung* or "transparent material under tension").

The trio of E III fighters appear to have seen limited combat. On July 9, 1916, the No. 16 Squadron of the British Royal Flying Corps reported that "a transparent German aeroplane marked with red crosses was pursued by French machines in the Somme area." Several more German aircraft were tested with the Cellon coating. These included four observation-light bombers: an Albatros B II, an Aviatik B, an Aviatik C I, and a Rumpler C I. Two heavy bombers, a VGO I and R I, had their tails and rear fuselages covered with the material.

Very soon, however, it was apparent this first attempt at a stealth airplane was a failure. A report dated July 11, 1916, states: "In clear weather, the aircraft is more difficult to spot, but in cloudy weather, it appears just as dark as other aircraft. In sunshine, the pilot and observer are unpleasantly blinded by the reflections." The major problem was the Cellon itself: "During longer periods of rain or damp weather . . . the covering becomes so loose that it would be better not to fly such aircraft . . . The covering itself is strong, but should a shrapnel go through the wing, the whole sheet would tear to pieces."

It was far more effective simply to paint the aircraft in camouflage colors. This could not make the plane invisible, as the German planes attempted to be, but would make the plane less visible.[1]

With the invention of radar in the mid-1930s, a new approach was needed. A variety of countermeasures emerged during World War II. The simplest means was strips of aluminum. Called "chaff" in the United States or "window" in England, the strips would be released from a plane. They would reflect the radar signals and produce false echoes, which would hide the plane. A more active method was to interfere with the radar. Called "noise jamming," the target plane transmitted signals on the same frequency as the radar. As the echo from a plane was a tiny fraction of the radar's original signal strength, it was possible for the plane to drown out the echo, making it impossible to detect the target plane.

With development of jet bombers like the B-47 in the late 1940s, it was thought that they would fly too high and too fast to be detected. This soon proved false, and development of electronic countermeasures (ECM) continued.[2]

During the Cold War, both the ECM and the tactics of its use grew more sophisticated. The first step was to avoid the radar entirely. The Soviet Union was vast, and many areas had little or no radar coverage. The bomber's route would take it through these gaps in the radar. The plane would not transmit any jamming signals, as this would only advertise the plane's presence. As the bomber neared the target, the number of radars would increase, and it would no longer be possible to avoid them. The bomber would then start to drop chaff and jam the radars. A more subtle approach was to transmit carefully timed signals, which made the plane appear farther from the radars, or at a different bearing. This is referred to as "deception jamming." As a last resort, the air defense centers, radars, and SAM sites would be bombed.

All this was based on the idea of hiding a plane's echo. As long ago as the mid-1930s, Sir Robert Watson Watt, who designed the first British radar, realized that bombers could avoid the whole problem by having a reduced radar cross section.[3] The problem was in the details. The radar cross section of a plane depends on three factors: the shape of the plane, the frequency of the radar, and the "aspect angle" between the plane and the radar.

The prime source of a large radar cross section is two or three surfaces, such as a wing and fuselage or the floor, sides, and back of a cockpit, which meet at a right angle. The radar signal strikes one surface, is reflected to the other, then is bounced directly back to the radar. Nor were tubular shapes immune—radar signals striking a round fuselage can actually "creep" around the fuselage and back to the radar. Still other sources are sharp points on the wings or tails, wing fences, external weapons, intakes that allow the front of the engine to be "seen" by the radar, gaps formed by access panels, and antennae.

Frequency has a similar effect. A feature that has a strong radar return at one frequency may not be detectable at another. This is quite independent of size—a small vent or grill may produce a major part of the plane's radar cross section.

The final factor, aspect angle, is the most complex. The interactions between the reflections from each part of the plane cause huge changes in the radar cross section. In some cases, a one-third-degree change in the aspect angle can result in a thirty-twofold change in the radar cross section. When all these factors are taken into account, a plane's radar cross section may vary by a factor of 1 million. A 1947 text on radar design noted:

> Only for certain special cases can [radar cross-section] be calculated rigorously; for most targets [it] has to be inferred from the radar data. . . . Only a rough estimate of the cross-section of such targets as aircraft or ships can be obtained by calculation. Even if one could carry through the calculation for the actual target (usually one has to be content with considering a simplified model) the comparison of calculated and observed cross-section would be extremely difficult because of the strong dependency of the cross-section on aspect.[4]

By the mid-1950s, basic research was underway in the United States on understanding the sources of a plane's radar cross section. A team headed by Bill Bahret at the Wright Air Development Center did much of this work. A large anechoic chamber was built to test the radar return of different shapes.

By the late 1950s, Bahret and his team felt they understood the sources of large echoes. Once they knew this, the obvious next step was to reduce the echoes. This would have two advantages in terms of electronic countermeasures: the amount of power needed to hide the plane's echo would be reduced, and, for a given jammer, the effectiveness would be increased. As yet, there was no intent to build a plane invisible to radar.

A second part of this effort was development of radar-absorbing material (RAM). Since World War II, Dr. Rufus Wright and a team at the Naval Research Laboratory had been working on RAM. Together with Emerson and Cuming Incorporated, a plastics manufacturer, they had developed a practical RAM. The material was in the form of thin, tilelike sheets. It was pliable like rubber and could be cut and formed into any shape. The navy lost interest in the project, and Wright went to the air force.

The air force was very interested—the RAM was both thin and strong and, therefore, could be attached to the skin of an airplane. After tests with scale models, it was decided to cover a T-33 jet trainer with the RAM. This was to verify the echo reduction predicted by the scale tests. The project was code-named "Passport Visa," although the white-painted T-33 was better known as "Bahret's White Elephant."

The Passport Visa T-33 was completely covered with the RAM. This included the skin, wing tanks, and control surfaces. The plane was only an experiment, with no operational applications in mind. The air force test pilot selected for the project was Capt. Virgil "Gus" Grissom. (The following year he was selected as a member of the first group of astronauts; he would later die in the 1967 *Apollo 1* launchpad fire.)

Test flights began in late 1958. The results were mixed—many of the echo reductions were confirmed, but the T-33's flight characteristics were degraded by the added thickness of material. Grissom found the plane was hard to control; it slid in turns, overdived, and coming in for a landing it behaved like a roller coaster.[5]

Clearly, a plane's radar cross section could not be reduced simply by covering it with RAM. It would have to be designed in. Despite all these efforts, there was no simple way to calculate the radar cross section of a plane. With the computers and theoretical models of the time, too many factors entered into the calculation for it to be a practical possibility.

This meant designers would have to take a crude cut-and-try approach. When Kelly Johnson wanted to test the radar cross sections of the A-12 and D-21, he first used small models. Then full-scale mock-ups were built and tested. From this data, the final designs were developed. Still, it was not until the planes actually took flight that the true radar cross section could be determined.

Such efforts could be made for Black airplanes. Reduced radar cross section had little impact on the design of operational aircraft. Until Vietnam.

PROJECT HARVEY

The air defenses of North Vietnam required a fundamental change in tactics. A typical Rolling Thunder strike was composed of sixteen F-105D bombers. The force needed to protect them was made up of eight EF-105F "Wild Weasels," which attacked SAM sites, and six F-4D escorts against MiGs. Even though each F-105D carried individual ECM pods, two EB-66 jamming aircraft would also accompany the strike force. The EB-66s, in turn, each required two F-4Ds as protection against MiGs. Thus, to protect sixteen bombers, a total of twenty jamming and support aircraft were needed since the support aircraft themselves needed protection.[6] The net result was that most of the available aircraft were diverted from attack missions to defensive roles.

The revolution in air defense caused by SAMs would be underlined in the October 1973 Yom Kippur War. The Egyptian and Syrian armies that attacked Israel were equipped with the new SA-6 Gainful SAM. Mounted on a tanklike transporter, it could move with the frontline troops. The Israeli air force did not have the ECM pods needed to counter the SA-6 and suffered heavy initial losses. During a single strike against a Syrian SA-6 battery, six Israeli F-4Es were lost. The air defenses also prevented the Israeli air force from providing close air support to ground troops.[7]

Although the Israelis overcame the early setbacks, the SA-6 was a clear warning. As long as U.S. countermeasures and tactics were specifically tailored to enemy radars and SAMs, they would be vulnerable to technological surprise. The Soviets were then in the process of deploying a new generation of SAMs. In the event of a war in Europe, NATO forces could suffer the same huge losses as the Israelis had. Many academics theorized the end of manned aircraft was at hand. Technical advances in radar design, such as the traveling wave tube and computers, had increased power and the ability to defeat ECM. Any new technological advances in ECM would be countered by improved radars.

Others realized that a new set of assumptions was needed. Countermeasures had always been based on overpowering the radar. Even Black aircraft with reduced RCS—the A-12 and Model 147-154 drones—used ECM equipment for protection. The key was not more powerful ECM, but to make the RCS a primary design consideration. It would be eliminated, not simply reduced. With no echo, the radar would be blind. No radar would provide early warning as the aircraft approached; no radar would direct MiGs, antiaircraft guns, or SAMs. There would be no need for support aircraft. Air defenses would revert to the 1930s, against an enemy traveling at near supersonic speeds.

The problem was the amount of RCS reduction needed. A tenfold reduction would only shorten the range at which a plane could be detected. A hundredfold RCS reduction would merely degrade the effectiveness of radar. It would take a thousandfold reduction of a plane's RCS to make it undetectable to radar.[8]

Moreover, to be fully effective this reduction in RCS would have to be combined with other design features to reduce detectability. Just as the aircraft could not reflect any radar signals, it also could not emit any—no bombing radar or ECM transmissions. The infrared emissions from the engine would have to be hidden. The engine could not produce smoke. The airplane also would have to be quiet; the sound of a plane gives warning of its approach. The plane could not produce a contrail—this had been a major problem with the Model 147 drones. The final problem was visibility. Although true optical invisibility was not possible, efforts had to be made to reduce the distance at which the plane could be seen. One problem was "glints" from the canopy. A plane could be seen at a distance of five to ten miles; the reflection of the sun could be seen at a distance several times that.

The effort to make this possible became known as "Project Harvey," after the invisible rabbit in the play and film of the same name.[9]

In 1974, the Defense Advanced Research Projects Agency (DARPA) issued requests to five aircraft manufacturers to study the potential for developing aircraft based on a minimal RCS. They were to design a small, low-cost test aircraft to demonstrate the possibilities. It was called the "XST," for "experimental survivable testbed." The companies were General Dynamics, Northrop, McDonnell Douglas, Grumman, and Boeing.[10] All had recent experience with fighter design and manufacturing. Lockheed, which had not built a fighter since the F-104 program of the early 1960s, was not included.

By early 1975, Ben Rich had learned of the program. He had been involved with the work Lockheed had done on the Dirty Bird U-2s, the A-12, SR-71, and D-21, and knew it gave Lockheed the experience needed for the DARPA project. Rich obtained a letter from the CIA granting permission to discuss the reduced RCS work of the earlier projects. This was part of the request to DARPA for Lockheed to be included in the program. The effort was successful, and Lockheed joined the design competition.

The keys to Lockheed's efforts were Lockheed mathematician Bill Schroeder and Skunk Works software engineer Denys Overholser. They produced the conceptual breakthrough that allowed a stealth aircraft to be designed.

Schroeder went back to the basic equations derived by Scottish physicist James Clerk Maxwell a century before. These described how electromagnetic energy was reflected by a surface. Maxwell's equations were revised at the turn of the century by German electromagnetic expert Arnold Johannes Sommerfeld. For simple shapes, such as a cone, sphere, or flat plate, these formulas could predict how radar signals would be reflected. In the early 1960s, a Soviet scientist named Pyotr Ufimtsev developed a simplified approach which concentrated on electromagnetic currents set up in the edges of more complex shapes, such as disks.

The Maxwell, Sommerfeld, and Ufimtsev equations still could not predict the RCS for a complex shape like that of an airplane. Schroeder's conceptual breakthrough was to realize that the shape of an airplane could be reduced to a finite set of two-dimensional surfaces. This reduced the number of individual radar reflections that would have to be calculated to a manageable number. Rather than a surface made of smoothly curving surfaces, the whole airplane would be a collection of flat plates, which reflected the echo away from the radar. This system of flat, triangular panels became known as "faceting," because it resembled the shape of a diamond.

Schroeder asked Overholser to develop a computer program that could predict the RCS of a faceted aircraft shape. It took only five weeks for the Echo I program to be completed. Now, with the faceting concept and the Echo I program, it would be possible to predict the RCS of an aircraft. Possible designs could be tested and refined in the computer. The way was clear to build a truly invisible aircraft.[11]

The initial design was dubbed the "Hopeless Diamond." When Overholser presented a sketch of the design to Ben Rich on May 5, 1975, Rich did not quite grasp what had been achieved. Rich kept asking how big the radar return of a full-size aircraft would be—as large as a T-38, a Piper Cub, a condor, an eagle, an owl? Overholser gave him the unbelievable answer. "Ben, try as big as an eagle's *eyeball.*"

The Hopeless Diamond met with a frosty reception by Kelly Johnson, who was still working as a consultant to the Skunk Works. Having built some of the most graceful planes ever to take to the skies, he was not impressed with this alien design. Johnson's opinion was shared by many of the Skunk Work's senior engineers and aerodynamicists. They preferred a disk-shaped design—a real flying saucer.

A disk had the ultimate in low radar cross section. The convexed surface of the disk would scatter the radar signals away from the source. The problem with a disk-shaped aircraft was control. The WS-606 project of the mid-1950s was to have relied on a large spinning fan to provide gyroscopic stability, while directional control was to be provided by thrusters on the rim of the disk, fed by a complex network of ducts. A disk also has poor aerodynamic qualities, such as high subsonic drag. How to make a flying saucer fly was the problem, and, as Rich later noted, "The Martians wouldn't tell us."

Johnson thought the radar return from the Hopeless Diamond would be larger than that of the D-21. A ten-foot mock-up of the Hopeless Diamond was built. On September 14, 1975, it was tested against the original mock-up of the D-21. The Hopeless Diamond had a radar return one-one thousandth that of the D-21. This was exactly that predicted by the Echo I program.[12]

By October 1975, the DARPA competition had been reduced to the Lockheed and Northrop designs. The Northrop XST was a pure delta wing with a faceted fuselage. The rear of the wing was swept forward, giving it the appearance of a broad arrowhead. The single, large intake was located above the cockpit. To mask the inlet from radar, a fine mesh screen was used. Two tilted fins shielded the engine exhaust.[13]

The Lockheed design, in contrast, had sharply swept-back wings—72.5 degrees. The rear of the fuselage came to a point; with the swept-back wings, this gave it a W shape. The two intakes were placed on the sides of the aircraft and were covered with grills. This allowed a higher speed than the screen on the Northrop XST, which was not usable above Mach 0.65. Twin inward-canted fins shielded the exhaust. These were slotlike and were called "platypus" nozzles.

The XST design philosophy was to have the lowest possible radar return from the front and bottom of the aircraft. As the plane would fly at high altitude, the top was not considered as important.[14]

In December 1975, Lockheed and Northrop built one-third-scale models of their XST designs. These were shipped to the Gray Butte Microwave Measurement Range in New Mexico and mounted on poles for radar signature testing. A second series of tests was run in January 1976 after minor modifications had been made. This was followed by a full-scale RCS model, which was tested at the air force measurement range at White Sands, New Mexico.[15]

The results were a breakthrough in aircraft design. During an early outdoor test, the radar could not detect the model. The radar operator thought it had fallen off the pole. Then a reflection was picked up—from a crow that had perched on it. At the White Sands tests, the reflection from the pole was many times brighter than the model. It was also discovered that the model had to be kept clean. Bird droppings increased the return by 50 percent. The series of measurements showed that the Lockheed design had one-tenth the radar return of the Northrop model.[16]

In April 1976, Lockheed was named the winner. It was to build two XST aircraft for aerodynamic and RCS testing. The contract was for $32.6 million from DARPA and the air force. Lockheed had to add another $10.4 million of its own money. The latter represented a big gamble on Lockheed's part. The Skunk Works had spent much of the late 1960s and early 1970s in an unsuccessful effort to sell a series of fighter designs. At the same time, losses in the L-1011 airliner program had brought Lockheed to the edge of bankruptcy. Even with federal guaranteed loans, Lockheed was still near failure in 1975 and 1976. But the $10.4 million investment was to bring in several billion dollars.[17]

STEALTH GOES BLACK

To this point, Project Harvey was unclassified, and stealth was freely talked about. In June 1975, *Defense Daily* carried a report that the air force was developing a small stealth fighter.[18] In August 1976, *Aviation Week and Space Technology* carried a brief story that Lockheed had won the development contract for a stealth fighter demonstrator.[19] The 1977–78 edition of

Jane's All the World's Aircraft carried a one-paragraph item that a "small" stealth fighter was being built by Lockheed and was expected to fly in 1977.[20] A June 1977 issue of *Aviation Week and Space Technology* revealed that the "Stealth Fighter Demonstrator" used J85 engines, that Kelly Johnson had acted as a consultant on the project, and that it would make its first flight in 1977.[21]

Soon after work on the XST started, Jimmy Carter was elected president. The program attracted the attention of the defense undersecretary for research and engineering, William J. Perry. The results of the model RCS tests indicated that stealth had the prospect of a fundamental breakthrough. As a result, the XST became a Black airplane in early 1977. Control was transferred from the largely civilian-staffed DARPA to the Air Force Special Projects Office. The word "stealth" also disappeared; it could not be used in any public statement or in an unclassified context. The program was pushed, even as the defense budget underwent major cuts.[22]

The program also received a new two-word code name. Unlike Aquatone, Oxcart, and Tagboard, it was a computer selected designation. Because it was an aircraft technology development project, the prefix "have" was given to the program. This new Dark Eagle became the "Have Blue."

HAVE BLUE

Have Blue was the first airplane whose shape was determined by electrical engineering, rather than aerodynamics. Not surprisingly, it had the aerodynamics of a household appliance. The design was inherently unstable in all three axes—pitch (longitudinal stability), roll (lateral stability), and yaw (directional stability). *Every* aircraft *ever* built had curved wing surfaces. On the Have Blue, the wings were made of long, wedged-shaped flat plates, meeting at a sharp edge.

The first Have Blue prototype would be used for aerodynamic and control tests. It had a long (and unstealthy) nose boom for the air-speed system. Because of the design's instability, it used a fly-by-wire control system, built for the F-16A, that was modified to make the Have Blue stable in all three axes. (The F-16 was unstable only in the pitch axis.) Stability was critical if the design was to be developed into an attack aircraft; an unstable aircraft cannot bomb accurately.

The second Have Blue prototype would be used to demonstrate the design's stealth qualities. It had an operational air-speed system and lacked a drag chute. Development work was also done on improved RAM and better ways to apply it. The prototype would also test the practical details. Unlike an RCS model, a real airplane has landing gear doors, a canopy, a fuel-fill door, screws, and vents. Any of these could greatly affect the plane's RCS. On the second Have Blue, greater care would be taken to insure that all gaps were sealed.

The Have Blue aircraft were 38 feet long and had a wingspan of 22.5 feet. This was 60 percent of the size of the planned production aircraft. They would have a top speed of Mach 0.8 and were powered by a pair of J85 engines. These lacked afterburners to reduce the infrared signature. There was no weapons bay and no inflight refueling equipment. Weight of the Have Blue was 12,500 pounds, and it was limited to a one-hour flight time.

To keep the development time short, as many existing components as possible were used. In addition to the F-16 fly-by-wire control system, the Have Blue aircraft used F-5 ejector seats, landing gear, and cockpit instruments. The J85-GE-4A engines were supplied by the navy from its T-2B trainer program. The Have Blue aircraft were built by hand, without permanent jigs (like the XP-80 and U-2s). As each part was designed, the plans were sent to the shop for fabrication. The work was done in a cordoned-off section of the Burbank plant. The two planes did not receive any air force serial numbers or designation, so Lockheed gave them the numbers 1001 and 1002.[23]

Two test pilots were selected to fly the Have Blue. Lockheed test pilot William Park would make the first flights. Park was so highly regarded at the Skunk Works that Ben Rich obtained a special exemption from the air force so he could be chief test pilot for the Have Blue. (He was not a test pilot school graduate, nor did he have an advanced engineering degree.) Years later, he recalled his first impression of the Have Blue: "Aerodynamically, it didn't look like it could fly at all. . . . It really looked like something that flew in from outer space."[24] Lieutenant Colonel Norman Kenneth "Ken" Dyson would serve as the air force project pilot. As events turned out, he would do the RCS testing.

As with other Dark Eagles, the Have Blue personnel had their own patch. It showed the cartoon character Wile E. Coyote holding a blue lightning bolt, signifying control of the electromagnetic spectrum, and the colorful code name. (The Road Runner, Wile E. Coyote's uncatchable nemesis, had been used in the 1960s as the symbol of the A-12 Oxcart.)

The completion of Have Blue 1001 was complicated by a strike at Lockheed. When the strike began in late August 1977, the Have Blue was in final assembly, with no fuel or hydraulic systems, no electronics, no ejection seat or landing gear. A thirty-five-man shop crew was put together from managers and engineers to complete it and check out its systems. They put in twelve-hour days, seven days a week, for two months. The initial engine test runs were done on November 4. To hide the plane, 1001 was parked between two semitrailers and a camouflage net was draped over them. The tests were done at night, after Burbank airport had closed. The only attention the tests attracted was a noise complaint.

Following the tests, the wings were removed and the plane was loaded aboard a C-5A transport for the flight to Groom Lake. The delivery was made on the morning of November 16, 1977. This was the first time a C-5 had flown from Burbank, and quite a crowd gathered. After arrival, the plane was taken to one of the old Lockheed A-12 hangars at the south end of the Groom Lake complex. Have Blue 1001 was reassembled in short order and engine thrust checks were made. Three days before the first flight, these tests uncovered a serious overheating problem. The engines were removed and a heat shield was added. This was made from an old steel tool cabinet.

The Have Blue then underwent four low- and medium-speed taxi tests. During the third run, the brakes overheated. This was to be a nuisance throughout the program. The functioning of the computer-stability system was also checked out, and minor adjustments were made in the yaw gains. The drag chute was also tested, and the plane was judged ready for its first flight.[25]

HAVE BLUE TAKES FLIGHT

Shortly before 7:00 A.M. on December 1, 1977, Have Blue 1001 was taken from its hangar and taxied to the end of the runway. The extreme security measures continued even at Groom Lake. The Have Blue 1001's entire surface was painted in patterns of light gray, black, and tan. This was not camouflage in the traditional sense, but was meant to hide the shape of the aircraft. Anyone observing the plane, either from the ground or from the air, would have difficulty seeing the faceting.[26]

As a further step to limit the number of people who knew of the project, everyone at Groom Lake not connected with Have Blue had been herded into the cafeteria before the plane left the hangar. The test was so secret that Lockheed Chairman Roy Anderson could not attend. The flight was also timed so that no Soviet reconnaissance satellites were in position to photograph Groom Lake during the flight. Both the White House situation room and Tactical Air Command Headquarters were monitoring the activities.

With security in place and the plane ready, Park made a final check and ran up the engines. The engines were quiet compared to a normal jet, due to the radar-absorbing grids. Park went to full power and released the brakes. The plane began to slowly accelerate down the long runway. Without an afterburner, it took nearly the whole length to reach flying speed. The little angular airplane finally lifted off and slowly climbed into the winter sky. As it did, Kelly Johnson slapped Rich on the back and said, "Well, Ben, you got your first airplane."

On this first flight, the landing gear was left extended while Park checked the Have Blue's airworthiness. A T-38 chase plane watched over the Have Blue throughout the flight. When the flight was completed, Park made a fast landing on the runway. Due to the plane's semidelta wings and lack of flaps, the landing speed was a very high 240 knots.[27]

Park was elated with the Have Blue's performance. He recalled years later, "It flew great. It flew like a fighter should fly. It had nice response to the controls."[28] The fly-by-wire control system had transformed the unstable airplane.

Over the next five months, the first Have Blue made a total of thirty-six test flights. Park and Dyson covered most of the aircraft's speed and altitude envelope. Only a few RCS flights were made; the aircraft had not really been intended for such tests. Have Blue 1001 provided data on flight loads, flutter, performance, handling qualities, and stability and control. The plane was unstable in pitch at speeds below Mach 0.3, and static directional stability was less than predicted. The plane was directionally unstable at speeds above Mach 0.65. Side forces were half that predicted by wind-tunnel data. These problems were corrected simply by changing the gain in the flight control system.

The platypus nozzles also affected stability—changing the power setting caused uneven heating, which warped their surfaces. This, in turn, generated forces that were picked up by the stability control system. The computer interpreted this as a change in flight direction and moved the fins to counter it. This resulted in the plane flying "crabbed" slightly to one side. The pilot had to adjust the trim each time the flight conditions changed.[29]

The only major flaw in the design was the high sink rate on landing, which would be corrected in the production aircraft. It was to be the undoing of Have Blue 1001.[30]

On May 4, 1978, Park was about to complete the thirty-sixth test flight when the plane hit the runway hard. Rather than risk skidding off the runway, Park took off and went around again. As he did, he retracted the landing gear. Park did not know that the right landing gear had been bent by the impact. When he lowered the gear, the T-38 chase pilot, Col. Larry McCain (the base commander), radioed that the right gear was jammed. Park added power and climbed. Over the next several minutes, he made several attempts to get the gear to extend. This included making another hard landing to jar it loose.

The fuel supply was running low and there was no time for additional attempts. As Park climbed to 10,000 feet, one of the engines flamed out from fuel starvation. When the other engine quit, he would have only two seconds before the Have Blue went out of control. Park radioed, "I'm gonna bail out of here unless anyone has any better ideas."[31] He then pulled the ejector seat handle, the canopy blew off, and the seat rocketed him out of the plane. As it did, Park's head struck the seat's headrest and he was knocked unconscious. His parachute opened automatically, but he was still unconscious when his limp body hit the desert floor. Park's leg was broken, he suffered a concussion, and his mouth was filled with dirt as the parachute was dragged across the desert by a strong wind. By the time paramedics reached him, Park's heart had stopped. The paramedics were able to save him, but Park never flew again. He was named Lockheed's director of flight operations.[32] The wreckage was examined, then buried at Groom Lake.

The accident was the first indication of the program's existence since it went Black. News reports were headlined "Plane Crash Shrouded In Mystery." It was speculated that Park may have been injured in the crash of a TR-1, a modified version of the U-2R which was then about to enter production. Spokesmen refused to comment, citing "national security reasons." At the end of the articles, "some sources" were quoted as saying it was part of a "stealth program," which was "aimed at developing reconnaissance planes that would be significantly less vulnerable."[33] The true importance of stealth was missed.

RCS TESTING

The second Have Blue, 1002, was delivered in July 1978, two months after the loss of 1001. It made its first flight on July 20. The pilot for this and all its later flights was Lt. Col. Ken Dyson.

Have Blue 1002 was intended for the RCS tests. As such, it lacked the air-speed boom of the first plane. To provide air-speed data, which was critical to the stability system, six measuring points were located in the upper and lower surfaces, the nose, and windshield center frame. Building an air-speed system that was both accurate and stealthy proved difficult; the design of the airframe restricted where such probes could be located. Have Blue 1002 was painted light gray overall rather than the camouflage finish of the first plane. It also lacked a drag chute and was equipped with a steerable nose wheel, which improved ground handling. Most important, it was covered with the RAM coatings and other materials needed to reduce its RCS.

Following several air-speed calibration flights, the baseline in-flight RCS measurements began. Following completion of the tests, modifications were made based on the initial results. A second series of penetration tests was run against ground radars and infrared systems.[34] A cover story to "explain" how an airplane could be invisible to radar was also prepared. The people involved with the tests were told that an ordinary plane was carrying a "black box" in its nose. This emitted a powerful beam which deflected the radar signals.[35]

In the final series of tests, Have Blue 1002 was flown against a simulated Soviet air defense network. These included the SA-6 Straight Flush tracking radar, and the Bar Lock, Tall King, and Spoon Rest early warning radars. These were either actual Soviet radars captured by the Israelis during the 1967 and 1973 wars, or copies built from scratch or modified from U.S. equipment. Tests were also run against airborne radars.

The results were phenomenal—most SAM radars could not detect the Have Blue until it was within the missiles' minimum range. This made it impossible for the SAMs to intercept it. The best approach was to fly directly toward the radar. This exposed only the Have Blue's tiny head-on radar return. Against VHF early warning radars, such as the Spoon Rest and Tall King, the results were more limited. Even so, the faceted shape reduced the detection range to half that of a normal aircraft. Against these radars, the plane would have to remain out of range, but since these radars were few in number, it would be a simple matter to bypass them. The Have Blue was undetectable by any airborne radar including the E-3 AWACS (airborne warning and command system). Fighter pilots would have to pick it up visually, the same as their World War I counterparts.

The Have Blue showed that further advances in RAM would be needed for operational aircraft. On the RCS tests, special care had to be taken. Before each flight, doors and access panels had to be sealed with metallic tape and the landing gear doors had to be adjusted for a correct fit. Then after Dyson climbed into the aircraft, the gaps around the canopy and fuel-fill door were filled with a paint-type RAM material and allowed to dry before the plane took off.[36]

On one flight, the Have Blue was picked up at a range of fifty miles. After landing, the plane was given a close inspection. Three screws had not been fully tightened and were sticking up less than an eighth of an inch above the plane's skin. This was enough to compromise the plane's low RCS.[37] Special efforts such as these were acceptable for a test aircraft, but for the operational aircraft, a more routine kind of procedure would be necessary.

The second Have Blue, 1002, was lost on July 11, 1979, during its fifty-second flight, a test against an F-15's radar. A weld in a hydraulic line cracked, spraying fluid onto the hot section of an engine. The fluid caught fire, and the blaze soon became uncontrollable. Dyson tried to get back to Groom Lake but had lost hydraulic power and was cleared to bail out. Dyson ejected and parachuted to a safe landing.

The plane crashed near the Tonopah Test Range, in the northern part of Nellis Air Force Base. A tall column of smoke rose above the debris.

Seeing the smoke, a group of people at the test range boarded trucks and headed toward the crash site. To chase them off, the F-15 pilot buzzed the trucks at 600 knots. One truck drove off the road as the fighter blasted past. The curiosity of the drivers was "satisfied," and they turned around and headed back. A helicopter from Groom Lake arrived and picked up Dyson.[38] The loss of the aircraft did not affect the program, as it was the next-to-last flight planned, and most of the test data had already been acquired. The wreckage of Have Blue 1002 was also buried at Groom Lake.[39]

Park was philosophical about the loss of the two Have Blue aircraft (and the accident that nearly killed him and ended his own flying career). Years later he noted, "We knew we had a problem but we couldn't fix it without a long delay in the program, and it was vital that we get the information. I don't mean we were going haphazardly. We did [the development] fast with a minimum amount of money. We wrecked two airplanes, but they were prototypes and served their purpose. . . . I smile a lot because I am just happy to be here alive. I believe that circumstances can occur that you cannot overcome no matter how good you are."[40]

The shape of the Have Blue remained secret for fourteen years. The code name was revealed in an October 1981 article in *Aviation Week and Space Technology.* Its existence was officially confirmed in 1988, and Park talked about his crash the following year. It was not until April 1991 that two photos of Have Blue 1001 were finally released.[41] Ironically, it is understood that the photos were released by accident.

UNVEILING STEALTH

The Have Blue was a particular shade of Black. The concept of faceting and, more importantly, its accomplishments were the darkest shade of Black. With a single pair of prototypes, every radar ever built had been rendered blind. The Have Blue had turned SAMs into expensive fireworks. Strategic airpower had undergone a revolution as great as that brought about by nuclear weapons.

The *idea* of stealth, however, was known. Because Project Harvey had been unclassified, the existence of a stealth demonstrator was also known. As time passed, it was becoming harder to keep the program a secret.

A 1979 article in the *Las Vegas Review-Journal* was one example. Its coverage of stealth was little better than gossip. The article described "an airplane so secret that 'whenever it comes out of its hangar, or when it comes in for a landing, a siren goes off and all personnel (except a select few) have to lie face-down on their stomachs to make sure they don't look at it.'" The article also talked about "the super-secret Stealthfire spy plane, which is supposed to be 'invisible' to radar." The "Stealthfire" was built of materials that were "non-reflective" to radar and had "a technological breakthrough which 'disperses' engine heat." This was described as "the only other way it could be tracked on radar." The article said that three Stealthfires had been built, but two had crashed—"one last year near Las Vegas and one more recently, perhaps in the past few weeks."

It also described "a new top secret fighter-bomber, tentatively known as the F-20." It claimed, "This plane would be outlawed by the proposed SALT II Treaty, but is secretly being developed in case the pact fails, the source claimed. The F-20 would be an advancement on the old B-1 design, which was scrapped last year by President Carter."[42]

The article did more to confuse than inform about stealth. The F-20 was only an improved F-5 fighter, which had nothing to do with the SALT II Treaty, and radar can not pick up the hot exhaust of a jet. Even so, the article caused major damage on several fronts.

During the summer of 1980, the pace of stealth leaks picked up. In the week of August 10, *Aviation Week and Space Technology*, the *Washington Post*, and *ABC News* all carried stories on stealth. (Up to this point, the popular press had ignored stealth.) The stories said that stealth technology was being developed for several types of aircraft, including bombers. They reported that it used RAM and curved surfaces to reduce the radar return. (The latter was entirely inaccurate.)

President Carter and Defense Secretary Brown said later that they had considered three options to deal with the leaks. Saying "no comment" would only fuel speculation. Disinformation—attempting to discredit the stories through false information—was also ruled out. It was seen as counter to the post-Watergate attitudes. The final option was to confirm the reports, in order to create a "firebreak" to additional leaks.[43]

And 1980 was also an election year.

Carter had been elected on the promise to cut defense spending, and during his presidency there was a major decline in U.S. military power. Funding shortages had caused over 7 percent of air force aircraft to be grounded due to a lack of spare parts. Air force crews wore flight suits that were so old the flame retardant had been washed out, but there was no money to buy new ones. The navy was particularly hard hit; it had half the number of ships of ten years before and could not fully man them. Nor was there enough ammunition to fill every ship's magazine. Enlisted personnel were so poorly paid—below minimum wage—that they had to put their families on food stamps or work second or third jobs.[44]

By 1980, the American public had become convinced the nation was in decline, pushed around by Third-World countries like Iran and facing a growing Soviet threat. The breaking point came with the failed attempt to rescue U.S. hostages held by the Iranians. The scenes of abandoned helicopters and burned bodies at Desert 1 damaged American confidence in a way even Vietnam did not. Unveiling the revolutionary possibilities of stealth seemed to be a way to counter Republican charges that President Carter had neglected defense.

On August 22, 1980, Defense Secretary Brown held a press conference:

> I am announcing today a major technological advance of great military significance.
>
> This so-called 'stealth' technology enables the United States to build manned and unmanned aircraft that cannot be successfully intercepted with existing air defense systems. We have demonstrated to our satisfaction that the technology works.

Brown noted that the effort had been kept secret for three years due to the efforts of the few people in government who had been briefed on the project and the contractors involved.

> However, in the last few months, the circle of people knowledgeable about the program has widened, partly because of the increased size of the effort, and partly because of the debate under way in the Congress on new bomber proposals. Regrettably, there have been several leaks about the stealth program in the last few days in the press and television news coverage.
>
> In the face of these leaks, I believe that it is not appropriate or credible for us to deny the existence of this program. And it is now important to correct some of the leaked information that misrepresented the Administration's position on a new bomber program. . . .
>
> I am gratified that, as yet, none of the most sensitive and significant classified information about the characteristics of this program has been disclosed. An important objective of the announcement today is to make clear the kinds of information that we intend scrupulously to protect at the highest security level.

Also at the press conference was William Perry. He explained that, "even as we acknowledge the existence of a stealth program, we will draw a new security line." The information to be guarded was the specific techniques used, how effective they were, the characteristics of thc aircraft under development, and the funding and schedules of the programs. (In retrospect, it is clear the secret they were trying to protect was faceting.) Perry also noted that "stealth technology does not involve a single technical approach, but rather a complex synthesis of many. Even if I were willing to describe it to you, I could not do it in a sentence or even a paragraph."[45]

The press coverage that followed the Brown statement indicated how little the popular press understood about stealth. A *Newsweek* article claimed stealth aircraft were equipped with "electronic jamming devices to reduce 'radar echo' aircraft normally give off." In fact, any electronic emissions would give the plane's location away. The article was illustrated with a CBS news drawing of a "stealth" airplane. It had nothing in common with what engineers thought a stealth aircraft would look like, nor did it look like the Have Blue. Instead, it resembled a navy F-8 Crusader with an inlet over the cockpit, two oddly bent curved wings, and a flat-tipped nose.[46]

In the wake of the press conference, the Republicans charged that the White House had released classified information for political gain. President Carter responded by blaming the Ford administration for not classifying stealth from the start.

Congress also became involved. On August 20, 1980, the House Armed Services Committee had been briefed on stealth. They were told that the subject was highly secret. Then, two days later, the press conference was held, which provided more information than they had been given. The committee held hearings, which cast doubts on the explanation for the disclosure. Benjamin Schemmer, editor of *Armed Forces Journal,* testified that in 1978 the magazine had withheld an article on stealth at the request of the Department of Defense; then in August 1980, he had been approached by William Perry, who encouraged him to publish a modified version of the original article. It was to be published no later than August 21—the day before the press conference.

More damaging was the testimony given by Adm. Elmo R. Zumwalt Jr., the former chief of naval operations. He testified that President Carter had decided to deliberately leak stealth information. This would be used as an excuse to announce the program's existence, so the administration could take credit for it. Zumwalt named as the alleged leaker of the information the deputy assistant to the president for national security affairs, David L. Aaron.

Aaron submitted an affidavit with the committee denying Zumwalt's charges. However, he refused to testify under oath due to a dispute between the committee and the White House over executive privilege.

The committee found Defense Secretary Brown's explanation for the press conference flimsy. In a report issued in February 1981, they stated they could not understand how the "damage-limiting tactic" was supposed to work. An official announcement was sure to attract more attention to the program than "no comment." Based on the testimony of Schemmer and Zumwalt, along with Aaron's refusal to testify and Brown's weak explanation, the committee concluded that the disclosure had been made for political ends. The committee also stated that the announcement of the stealth program had done "serious damage . . . to the security of the United States and our ability to deter or to contain a potential Soviet threat."[47]

Any attempt to use stealth as an election year ploy by the Carter administration had backfired. The concept and possibilities of stealth, which few in the public and press could understand, did not mitigate the failure at Desert 1. Ronald Reagan was elected president in a landslide.

The unveiling of stealth had another effect. It was to warn both Republicans and Democrats that misfortune awaits those who disclose Black projects. Not surprisingly, the new administration had very different ideas on how to handle Black airplanes.

They disappeared from sight.

THE GROOM MOUNTAINS LAND SEIZURE

When the Groom Lake test site was established in 1955, the location's isolation was sufficient protection. It was about twenty miles from the nearest highway and thirty miles from any town. "Ozzie" Ritland, who selected the site, said they were looking "for the most isolated part of the United States of America" to test fly the U-2. During the U-2 operations, there were a few sightings, but as Ritland recalled, "they would see the airplane, but they weren't so curious and it was far enough into the desert."[48]

Although there were a number of A-12 sightings, Groom Lake remained in isolation. Only crashes, such as Have Blue 1001, brought attention to the site. Despite this, no published account had mentioned Groom Lake. It was always "a base in Nevada."

The 1979 *Las Vegas Review-Journal* article changed this; it named Groom Lake as the location of the test site. The article noted that while older maps showed Groom Lake, more recent ones did not. The article stripped Groom Lake of the anonymity that had protected the site for nearly twenty-five years. This was to have major consequences in the decade to follow.

With the increased press attention, several code names for the test site soon became known. These included the Ranch, Area 51, and what was actually the call sign for Groom Lake's tower. The name's evocative and sinister sound ensured it caught the public's fancy—"Dreamland."

With the publication of the name Groom Lake, the flaw in Dreamland's security became apparent. The site was bordered by public land. It was possible to hike into the area and observe the site, without trespassing. There was an obvious solution.

The withdrawal of public land for military use must be periodically renewed. When the air force submitted a renewal to Congress, it added 89,600 acres of the Groom Mountains to the 3.3 million of the Nellis Air Force Base range.[49] For a full two years, Congress did nothing about the request.[50]

In the meantime, the Reagan administration had begun a major buildup of strategic nuclear forces. This sparked opposition by "peace activists," such as Greenpeace. A standard tactic was attempting to disrupt military activities, particularly those dealing with nuclear weapons. On April 18, 1983, four Greenpeace demonstrators entered the nuclear test site via the dirt road that led to Groom Lake. For five days, they hid out in the mountainous terrain, before finally surrendering to the test site guards. It was widely believed that this incident caused the government to take action.

In March 1984, acting on the orders of Secretary of Defense Casper Weinberger and/or President Reagan, armed guards were posted on the land. When hunters and hikers approached Groom Lake, the guards requested them not to enter the area.

A reporter who went out to the site in May found the guards were "especially polite as they tell visitors they cannot drive farther along the dirt road" that led to Groom Lake. The reporter chatted with the guards and watched television while they waited for the supervisor to arrive. At the same time, orders were issued to ground all aircraft while strangers were "within earshot." The reporter repeatedly asked for the legal justification for the air force denying public access. The only reason given was "national security."[51]

Throughout the west, resentment was building over federal land policy. About 87 percent of the land in the state of Nevada was not under state control; it was federal land. The Groom Mountain land seizure quickly became part of this "sagebrush rebellion." Local members of Congress were quick to become involved. Representative Harry Reid (D-Nevada) said, "People have a right to be upset. There has been no land withdrawal. They simply have closed land off for national security reasons." Representative Barbara Vucanovich (R-Nevada) requested a hearing "to bring it into the open."[52]

It would be August 6, 1984, before Congress could get around to holding hearings on the Groom Mountains. The hearings before the House Subcommittee on Lands and National Parks saw a parade of Nevada officials, hunting and mining interests, and environmental groups such as the Sierra Club and Audubon Society. Nevada Governor Richard Bryan attacked the air force, saying that it had tried to "hoodwink the Congress and the state of Nevada." He continued: "For years, Nevadans have acquiesced to defense-related land withdrawals, but the time has come to draw the line. I strongly suggest to you that the day is past when the federal government can look at Nevada . . . as an unpopulated wasteland to be cordoned off for whatever national purpose seems to require it."[53]

Governor Bryan did not object to "the legitimate security needs of our country," but said, "if the federal government withdraws the land, then Nevada must be compensated."[54]

The subcommittee chairman, John Seiberling (D-Ohio), also attacked the air force. He told John Rittenhouse, the air force representative, "There is no higher level than the laws of the United States." When Rittenhouse said he could explain the reasons only in a closed briefing, Seiberling exploded: "Shades of Watergate. All I am asking you is under what legal authority this was done. I am not asking you the technical reasons. That certainly is not classified."

Rittenhouse responded, "We had no legal authority, but we asserted the right to request people not to enter that area."[55] Newspaper headlines read, "AF admit to illegality."

The Groom Mountains land issue also became involved with wilderness policy. Representative Sieberling proposed a trade-off—the air force could have the Nellis Air Force Base and Groom Lake land *if* 1,408,900 acres of the National Wildlife Range was designated a wilderness area. This would close it to any development and restrict access to backpackers. The deal would also restrict the tests the air force could conduct and limit the land withdrawal to December 31, 1987, pending an environmental report. He then offered to withdraw the wilderness provisions, *if* Senator Paul Laxalt (R-Nevada) would provide assurances he would get the Nevada congressional delegation to act on a wilderness bill that year. (Nevada had yet to produce a wilderness plan ten years after it was required.)[56] Representative Reid said the provision had been put into the bill to force the Nevada delegation to come up with a wilderness bill.[57]

The legal maneuvering continued for the next three years, and involved "compensation" for the loss of recreation, grazing, and mining claims on the land. Many of the land use–wilderness issues, such as whether snowmobiling would be allowed in some areas and the building of a paved road from Rachel, Nevada, into the Nevada Test Site, had nothing to do with the Groom Mountains, but they blocked passage.[58]

By March 1988, the issue had not been resolved, and the temporary land withdraw would soon expire. John Rittenhouse told the Senate public lands subcommittee: "We have operations which would have to cease if the public were allowed to be [there]. It would be extremely detrimental to our national defense effort. . . . Our concern is for any visual sightings by anyone."[59]

The extension to the land withdraw the air force sought was itself part of the political power plays—Reid wanted only a ten-week extension, in order to pressure Senator Chic Hecht (R-Nevada) to act on the wilderness bill.[60] Environmentalists also continued to complain they were not getting enough. The groups Citizen Alert and the Rural Coalition tried to use two mining claims as "bargaining chips." They would be given up in exchange for the groups having a "say" in the writing of a report on military activities in

Nevada, action on land claims by the Western Shoshone Indians, and return of one member's pilot license.[61]

The day before the extension was to expire, the House separated the Groom Mountain issue from the wilderness bill. The withdrawal was approved on a voice vote and sent to the Senate.[62] Approval was given and it was sent to President Reagan.

This brought the Groom Mountain land seizure controversy to a close. It had taken a total of six years—twice the time needed to develop, build, and conduct the flight and RCS tests of the Have Blue. The new boundaries of the Dreamland restricted area were laid out in straight lines. It was not realized at the time that a few spots had been missed, but that did not matter—for the moment.

While this controversy dragged on, the descendent of Have Blue had made its first flight, undergone systems development, and reached operational status behind the shield of the mountains. A few months later, this Dark Eagle would be publicly unveiled to questions about its cost and whether stealth would work.

Two years later, it would make history.

Chapter 8

The Black Jet of Groom Lake
The F-117A Senior Trend

Subtle and insubstantial, the expert leaves no trace;
divinely mysterious, he is inaudible.
Thus he is master of his enemy's fate.

Sun Tzu
ca. 400 B.C.

By mid-1978, the Have Blue 1001 had proven the basic concept of stealth. Lockheed proposed two different operational stealth aircraft. One was a medium bomber about the size of the B-58 Hustler. It had a two-man crew and four engines. The other was a fighter-sized aircraft with a single-man crew, two engines, and a payload of a pair of bombs.[1]

The air force chose the stealth fighter design, and on November 16, 1978, Lockheed was given a contract to begin preliminary design work. Extreme secrecy enveloped the program, code named "Senior Trend." At the start, only twenty people were authorized to know of this Dark Eagle's existence.[2]

SENIOR TREND

The Have Blue aircraft had been designed solely to test faceting, with no allowances for tactical systems or weapons. The little experimental plane would have to be transformed into an operational aircraft. This meant more than simply adding these systems; the aerodynamic and RCS testing had also revealed the need for other design changes.

The most obvious change to emerge during the redesign was the tail. The Have Blue's twin fins were canted inward to shield the platypus exhausts from infrared detectors above the aircraft. In practice, however, the fins reflected the heat toward the ground, making the plane more visible from below. The twin fins were also mounted on a pair of booms, which proved structurally inefficient. In the stealth fighter, the fins were moved farther aft and canted outward, in a V shape (similar to the V-tail of the Beech Bonanza light plane). This also improved control effectiveness. The fins were attached to a central spine that also carried the weight of the weapons.

The Have Blue's wing sweep was an extreme 72.5-degree angle. This resulted in a poor lift-drag ratio, which cut into payload and range performance. Highly swept, low-aspect ratio wings also lose airspeed rapidly in a sustained high-g turn. The sweep angle was reduced to 67.5 degrees, and the wings were extended as far back as possible to improve performance.

Operational requirements also resulted in a change to the design of the windshield and nose. The pilot would need a heads-up display (HUD) for flight information. The plane would also carry two infrared imaging systems—one looking down, and the other looking forward. Neither the HUD nor the forward-looking system could be fitted into the Have Blue's nose shape. This gave the new design a distinctive appearance, over the more conventional shape of the Have Blue's nose section. Although operationally required, the change did slightly increase the plane's RCS.[3]

A major concern was maintenance: extreme care had to be taken with the Have Blue to preserve its stealth. With the operational aircraft, the total number of maintenance hours per hour of flight time was to be similar to that of conventional twin-engine fighters. The portion related to the stealth design was to be limited to a small fraction of the total. To meet the requirements, servicing accesses for aircraft subsystems were located in the wheel wells and weapons bays. All the aircraft's avionics were located in a single bay. This minimized the need to remove and replace RAM coating during maintenance.[4]

Most of the changes from the Have Blue were internal—a reengineered cockpit, revised inlets and exhaust system, tactical systems, a braking parachute and arresting hook, an anti-icing system for the inlet grid, fuel tanks in the wings, retractable antennae, formation and anticollision lights, an in-flight refueling receptacle, and, finally, two weapons bays.[5] Each bay would hold a single 2,000-pound bomb. Those bombs would be as remarkable as the aircraft itself.

During the Vietnam War, the United States had developed laser guided bombs (LGB), better known as "smart bombs." The stealth fighter would be equipped with a laser. The pilot would put the laser beam on the aim point, and the bombs would home in on the laser light reflected from the target. The guidance system would compensate for shifting winds: all the pilot had to do was hold the beam on the target. It was now possible to hit a target within inches of the aim point.

Stealth meant a single aircraft could penetrate the heaviest air defenses. LGBs meant this single plane could then destroy any target, no matter how small or hardened against attack. No longer was it necessary for massive formations to rain bombs on a target in hopes of destroying it. One plane, one bomb, one target. This was the attack profile the stealth fighter was to undertake. It would change airpower.

The aircraft which emerged from the redesign had a shape similar to that of the Have Blue, but the fuselage was wider and more squat. The Senior Trend was 65.9 feet long, with a wingspan of 43.25 feet. The high canopy trailed off to a very thin rear fuselage. Seen from the front, it resembled a pyramid; from the rear, the plane looked almost flat. By late 1979, a wooden mock-up was completed. This was used to check placement of equipment and systems. A full-scale Senior Trend pole model was also built for RCS testing. This posed a security problem—such testing was done outdoors where the model might be photographed by Soviet reconnaissance satellites. To prevent any sightings, the testing was done at night.[6]

In December 1979, a contract was awarded to Lockheed to build five full-scale development (FSD) test aircraft and fifteen production aircraft. This would provide a full squadron of the aircraft.[7] The first Senior Trend was given the aircraft number 780, for its scheduled first flight date of July 1980.

Because of the short time, existing systems were used. The General Electric F404-GE-F1D2 turbofan engines were from the navy F/A-18, without the afterburners of the fighter. The F/A-18 also provided the multifunction cathode-ray tubes, HUD, fuel controls, stick grip, and throttles. The sensor displays were from systems developed for the OV-10D and P-3C. The navigation system was from the B-52. Other systems came from just about every Lockheed aircraft built since the T-33; these included the SR-71, C-130, L-1011, and even the F-104.[8]

Of critical importance was the flight-control system. Like the Have Blue, the Senior Trend was aerodynamically unstable. Harold C. "Hal" Farley Jr., the Lockheed test pilot selected to make the first flight, later described the plane's "aerodynamic sins": "In fact, the unaugmented airframe exhibits just about every mode of unstable behavior for an aircraft; longitudinal and directional instability, pitch up, pitch down, dihedral reversal, and various other cross axis couplings. The only thing it doesn't do is tip back on its tail when parked."

It was obvious that a computer-controlled, fly-by-wire system was needed. There was no manual backup system because it was impossible for a pilot to control the plane without the computer. To reduce risks, it was decided to use a proven off-the-shelf system. The F-16's fly-by-wire system was selected; actuators, flight-control computer chassis, and power supply were modified slightly. New computer programs had to be developed.

The control system was designed so the Senior Trend would handle like "an ordinary plain vanilla aircraft." Programming was tested on an NT-33, "by real pilots flying in a real airplane in real turbulence." Some flight tests assumed the directional stability of the Senior Trend was even worse than predicted.[9] It would prove to be a wise precaution.

Despite the Skunk Work's best efforts, by the summer of 1980 the project was behind schedule, and the first flight was nowhere in sight. Each day seemed to bring new problems and no solutions. Ben Rich said years later that this was the low point of his life. The meetings went from before dawn and continued long after dark. In the midst of this, Ben Rich's wife, Faye, died of a heart attack, leaving him emotionally devastated. When he returned to work, Alan Brown, the Senior Trend program director gave him a note. Written on it was Rich's next birthday "June 18, 1981." When Rich asked him about it, Brown said, "That's the date we test-fly the airplane." He continued, "The date is firm. In granite. Count on it."[10]

SENIOR TREND TAKES FLIGHT

By the fall of 1980, aircraft 780 was beginning final assembly at Burbank. By early June 1981, final checkout was completed. The wings were removed and crated. The fuselage was covered in a shroud, and a wooden framework was added to the nose to further hide its shape. Under cover of darkness, number 780 was loaded aboard a C-5 and flown to Groom Lake.

Upon arrival, 780 was taken to a hangar at the south end of the flight line and reassembly began. Even here, security remained paramount: a camouflage net was stretched across the open hangar door. Once assembly was complete, static engine test runs were made. The plane was kept inside the hangar, with the exhaust vented out the open door.[11]

When delivered, 780 was a dark gray color. Before the first flight, patterns of light blue and light brown were painted on the aircraft. As with Have Blue 1001, this three-color finish was meant to hide the faceting. (780 was the only one of the FSD aircraft to be camouflaged.) The paint finish was ragged looking and appeared to have been hurriedly done.[12]

On June 18, 1981, just as Brown had predicted during the dark days of nearly a year before, everything was ready. As dawn broke, 780 was rolled out of its hangar for the first time. Hal Farley ran up the engines, and 780 started down the runway. As with the first hops of both the U-2 and A-12, it was to be an eventful flight.

It was a difficult task to design an air-data system that was stealthy. Four probes extended from the nose, along with a conventional boom. Because the air-data probes had shown erratic readings during ground vibration testing, it was decided to ballast the aircraft to a far forward center of gravity point, turn off the angle of attack, and sideslip measurements to the flight-control system.

Immediately after 780 lifted off the runway, it became apparent to Farley that the directional stability was much worse than predicted. Farley immediately switched on the sideslip feedback to the flight-control computer. The plane's handling "stiffened up," and the rest of the flight was routine. Subsequent analyses indicated the Senior Trend's directional stability and directional-control power were less than predicted. The solution was to increase the area of the fins by 50 percent. The new fins were installed by the fall of 1981. This cured the instability but would cause other problems later in the test program.[13]

Two more test pilots soon joined the program—Skip Anderson (air force) and Dave Ferguson (Lockheed). After a few months, 780's desert camouflage was removed, and it was repainted light gray. The plane had no national markings (in common with most of the other Groom Lake aircraft), but "Hal," "Skip," and "Dave" were painted on the canopy rail.[14]

Between mid-1981 and early 1982, the other four FSD Senior Trend aircraft were delivered to Groom Lake. While the first two (780 and 781) were aerodynamic test aircraft, the other three (782, 783, and 784) were systems aircraft. As such, they had the full set of cockpit displays, just as on the operational aircraft. It has been reported that 782 and 783 had the Skunk Works emblem on their tails. The fifth FSD aircraft (784) reportedly sported a full-color painting of Elliott, the dragon from the Disney film *Pete's Dragon.* This was in honor of Col. Pete Winter, air force commander at Groom Lake. (For the uninitiated, Elliott was invisible to everyone except Pete.)[15]

Initially, the FSD aircraft were painted gray. The commanding general of the Tactical Air Command then ordered they be painted black. (Although most of the test flights were done in daylight, the aircraft's operational missions were flown at night.) Markings became more formal: national insignia, "USAF," and the aircraft number on the tail.[16]

FLYING ON THE EDGE

With the full complement of FSD aircraft on hand, along with additional test pilots, the test program got under way in earnest. The flight-test program explored a number of unknowns related to the Senior Trend's stealth design.

Many test hours were required to calibrate the air-data system for angle of attack, sideslip, airspeed, and altitude. A critical part of the flight control system was the angle of attack limiter. Angle of attack (AOA) refers to the angle between a plane's wing and the direction of the airflow. Wind-tunnel tests and free flights of unpowered models indicated the Senior Trend would pitch up at high angles of attack. (The nose would abruptly and uncontrollably rear up.) The aircraft would then enter a "deep stall" and would not be recoverable. The AOA limiter would have to automatically move the control surfaces to prevent the aircraft from exceeding the critical value. Because of the risk, the AOA testing was done in slow steps. There were literally hundreds of individual tests run.

Validation of this approach came on May 23, 1983. One of the FSD aircraft was on final, with its left wing low, when a strong wind gust hit it. This caused the AOA and sideslip to instantaneously reach levels higher than any tested—higher, in fact, than could be tested in the wind tunnel. The AOA limiter countered with full down elevon in less than 0.4 seconds and moved the fins 90 percent of their full travel. The plane successfully recovered.[17]

Flutter testing was also prolonged. The early tests showed no problems, but during a weapons compatibility test, an air force test pilot put the aircraft into a sideslip while flying at near maximum speed. The left fin underwent "explosive flutter" and disintegrated. The pilot made it back to a successful landing despite very poor stability. Farley called it, "a very professional response by a real pro."

The problem was traced to the redesign of the fins—the added area had reduced the fin's stiffness. The problem had been hidden during the earlier tests by the friction of the fin bearing.[18]

One of the more unusual problems was testing the Senior Trend's inlet grids. There was some concern they could distort the air flow to the engines. In fact, they acted like "flow straighteners," giving the engines a constant flow of air.

More serious was grid icing. Tunnel tests indicated that, in Farley's words, "the inlet grids not only looked like a giant ice cube tray, but acted like one as well." A wiper system and alcohol dispenser was developed. Ironically, airframe icing was not judged to be a problem; chief aerodynamicist Dick Cantrell said that any ice buildup would only help the plane's aerodynamics.[19]

If the Senior Trend was to be an effective bombing platform, the avionics systems would have to show capabilities never before achieved. The pilot would have to find the target, which was not an area or a wide-spread factory but rather a specific part of one specific building, then direct the LGB to the aim point—all in the dark.

The heart of the Senior Trend's bombing system was a pair of infrared turrets—the forward-looking infrared (FLIR), located in the front of the plane, and the downward-looking infrared (DLIR) on the plane's underside. Each turret was mounted in a well that was covered by a fine-mesh, radar-absorbing screen. The two-turret design was able to scan from just above the horizon to below and behind the aircraft. The image from the system was displayed on the instrument panel's central cathode-ray tube.

The design posed many problems. To give one example, the FLIR turret would have to pick up the target at long range, then track it as the plane approached. The FLIR would then have to "hand off" the target to the DLIR without losing the target lock. To create this seamless display, the two turrets had to be exactly aligned (called boresighting).

Flight testing of the system revealed numerous problems. It proved impossible to electronically boresight the two turrets, which created problems in the handoff. Problems with the video display included "windshield wiper noise," "jello," "shimmering," "picket fence noise," and "horizon shadowing." Added difficulties included problems with level and gain controls, turret slew rates, and target acquisition and illumination.

Some of the problems were purely subjective, which made it even more difficult. There were three test aircraft (FSD 3, 4, and 5) and six test pilots; what one pilot judged unacceptable on one plane was called good by another pilot. A "Tiger Team" was organized to sort out the systems problems. It was headed by the Skunk Works chief scientist and drew man power from other Lockheed divisions.

Like the engineers who had worked on earlier Black airplanes, the team came up with innovative, simple solutions to the complex problems. They abandoned efforts to electronically boresight the turrets and used a mechanical procedure. The aircraft was rolled up a thirty-inch-high ramp. This raised the nose and allowed both turrets to view the same target board simultaneously. They could then be aligned. To prevent the control surfaces from scraping the ground, mattresses were placed under them when the hydraulic system was shut down. A portable boresight fixture was developed, the turret mounts were fixed in place with epoxy, and tolerances were tightened. Reliable handoffs could then be accomplished even in "dive-toss" drops: this involved the plane going from level flight into a dive, then pulling up and releasing the bomb. Three FLIR-DLIR handoffs were required for this maneuver. Another difficult flight maneuver was loft bombing, where the plane goes from level flight into a steep 4-g pull up. The bomb is released during the pull up and is "thrown" toward the target. The video display problems were traced to electromagnetic interference due to poor shielding. The target lock-on problems were corrected with new software.

For flight testing of the modifications, the team developed what was described as "a broadband, wide-spectrum, inexpensive, expendable, point-source IR target." This was a barrel filled with glowing coals—a backyard barbecue.

In all, it took a year and some 100 test flights to correct all the problems. When the work was completed, one aircraft dropped a 2,000-pound, inert GBU-27 laser-guided bomb, which scored a direct hit on the barrel.[20]

THE 4450TH TACTICAL GROUP

Major Alton C. "Al" Whitley had flown two tours in Vietnam, one in F-100s and the other as a search and rescue A-7 pilot, but this was a new and novel experience. He had been called to a small interview room at Nellis Air Force Base. When he knocked on the door, a man opened it an inch and asked if he was Whitley. He said yes, and the man asked for his ID card. The man took it and closed the door. A minute later, he reopened it and said, "Yeah, you're Whitley."

Once Whitley was in the room, he was offered a chance to fly with a top-secret unit. The duty would require constant separation from his family, and he could not be told much more. He had five minutes to decide, and when he left the room, the decision stood. Whitley did not know what airplane he would be flying, or what the unit would be doing. He responded, "Sign me up. I'll do it."

Whitley thus became a member of a secret brotherhood composed initially of about ten officers and a dozen enlisted men—the 4450th Tactical Group. The unit commander was Col. Robert A. Jackson. He was to select and train the initial group of pilots. He was looking for pilots who were both experienced and mature—majors and senior captains with a thousand hours of flight time and air-to-air and/or air-to-ground backgrounds (F-4s, F-15s, F-111s, A-7s, and A-10s). Colonel Jackson met with each candidate. Once they agreed, the pilots were told to go back to their units and wait for a call.

Although the 4450th Tactical Group was formally established on October 15, 1979, training did not begin until June 1981. Whitley and the other pilots spent time at Lockheed in the cockpit procedures trainer. This was a model of the cockpit on which the pilot could learn the layout of the various consoles, but not how the plane would fly. On seeing the FSD aircraft under construction, the pilots' initial reactions were to wonder if that strange-looking airplane could really fly.

Ironically, the unit that was to fly the world's first stealth aircraft had no airplanes. All the FSD aircraft were used for testing. Soon after the unit began training, they received A-7s. These were organized into the P-Unit (later called the 4451st Test Squadron). The A-7s were to provide both pilot-proficiency training and cover. The 4450th Tactical Group's cover story was that it was an A-7 avionics and evaluation unit.[21]

At the same time, the base the unit was to use was under construction—the Tonopah Test Range (TTR). It had originally been built to support drop tests of nuclear weapons. The TTR was on the north side of the Nellis range, 140 miles from Las Vegas and northwest of Groom Lake. The nearest town was Tonopah, Nevada. The area was open range, with wild horses running free.

The base was immediately staffed with air force security police. The flight line was walled off with a double fence; the only access to the runway was through gates. The area between the fences was lighted at night and had intruder detectors. At first, the facilities were limited to a few buildings, a small mess hall, and sixteen winterized trailers. These were soon replaced by dormitories and hotel-style rooms for the pilots and support personnel. Because all the flights were done at night, the rooms featured blackout curtains to keep out the sun during the daylight sleeping hours. The runway, taxiways, and aprons were all improved, while maintenance facilities, fuel and water tanks, fire stations, and a dining hall were built. Individual hangars were also constructed for each aircraft. In addition to providing protection from the weather, these hangars also hid the planes from prying eyes during the day.[22]

Before the 4450th Tactical Group could become a fully functional unit, it would need the production Senior Trend aircraft. It was the spring of 1982 before the first aircraft, 785, was ready. Its first flight would end in a near fatal crash.

The accident occurred on April 20, 1982. With Lockheed test pilot Robert L. "Bob" Riedenauer at the controls, 785 began its takeoff roll. The plane lifted off correctly, but moments after the main wheels left the runway, it yawed violently and went out of control. Within seconds, the plane went inverted going backward and slammed into the ground on the shore of the lake bed. It took rescue crews some twenty minutes to pull Riedenauer from the wreckage. He had suffered major injuries. He recovered, but never flew again.[23]

The cause of the crash was traced to incorrect installation of several wires to the flight-control system. The computer read the pitch-up as an uncommanded yaw movement and "corrected" for it. Riedenauer never had a chance to get the plane under control.[24]

Soon after, the 4450th Tactical Group received a new commander. Colonel Jackson was replaced on May 16, 1982, by Col. James S. Allen. Although Jackson had organized the unit and started construction of TTR, Allen would oversee flight training and the move to operational status.

In June 1982, aircraft 786 was delivered to Groom Lake, but was used for flight testing. Senior Trend 787 thus became the first plane delivered for the 4450th Tactical Group. In September, this single aircraft became the core of the Q-Unit, nicknamed the "Goatsuckers" (later renamed the 4452d

Test Squadron). Major Alton Whitley was picked to make the unit's first operational flight. This was successfully completed on October 15. As with the A-12/SR-71 pilots, he was given a personal designation—"Bandit 150." As each new pilot made his first flight, he was given his own Bandit number.

Whitley was later given a plaque marking that first flight. It would be another six years before he was allowed to tell his family what the inscription meant. All it said was: "In Recognition of a Significant Event, Oct. 15, 1982."[25]

LIFE AT TTR

Before Christmas 1982, Senior Trends 790, 791, and 793 had been flown to TTR and flight operations had begun in earnest.[26] Unlike Groom Lake, all flight operations at TTR were conducted at night. The pilots would leave Nellis Air Force Base on Monday afternoon and fly to TTR on Key Airlines, which operated a shuttle service to the base. Before each night's flights, there would be a mass briefing of the pilots, followed by target and route study.[27] The hangar doors were not opened until one hour after sunset.[28] This meant the first takeoff would not be made until about 7:00 P.M. in winter and 9:30 P.M. in the summer.

For the first year, flights were restricted to the Nellis range. This continued until sufficient confidence had been gained in the aircraft. Even so, it took a presidential authorization to begin off-range flights. In the event of an unscheduled landing, the pilots carried a signed letter from a senior air force general ordering the base or wing commander to protect the aircraft.

Once sufficient aircraft had been delivered, two waves were flown per night. This involved eight primary aircraft and two spares, for a total of eighteen sorties. The aircraft would fly the first wave (called the "early-go"), then return to TTR and be serviced. A second group of pilots would then fly the second wave (the "late-go").

Typically, the training flights simulated actual missions. A normal mission would have two targets and several turn points. On other nights, there would be a "turkey shoot" with some fourteen targets. The pilots would get points for each one; at the end of the night, they would be added up to see who "won." The missions ranged across the southwest, and the targets were changed each time, to make it more challenging.

The targets themselves were also challenging. The infrared system made picking up buildings too easy. Rather, the targets would be such things as a fire warden's shack in a forest, or the intersection of two dirt roads. When it snowed, it was even harder to pick them up, as there was little temperature difference between the targets and the ground. The hardest target was a dock at the Lake Tahoe marina. It was not visible against the cold water of the lake, and none of the pilots found it.

The second wave was completed by about 2:30 or 3:00 A.M. in the winter, a few hours later in the summer. The planes had to be in their hangars and the doors closed one hour before sunrise. After landing, the pilots would be debriefed.

The pilots then began a race with the sun. It had been found that sleep is disrupted if a person tries to go to bed after seeing the sunrise. Like vampires, they had to be indoors before the sun rose.[29] They would sleep six or seven hours, then begin their twelve-hour "day" again. Each pilot would make two or three flights during each four-day period at TTR. One of these flights would involve an in-flight refueling. During a month, each pilot would make ten to twelve flights in the Senior Trend and another five or six A-7 flights.[30] This took its toll—by Thursday night they were "a wreck."

Friday afternoon, the pilots would pack up and fly back to Las Vegas to their families and a normal day-night cycle. They would spend the weekend at home, then start it all over again Monday afternoon.[31]

Security affected everything the pilots and ground crews did. The pilots could call home from TTR every day but could not say where they were. Nor did the families know what they were doing while they were gone. One pilot's wife told her children that their father was "at work." The whole situation took its toll on the pilots and their families.[32] One consolation was a sign in the ready room—"Someday They'll Know."[33]

The area around TTR was closely monitored. If a truck was seen in the hills around the base, it would be checked out, as were airplanes flying near the base's restricted airspace. Trips into Tonopah were also discouraged—security did not want a lot of air force uniforms visible.

Internal security at the base was extremely important. The operations building had no windows; it was a giant vault. Within the building was another vault room where the aircraft flight manuals were stored. When in use, the manuals always had to be in the pilot's physical possession—if a pilot had to go to the bathroom, his manuals were given to someone or returned to the vault.[34]

Before personnel were allowed access to the flight line, they underwent an electronic palm print scan.[35] During training flights, security also had to be maintained. On off-range flights, the pilots talked to the air traffic controllers as if they were in an A-7. Each plane also carried a transponder that indicated to radar operators that it was an A-7. Even though the planes flew only at night, special care was taken to avoid sightings. The routes avoided big cities. If a plane flew under a high overcast, the reflected city lights would silhouette it against the clouds. The phase of the moon also affected flight operations. Several routes were not flown if the moon was more than 50 percent full.[36]

SENIOR TREND TACTICS AND ROLES

The Senior Trend aircraft represented a complete break with past attack aircraft and, accordingly, needed a whole new set of tactics. Conventional tactics were intended to prevent the plane from being shot down—the attack was secondary. The plane would hug the ground to escape radar detection. When the target was reached, the plane would have to pop up to a higher altitude to release the bombs. Such low-altitude, high-speed attack profiles made LGB drops difficult.

The Senior Trend, on the other hand, could go in at high altitude. This allowed the pilot to concentrate on the attack, rather than on avoiding hitting the ground. The high-altitude flight also permitted the target to be picked up at much longer ranges. In addition, the drop was made while flying straight and level. This meant the bomb would hit the target vertically, improving accuracy and penetration.[37]

The pilots talked tactics every day. In a sense, they were trying to define the role of the airplane. A new technology had been developed and they had to discover how best to use it, much as airpower theorists had done in the 1920s and 1930s.

Originally, the air force envisioned only a single squadron of eighteen Senior Trend aircraft. These would be used for Delta-Force-type missions. One or two planes could attack a single, high-value target without being detected. The early success of the program, however, convinced the air force and Congress that a full wing was needed. This involved three eighteen-plane squadrons. Orders were placed with Lockheed for a total of fifty-nine production aircraft. The first of these added squadrons, the I-Unit "Nightstalkers," was activated in July 1983, followed in October 1985 by the Z-Unit "Grim Reapers" (later redesignated the 4450th Test Squadron and the 4453d Test and Evaluation Squadron respectively).

The pilots doubted that there was a role for even a two-squadron unit: this was considered too large a force for limited clandestine missions, and participation in any general war scenarios was not foreseen. Part of the problem was the planes' extreme secrecy. A plan for using the full wing could not be developed when even the senior air force commanders in Europe and the Pacific had not been told about the aircraft. Nor was it possible to use the plane in Red Flag exercises.[38]

TO THE BRINK

A year after Whitley's first flight, the Senior Trend was prepared to go to war.

On October 23, 1983, terrorists launched a truck-bomb attack on the U.S. Marine barracks in Beirut, Lebanon, killing 241 and wounding 100 others.[39] Five days later, the 4450th Tactical Group was declared to have achieved an initial operational capability. Reportedly, orders were also received to prepare for attacks on PLO camps in southern Lebanon, in retaliation for the Beirut bombing. The unit was alerted and five to seven aircraft were armed.

According to these reports, the aircraft flew from TTR to Myrtle Beach, South Carolina. The planes were placed in hangars, and the pilots rested for forty-eight hours. The pilots then began their final preparations before takeoff. They would fly nonstop to southern Lebanon and strike terrorist targets. Only forty-five minutes before takeoff, they received word that Defense Secretary Casper Weinberger had canceled the strike.[40] Over two years would pass before Senior Trend again went to the brink.

In the years following the aborted attack, the 4450th Tactical Group saw changes in command. Colonel Allen was replaced by Col. Howell M. Estes III on June 15, 1984. Colonel Estes led the unit through its first operational readiness inspection, earning a rating of excellent. On December 6, 1985, Col. Michael W. Harris was named commander. He oversaw the expansion to three squadrons and was the first operational pilot to reach three hundred hours in the Senior Trend.[41]

Soon after, the unit was reportedly alerted for a second possible combat mission. The erratic leader of Libya, Col. Mu'ammar al-Gadhafi, had long been suspected of backing terrorist attacks. Proof was lacking, however. During the evening of April 4, 1986, a message from the Libyan embassy in East Berlin was intercepted and decoded by British intelligence saying a bombing was about to take place and that American soldiers would be hit. Just before 2:00 A.M. on April 5, a bomb exploded in the La Bella Discotheque, killing two GIs and a Turkish woman. Minutes later, the Libyan embassy sent a coded message that the operation had succeeded and could not be traced to them. These decoded messages were the "smoking gun" of Libyan involvement. Approval was given four days later to attack Libya.[42]

Libyan air defenses were more numerous than those of North Vietnam more than a decade before. Indeed, only three targets in the Soviet Union and Warsaw Pact countries were better defended than Tripoli and Benghazi, Libya. The Senior Trend aircraft was judged ideal for such a mission and was reportedly included in the attack plan. It has been stated that less than an hour before takeoff, Weinberger again canceled their participation, on the grounds that the targets were not worth risking the planes.

The rest of the attack, code-named "Operation El Dorado Canyon," went forward on April 15, 1986, using F-111s from England and A-6s, A-7s, and F/A-18s from two carriers. All five target areas were hit, with the loss of one F-111.[43] It would be another three years before the Senior Trend would see action.

NAMES AND PATCHES

Secret symbols have long been associated with the Dark Eagles, but the tradition reached new heights with the Senior Trend program. "Senior Trend" was a computer-generated code name with no meaning or style. It cried out for a nickname.

The plane's first one came during the flight-test program. After finding a "huge" scorpion in their office area, the test team adopted this as their symbol. The FSD aircraft were dubbed Scorpion 1 through 5.[44] The scorpion symbol also found its way onto patches. One showed a black scorpion and the words "Baja Scorpions" (a reference to their location in "Baja" [southern] Groom Lake). Another showed a scorpion and a black T-38 trainer and "Scorpion FTE" (flight test and evaluation). The FSD 4 aircraft had its own patch—a red delta shape with the number "4" and a black scorpion superimposed over it. The shape was based on the wing of the plane.

The "Pete's Dragon" aircraft also had its own patch—a black shield with a green dragon and "Pete's Dragon" in red. Another patch showed a green dragon and "Dragon Test Team." Such patches were not seen as a security problem. None of them showed the aircraft, and one had to be part of the program to understand the symbolism.[45]

Other than the "Pete's Dragon" design, artwork on the FSD aircraft was limited. The most spectacular exception occurred in 1984, when Weinberger went to Groom Lake. As part of the display of Black airplane activities, a Senior Trend made a flyby. The plane approached the reviewing stand from the south and banked to show its top surface. As it reached the center of the crowd, the plane banked again and showed a flag design painted on its underside. The crowd went wild.[46]

The 4450th Tactical Group had its own names and patches. Upon seeing the odd-looking plane for the first time, the air force pilots dubbed it the "Cockroach." Later, reflecting its role as a nocturnal predator, the air force pilots dubbed the plane the "Nighthawk." It was also called "the Black Jet," to differentiate it from the camouflaged A-7.[47]

Squadron patches had been an air force tradition dating from World War I. This continued with the 4450th Tactical Group. The patch for the Q-Unit showed an A-7 chasing a goat and the words "Goat Suckers." (As any bird watcher would know, the North American Nighthawk is also called the

Goatsucker.) The I-Unit patch showed a hawk swooping out of the night sky and the word "Nightstalkers." The Z-Unit patch showed a hooded figure pointing a bony hand at the viewer and the title "Grim Reapers."[48]

In all, about forty patches are known to be related to the aircraft. These included patches related to individual test programs. Even the C-5 flight crews that picked up the completed aircraft at Burbank had their patch—a black circle with a white crescent moon and a large question mark. On a tab at the top of the patch was "DON'T ASK!," while another tab at the bottom carried the enigmatic letters "NOYFB."[49]

And then there was the plane's designation. Pilots flying Black airplanes at Groom Lake logged their flight time with the code "117." When the Senior Trend began to fly, Lockheed started referring to it as "117" until the actual designation could be given. When Lockheed printed the first copies of the Dash One Pilot's Manual, F-117A was printed on the cover.[50] For year after year, the "F-117A" remained secret. But it was a secret that was proving harder and harder to keep.

LEAKS

The incoming Reagan administration increased the secrecy surrounding the stealth program. Although several projects would remain unknown for a decade and more, the effort was not entirely successful with the stealth fighter.

In June 1981, an article in *Aviation Week and Space Technology* said the Lockheed demonstrator aircraft was undergoing tests against Soviet equipment. It described the aircraft as "rounded" in shape.[51] In October 1981, *Aviation Week* carried another article, which said the Lockheed stealth fighter "will fly this year." (In fact, it had flown nearly four months before.) It also stated that the Fiscal 1983 budget was about $1 billion, that twenty aircraft were on order, and that they would be delivered within two years (essentially correct). The aircraft was described as resembling the space shuttle's wing platform (very wrong).[52]

The plane also acquired a "designation." Since the Northrop F-5G had been redesignated "F-20" in 1982 and the previous fighter was the F/A-18, it was assumed that "F-19" was the (secret) designation of the stealth fighter.[53]

By 1983, artists' conceptions of the F-19 began to appear. The general pattern was a long SR-71–like fuselage, elliptical wings at the rear, a bubble canopy, canards, and twin inward-canted fins. As it was now known the SR-71 had a reduced RCS, it was assumed the "F-19" was similar.[54]

In November 1983, *Defense Week* carried an article indicating two squadrons would be built. Again, stealth was described as relying on "curved airframe surfaces, inset engines, [and] rounded inlets." The article also revealed a stealth aircraft had crashed in "mid-April 1982" (the loss of the first production aircraft). At the same time, it noted that "some experts scoff at the suggestion that the Air Force could deploy 40 stealth fighter planes in the western deserts without public knowledge. Others speculate that these first stealth fighters rely on techniques to absorb or distort radar signals, and to the untrained eye do not appear radically different from other fighter planes, despite their somewhat smaller size."[55]

These accounts were in the technical press and so had little impact on the public. This changed in May 1986, when the F-19 arrived at the local hobby shop.

THE TESTORS F-19 STEALTH FIGHTER

Among those following the stealth story was the Testors Corporation. In 1985, they began work on a conceptual model of the F-19. The design was based on technical data, such as the *Radar Cross Section Handbook,* on a description from an airliner pilot of a black airplane seen over Mono Lake in 1983, and on the various published reports.

The design was well thought-out, looking like "a high-tech water beetle." It had inwardly canted rudders, curved surfaces, and blended air inlets. Testors was even able to test it in a San Diego defense contractor's RCS test range. This indicated problems with the intakes, which were corrected in the final design.[56]

The basic flaw was that it followed the SR-71 idea of stealth, and the report that the F-19 had a double delta wing platform like the space shuttle. Although it was described as being about "80 percent accurate," only two features were correct—the pitot tube and the platypus exhaust.[57] Yet only Lockheed and the F-117 pilots knew. With the rest of the world, the F-19 kit was an immediate best seller. In the next eighteen months, nearly 700,000 copies were sold.[58]

The model also figured in congressional hearings. In late June 1986, two Lockheed engineers made accusations that hundreds of documents, photos, films, and tapes were missing. Representative John D. Dingell (D-Michigan) accused Lockheed of falsifying document audits to cover up the problem.[59] Dingell, as chairman of the oversight and investigation subcommittee of the House Energy and Commerce Committee, held hearings on the alleged security leaks on the F-19.

During the hearings, a model of the Testors F-19 was passed around. Representative Ronald Wyden (D-Oregon) righteously complained, "It's bizarre. What I, as a member of Congress, am not even allowed to see is ending up in model packages."[60] Some officials complained that Dingell's hearings and the resulting publicity had unnecessarily compromised the program.[61] In the meantime, a greater, more tragic compromise had taken place.

LOSSES

The night operations of the 4450th Tactical Group continued to have an adverse effect on its pilots. One F-117 pilot later noted, "Each and every pilot in this group deals with being tired." This was made worse with the summer and the shorter flying hours. Part of the problem was the *Right Stuff* attitude—a pilot would never admit he was too tired to fly.

Lieutenant Colonel John F. Miller, one of the three squadron commanders, wrote a memo on Thursday, July 10, 1986, (the last flying night of the weekly cycle). He noted, "I believe that these extended hours are taking their toll on overall pilot performance. I have detected more and more instances of poor judgment that weren't evident 2–3 months ago." He cited unpredictable physical reactions to the continued stress, and "a major problem with fatigue-induced burnout that is getting worse with time." He added, "if we liken our usual late-go to a time-bomb waiting to go off, then our extended summer hours are accelerating the countdown to zero. I believe we are on a collision course with a mishap." Lieutenant Colonel Miller recommended that the pilots be forced to take "extra time off every two or three weeks."

As Miller was writing his memo, Maj. Ross E. Mulhare was preparing to fly a late-go mission. Mulhare had been declared mission ready in the F-117A on March 18, 1986. He was developing a new tactics training concept for the aircraft. Although an experienced pilot, he had a total of only fifty-three and a half hours in the plane. As he got ready, Mulhare told a colleague that he was tired and "just couldn't shake it."

Mulhare took off from TTR at 1:13 A.M. PDT, July 11, 1986, in F-117A number 792. He flew northwest to the town of Tonopah, then headed southwest and climbed to 20,000 feet. The night was clear and dark, with no moon. After crossing the Sierra Nevadas, Mulhare turned south along the eastern edge of the San Joaquin Valley. During the flight, Mulhare was in contact with the Los Angeles and Oakland Centers. He received permission to descend to 19,000 feet. As he neared Bakersfield, Mulhare turned southeast, requested a descent to 17,000 feet, and canceled his instrument flight plan at 1:44 A.M.

On the ground, Andy Hoyt, his sister Lisa, and her sixteen-year-old son, Joey, had pulled over at a rest stop. Hoyt saw "three red lights and a dark image behind them like an upside-down triangle." Hoyt got out his camera and took two or three photos before the object disappeared behind a hill. Suddenly, a pair of explosions "lit up the sky like it was daylight out."[62]

F-117A number 792 had slammed into a hillside in the Sequoia National Forest, about fifteen nautical miles from Bakersfield. Major Ross E. Mulhare was killed in the crash.[63]

An air force search party soon arrived and ordered all civilians out of the area.[64] The crash site was declared a national security area—no unauthorized people could enter the site, and no planes could fly within five miles of the crash site at altitudes below 8,500 feet. When Hoyt called Edwards Air Force Base to report what he had seen, the air force brought the three of them to a command post near the crash site. The film was developed, and two sets of prints were returned, minus the shots of the aircraft.[65]

The air force said only that a plane had crashed and the pilot had been killed. They would not say what type of aircraft, where it had taken off, its mission, or where it was going. There was no doubt among the press about what had crashed; an investigator with Dingell's subcommittee was quoted as saying, "It is clearly the F-19 that crashed." Dingell requested a briefing on the crash, but was turned down.[66]

The reports that followed the crash were a mixture of guesswork and speculation. *Newsweek* ran an article that claimed over seventy-two stealth fighters were operational; it speculated that the crash site would have to be cordoned off "forever" to prevent the Soviets from recovering debris. (In fact, the site was reopened after several weeks.) The *New York Times* claimed that the F-19 cost $150 million each. (The actual fly-away cost of the F-117A was $42.6 million, nearly identical to the $40 million cost of the F-15E Strike Eagle.)

Other articles were more accurate. An August 22, 1986, *Washington Post* story said that about fifty aircraft were operational, that the F-19 designation was incorrect, and that the plane was described as "ugly" due to its bulging, nontraditional shape. The plane's base was also identified as being Tonopah. Other reports described the daily flights to and from the base.[67]

While the press chased rumors and shadows, the air force tried to find the cause of the crash. This was made difficult by the condition of the debris. A report by Robert M. McGregor, an engineer at the Air Force Sacramento Air Logistics Center, stated: "Without exception, in terms of physical damage to the aircraft, this is the worst crash that I have worked. Structural breakup was almost absolute. 'Shattered' may best describe the aircraft after impact. . . . The right engine compressor drum . . . was crushed to half its normal length." The F-117A had hit the ground in a steep dive, between 20 and 60 degrees. There had been no in-flight fire, the engines were at a high-power setting at the time of impact, and Mulhare had not attempted to eject.[68]

The most probable reason was pilot disorientation. At night, without the normal visual clues, a pilot cannot tell if he is flying straight or is in a turn.

Lights on the ground can also be mistaken for stars. The problem was compounded by the F-117A's instrument panel design. Normally, the artificial horizon and other instruments are in the center of the panel, so the pilot can read them without moving his head. On the F-117A, the center of the panel is occupied by the FLIR-DLIR screen. To look at the instruments, the pilot had to turn his head. This took some getting used to, and more important, it could cause vertigo. An F-117 pilot later noted that most cases of disorientation were caused by a pilot moving his head while flying on instruments.[69]

The death of Mulhare was a wake-up call for the 4450th Tactical Group. Although the training schedule remained demanding, pilots were more closely monitored for signs of fatigue and were better trained to resist disorientation. The attitude also began to change: admitting you were not fit to fly a mission was seen as a sign of strength, not weakness. Pilots would watch each other; if someone showed signs of fatigue, a buddy would pull him aside for a private chat.

One of those who worked to instill this attitude was Maj. Michael C. Stewart. Several times, he had spoken about the need to avoid unnecessary risks. At 7:53 P.M. on October 14, 1987, he took off in F-117A number 815. The flight was a single-plane mission that would remain within the Nellis Air Force Base range. The night was clear and dark, with no moon. The mission was under visual flight rules. At 8:33 P.M., about three-fourths of the way through the mission, radar controllers noticed the plane had strayed to the left of its planned ground track. It then disappeared from their radar.[70]

Shortly thereafter, a large fire on the Nellis range was reported to the Bureau of Land Management (BLM), which relayed word to the air force. The air force asked the BLM not to say where or how big the fire was.[71]

The F-117A hit the ground in an area of gently sloping desert, digging a crater six to seven feet deep. The plane was 28-degrees nose down and in a 55-degree right bank at impact. There was no in-flight fire, and the F-117A was intact before impact. The engines were at a low-power setting. Major Stewart never attempted to eject.

There were a number of similarities to the loss of Mulhare the year before. Both were experienced pilots but had limited time in the F-117A—53.5 hours for Mulhare, 76.7 hours for Stewart. Both accidents occurred on dark, moonless nights. Again, disorientation was blamed.[72]

As with the earlier crash, the air force released minimal information, but the press had no doubt that another "stealth fighter" had crashed. Because the crash had occurred within the Nellis range, however, there was not the publicity of the first loss.

This changed less than a week after Stewart died. On October 20, an A-7D on a cross-country flight suffered a flameout. Its pilot, Maj. Bruce L. Teagarden, attempted to make an emergency landing at Indianapolis Airport, Indiana. The attempt was unsuccessful due to weather, and Teagarden was forced to eject. He landed safely, but the A-7D crashed into the lobby of a Ramada Inn, killing nine people. Press interest grew when it was learned that Major Teagarden was a member of the 4450th Tactical Group—the same unit Mulhare and Stewart had flown with.

Yet again, the press speculated. It was suggested that the A-7s had been modified with stealth systems or were playing the role of Soviet aircraft for stealth fighter practice missions.[73]

OUT OF THE BLACK

Although the erroneous reports effectively hid the true information, it was clear that the wall of secrecy around the F-117A was breaking down. In January 1988, *Armed Forces Journal* revealed the aircraft's actual designation was the F-117 Nighthawk.[74]

More important, the "big secret" of stealth, faceting, was starting to leak out. In 1986, there were reports that the F-19 was not smooth, but rather had "a multi-faceted outer-body surface" and a "cut-diamond exterior." This was described as being thousands of flat surfaces, none more than eight square inches in size, which did not share the same "reflectivity angle."[75] The F-117A actually used large panels, but the basic principle was the same.

A second major disclosure came in September 1987 with the release of a second Testors kit. This was a model of a (hypothetical) Soviet stealth aircraft, the "MiG 37B Ferret-E." The aircraft was made up of large, flat panels—faceting.[76] Reporters were soon being told by "reliable sources" that "if you want to see what the F-117 looks like, look at that MiG 37 model."[77]

The stealth fighter was such an open secret by the fall of 1988, that even the air force could joke about it. At that year's Edwards Air Force Base Air Show a large area was roped off. It contained a ladder, wheel chocks, and an official display sign labeled "F-19 'Flying Frisbee.'" Of course, this was an invisible airplane, so no one could actually see it.[78]

Bringing the F-117A out of the Black would have a number of advantages. The plane could be used in Red Flag exercises and could become part of standard war planning. There were benefits for the pilots too. With daylight flights possible, the fatigue from the late hours would be lessened. Night flights could start earlier and not have to race with the dawn. This would make the pilots' home lives easier. The problem was, ironically, that 1988 was also a presidential election year.[79]

The air force had planned to announce the F-117A's existence in early October 1988. This ran into congressional problems. Senators Sam Nunn

(D-Georgia) and John Warner (R-Virginia), the chairman and ranking minority member of the Armed Services Committee, warned that any release so close to the election could be seen as using classified information for political ends (harking back to the 1980 stealth announcement). They also complained that they had not been "adequately consulted." It was argued that if Congress had funded a Black program, then Congress should be consulted in any decision to declassify the program. A congressional staffer complained, "They can't just unilaterally release information at their insistence," and referred to the air force's "irresponsible handling" of the matter.[80]

On November 10, 1988, the announcement was made. Pentagon spokesman Dan Howard admitted that the stealth fighter did exist and that its official designation was the F-117A. He stated: "It has been operational since October 1983 and is assigned to the 4450th Tactical Group at Nellis Air Force Base, Nev. The aircraft is based at the Tonopah Test Range in Nevada." The press release contained the facts that the first flight had been made in June 1981, that three had crashed, and that fifty-two had been delivered out of a total of fifty-nine ordered from Lockheed. A single photo was also released. The angular shape came as a surprise, as did the out-of-sequence "century-series" designation. The photo showed the plane in a slight turn and gave no clue as to size. (In fact, the photo had been electronically altered.) No information was given out as to the F-117A's dimensions, cost, range, or speed.[81]

Where facts were lacking, the press was quick to speculate. Three-view drawings were published, showing a plane that was shorter, with much less wing sweep than the actual F-117A. Photocopies of the drawing were handed out to F-117 pilots. Some of the pilots, "laughed so hard that they started to cry."[82] It was suggested that the plane was supersonic, and that the air speed probes might be "gun barrels."[83] *U.S. News and World Report* claimed it had been flown near the Soviet border on reconnaissance missions.[84]

The F-117A now began daylight flights, and the sightings also began. On April 12, 1989, ten F-117As were seen flying near Mojave, California, at about 5:30 P.M. They were heading east, flying about eight minutes apart. Between 10:30 and 11:30 P.M. that night, another group of at least six planes was spotted flying the same route. On April 18, two similar waves of F-117As were also seen. Other sightings were made at TTR, where camera-equipped observers photographed the planes as they made takeoffs and landings.

Despite the F-117As coming out of the Black, speculation about the aircraft continued. It was reported, for example, that the F-117A had "a distinctive, although faint engine whine"—a claim that amazed Tom Morgenfeld, a Lockheed test pilot who had flown the F-117A since the early

1980s. He had "never heard anything more than the standard GE F404 engine noise."[85] Two different "sizes" of F-117As were also reported. People began watching the skies, looking and listening for "other" Black airplanes—the ones that were still being kept secret.[86]

With the disclosure of the F-117A, the 5540th Tactical Group underwent a name change. It became the 37th Tactical Fighter Wing. This was the former designation of the F-4G Wild Weasel unit. The 4450th, 4453d, and 4452d Squadrons became the 415th "Nightstalkers," the 416th "Ghostriders," and 417th "Bandits" Tactical Fighter Squadrons (TFS). These had been the designations of the first U.S. night-fighter squadrons during World War II.[87]

Thirteen months after it came "out of the Black," the F-117A flew its first combat mission.

PANAMA

During this time, relations with Panama were deteriorating. In early 1988, Panama's military dictator, Gen. Manuel Noriega, had been indicted by two Florida grand juries on charges of laundering drug money. He laughed off the charges and dismissed Panama's president in February. During the May 1989 presidential election campaign, Noriega's "Dignity Battalion" goon squad beat up opposition candidate Guillermo Endara. Endara won the election, but on October 1, Noriega prevented him from taking office. Two days later, a coup attempt was made but collapsed when loyalist Panamanian Defense Forces (PDF) rescued Noriega. The coup leaders were executed the following day. On December 15, Noriega declared a state of war between the United States and Panama. The following evening, PDF soldiers killed a marine lieutenant and arrested a navy lieutenant and his wife who had witnessed the shooting. The officer was beaten and his wife was threatened with sexual abuse.[88]

In response to these events, President George Bush issued orders to invade Panama. The attack was to strike PDF forces, capture Noriega, and rescue political prisoners. One of the targets was the Battalian 2000 barracks at Rio Hato. United States Army Rangers were to be dropped at the adjoining airfield. The PDF troops would have to be neutralized before the airdrop. Army Lt. Gen. Carl W. Stiner, the XVIII Airborne Corps commander, requested F-117As be used. They would not bomb the two barracks, but rather the 2,000-pound LBGs with time-delay fuzes would be directed to aim points near the buildings. They would act as "a giant stun grenade," to confuse the PDF troops without killing them. The use of F-117As was based on their night-bombing accuracy, rather than stealth, as the PDF lacked heavy air defenses.[89]

On the night of December 19, 1989, six F-117As from the 415th TFS took off from TTR. The flight would require five in-flight refuelings. Two of the planes were targeted on Rio Hato, two more were to provide support for an attempt to capture Noriega, and the final pair were in-flight spares should any of the others suffer malfunctions.

As the planes neared Panama, the attempt to capture Noriega was called off because he was not at any of the potential targets. Two of the F-117As continued on to Rio Hato. While they were in flight, the first problem occurred. Three hours before the invasion was to start, the PDF was tipped off to the coming U.S. attack (possibly due to American press reports). By H hour, 1:00 A.M. December 20, they had already occupied the airfield.[90]

As the two F-117As approached the release point, a moment of confusion occurred that would mar their debut. The original plan was for the lead plane to drop its bomb in a field near the barracks on the left, while his wingman would drop his bomb in a field near a barracks on the right. Just before the drop, the wind direction changed. The lead pilot, Maj. Gregory A. Feest, responded by telling his wingman to switch targets with him. At the drop point, however, the lead pilot bombed his original aim point. The wingman adjusted his aim point even farther to the left, following the changed plan.[91] One bomb, intended to land about 100 yards from the 7th Company barracks, actually landed 260 yards away. This was only 18 yards from the 6th Company barracks, which was too close. The other bomb impacted near a basketball court, about 40 yards farther from the barracks than intended.[92]

Despite these problems, the explosions caused the desired confusion. Initial reports spoke of PDF soldiers running around in their underwear, while others threw down their weapons. Several Rangers were killed in the subsequent firefight, but the airfield was taken and U.S. aircraft were landing within two hours.[93] In the confusion the miss was not immediately noticed. Defense Secretary Richard B. Cheney was advised both bombs hit their targets. He later spoke of the attack's pinpoint accuracy.

"Operation Just Cause," as the invasion was code-named, was effectively completed by the afternoon of December 20. The following days saw the running down of scattered snipers and a prolonged hunt for Noriega. The controversy over the invasion was more prolonged. Representative Charles B. Rangel (D-New York) said, "I strongly believe the invasion was totally illegal."[94] Former Attorney General Ramsey Clark claimed two thousand to four thousand Panamanians had been killed and secretly buried. His unsubstantiated claims were later repeated on *60 Minutes*.[95]

Particularly vitriolic attacks were directed against the use of the F-117A. *Time* carried a diatribe titled "Bombing Run on Congress." It claimed the "supersecret" F-117A had only been used "to wage a public relations assault on the U.S. Congress." It quoted a "congressional defense expert" as calling the mission "pure pap—a gimmick." He said the mission "could have been flown with an Aero Commander, or let Mathias Rust [a West German teenager who landed a Cessna in Red Square] do it." The article dismissed the plane's accuracy by saying, "Some Air Force pilots consider the plane so unstable in flight that they call it the Wobbly Goblin." It concluded, "The real objective was to save Stealth technology from the congressional budget ax. . . . The Air Force unleashed its F-117As not to scare Manuel Noriega but to build a case that high-tech aircraft have a role even in a low-tech war."[96]

It was not until April 1990 that word was published about the miss. Headlines such as "Stealth error kept under wraps" and "General didn't report Stealth flaws in Panama" were used.[97] The press had its "cover-up" story, and the usefulness of the F-117A was further questioned.

"THE END OF HISTORY"

As the F-117A was coming out of the Black, the world was emerging from another kind of darkness. During 1989, one by one, like dominoes, Eastern European countries cast off their Communist governments. The Soviet Union became a multiparty democracy. On November 9, 1989, the Berlin Wall fell. The world had changed—whatever followed would be far different than the past forty years. Questions arose about what role, if any, the U.S. military would play in this brave new world.

At a press conference, a reporter pointedly asked President Bush, "Who's the enemy?" As in 1919 and 1945, it was assumed that having defeated one enemy, there would be no more.[98] *Time* asked "Who Needs the Marines?" *Newsweek* predicted U.S. military forces would be cut in half by the turn of the century.[99] Academics began to talk about "the End of History." "National security" would be defined in terms of education, cultural enrichment, and environmental enhancement. In parallel with "demilitarization" was "denationalization"—in an "interdependent world" nothing could be achieved along only national lines.[100] Mankind had evolved; in the bright future, there would be no need for the use of force. But Communism was only one form of evil. Evil remained.

August 1, 1990, was the thirty-fifth anniversary of the U-2's first "hop." Now its descendant, the F-117A, was being dismissed as a useless relic of an era never to return. At Groom Lake, as the afternoon passed, the shadows from the mountains lengthened toward darkness. In the Mideast, it was now 2:00 A.M., August 2, 1990.

Suddenly, three Iraqi armored divisions, backed up with MiGs and helicopters, attacked Kuwait.

LINE IN THE SAND

Within hours, resistance had collapsed and Kuwait became Iraq's "19th province." It seemed that Iraqi dictator Saddam Hussein's ambition would not end there. Soon after the invasion, seven more Iraqi divisions took up positions along the Saudi-Arabian border. This was followed by a series of border incursions. The Saudis concluded that an Iraqi invasion was imminent. The Iraqis could take the Eastern Province in six to twelve hours, and the whole country in three days. This would give Saddam effective control of the world's oil supply and the world's economies.[101]

On August 6, King Fahdibn Abd al-Aziz Al Saud of Saudi Arabia invited U.S. troops into the country. Within two days, F-15s and the first elements had arrived to draw "a line in the sand." The third great conflict of the twentieth century had begun.

On August 17, Alton Whitley, now a colonel, was named commander of the 37th Fighter Wing. Four hours later, he was ordered to deploy the 415th TFS to Saudi Arabia. On August 20, eighteen F-117As were on their way. They landed at King Khalid Air Base at noon the next day. The brand-new base was located at the southern tip of Saudi Arabia, outside the range of Iraqi Scud missiles. It had state-of-the-art hardened aircraft shelters and even hardened crew quarters. The base was soon dubbed "Tonopah East."[102]

The unit began an intensive training program. Only four of the sixty-five stealth pilots had flown combat, one of them in Panama. The flights exactly simulated the operational missions, right up to the point that the F-117A would head into Iraqi airspace. Three exercises were also held to test the readiness of "Team Stealth," as the unit was now called.[103]

To keep up morale, a longtime tradition was revived—nose art. To remain stealthy, it was applied to the bomb bay doors. There were names such as "Unexpected Guest," "Dark Angel," "The Toxic Avenger," "Habu II," "The Overachiever," "Once Bitten," and "Christine." The Saudis nicknamed the plane *Shaba,* Arabic for "ghost."[104]

On November 8, 1990, President Bush ordered a major increase in U.S. forces in the Gulf. As part of this, another twenty F-117As from the 416th TFS flew to Tonopah East, arriving on December 4. The unit was redesignated the 37th Tactical Fighter Wing (Provisional). Both squadrons of combat F-117As had now been deployed. (The 417th TFS was the training unit.)[105]

As the Allied buildup continued in the Gulf, doubts were expressed that the effort would be successful. Since the 1970s, a network of "military critics" had developed; they depicted the U.S. military as incompetent, as building weapons that were not needed and did not work, and as "fighting the last war." A central theme was that airpower was doomed to failure.

Bombing was indiscriminate, they said, hitting civilians, schools, and hospitals, which would only stiffen Iraqi resolve. Dug-in troops could not be dislodged by bombing, nor could airpower cut off supplies to Iraqi troops. Harvard economist John Kenneth Galbraith said Americans "should react with a healthy skepticism to the notion that airpower will decide the outcome of a war in Kuwait and Iraq." Another voice added, "The United States relies on the Air Force and the Air Force has never been the decisive factor in the history of wars." The voice was that of Saddam Hussein.

The F-117A was specifically criticized: stealth could be defeated by multiple radars, stealth required too much maintenance time, "delicate" and "complex" high-technology systems could not withstand the demands of sustained combat or the desert heat and dust.[106] (In fact, the F-117A had readiness rates in the Gulf *higher* than the peacetime standard.) The Iraqis tried to encourage such beliefs, with such statements by Saddam as, "[The F-117A] will be seen by a shepherd in the desert as well as by Iraqi technology, and they will see how their Stealth falls just like . . . any [other] aggressor aircraft."[107]

The war for Kuwait, it was argued, would not be decided by airpower, but by ground combat with the "battle-hardened" Iraqi army. The Iraqi use of poison gas in the war with Iran brought back echoes of the mass slaughter of World War I. Estimates of U.S. casualties from such a ground war ran as high as forty-thousand-plus. Politicians warned such casualties would fracture the nation, just as Vietnam had.[108] An "antiwar" movement had already organized under such slogans as, "No Blood For Oil," "Protest The Oil War," "Bury Your Car," and the ever popular "Yankee Go Home."[109] In a real sense, the United States had to fight not only the Iraqis, but also the ghosts of its Vietnam experience.

On January 12, 1991, the Congress approved the use of force to back up a United Nations (UN) resolution calling on the Iraqis to withdraw from Kuwait. It was, in every sense of the term, a declaration of war. The UN deadline expired on January 15, and President Bush ordered combat operations to begin.

On January 16, 1991, the F-117 pilots were told to get a good meal. They began to suspect something was afoot. The maintenance and weapons personnel were ordered to make one simple change in the bomb loading procedures—the arming lanyards were attached to the bombs.[110]

The pilots reported for duty at 3:00 P.M. and were told they would attack Iraq that night. Each pilot was then given his target data. This war would begin over Baghdad and would strike at the heart of Iraqi air defenses and communications facilities. The F-117As would strike the National Air Defense Operations Center in Baghdad, the regional Sector Operation Centers

(SOCs), and the local Intercept Operation Centers (IOCs). This air-defense network controlled some five hundred radars, the SA-2, SA-3, SA-6, SA-8, and Roland SAMs, and some eight thousand antiaircraft guns. Baghdad alone was protected by about four thousand antiaircraft guns and SAM launchers.[111] The complete system provided a thicker air defense than any in Vietnam or Eastern Europe, while the defenses of Baghdad rivaled that of Moscow or Vladivostok.[112] And the F-117 pilots would have to face it all alone.

Ironically, the senior commanders and the F-117 pilots had very different images of the plane. The commanders had great faith in stealth, but due to the flawed Panama attack, there were questions about the plane's bombing accuracy. The F-117 pilots, on the other hand, had absolute faith in their ability to hit the targets. The plane's stealthiness was the unknown factor to them. As they suited up for the first night's attacks, several pilots were heard to say under their breath, "I sure hope this stealth shit works!"[113]

At the briefing, Colonel Whitley tried to prepare them for what was ahead. He explained what it would be like when the whole world seemed to be firing at them. He recalled, "I told them there would be hormones that would flow that they'd never tapped before. I told them they would know what I meant after they came back."

The pilots arrived at their planes about 10:30 P.M. and began the preflight inspection. When this was complete, they boarded the aircraft. The ground crews then handed them the paperwork for the mission—target photos, maps, checklists, and locations of emergency airfields. Each pilot also carried a protective suit against chemical attack, a rescue radio beacon, a "blood chit" (in English and Arabic), which promised a large reward for helping a pilot escape, and a 9mm Berreta automatic.

The first wave was made up of 415th TFS pilots; they had been at Tonopah East since August, so Colonel Whitley felt they should have the honor of being first. One pilot almost missed his chance; Capt. Marcel Kerdavid discovered he could not start his plane's port engine. He grabbed his paperwork and the tape cartridge that held the mission data and was driven to the spare F-117A. He did a fast preflight and was ready to go.

Just before midnight the F-117As were towed out of the hangars and began moving down the taxiway. The day shift had just come off duty, and the taxiway was lined with maintenance personnel. They saluted as the planes went past. Just after midnight, the first F-117A took off; by 12:22 A.M., January 17, the last was gone.

The F-117As flew in pairs to the tankers. The first refueling occurred soon after takeoff. The second was completed thirty-five nautical miles from the Iraqi border. So far, everything was exactly the same as the training missions.

The first pair completed their refueling, left the tankers, and slipped undetected into Iraqi airspace, and the unknown.[114]

A NIGHT OF THUNDER

At home, the day of January 16, 1991, had passed slowly. It was clear that war was inevitable. People gathered around their televisions, waiting for news. At 6:35 P.M. EST (2:35 A.M. in Baghdad), CNN's David French was interviewing former defense secretary Casper Weinberger. He stopped and said, "We're going to Bernard Shaw in Baghdad." Shaw began his report: "This is—something is happening outside. . . . The skies over Baghdad have been illuminated. We're seeing bright flashes going off all over the sky."[115]

The sky above Baghdad had erupted with antiaircraft fire, but, as yet, there were no U.S. aircraft over Baghdad. At 2:39 A.M., only minutes after CNN began broadcasting from Baghdad, army Apache helicopters blasted two Iraqi early warning radar sites. This opened a gap in radar coverage, and F-15Es flew through it to strike Scud missile sites in western Iraq. Two F-117As had already crossed into Iraq. They were followed by six more. Unlike the F-15Es, they did not have support from EF-111A jamming aircraft. It was one of these follow-on F-117As that opened the Black Jet's war.

The target was the Nukhayb IOC in western Iraq. Located in a hardened bunker, it could coordinate attacks on the incoming F-15Es and the follow-on strikes. The pilot was Major Feest, the lead pilot for the Panama strike. He located the target and released the bomb at 2:51 A.M. He saw the bomb penetrate the bunker's roof and blow off its doors. He turned toward his second target, an SOC at the H2 Air Base. When he looked back, Feest saw the night sky was filled with antiaircraft fire, triggered by the bomb's explosion. When he looked toward the second target, he saw the whole sky was alive with ground fire.

As the other F-117As closed on Baghdad, antiaircraft fire seemed suspended above the city. Lieutenant Colonel Ralph Getchell, 415th TFS commander and leader of the first wave, likened it to Washington, D.C., on the Fourth of July. The firing at the empty sky had been going on for a full twenty minutes, but at 2:56 A.M., a cease-fire order was issued. A stillness fell over the city. From their cockpits, the pilots could see the eerie glow suddenly disappear. Through the IR displays, individual buildings took shape. Baghdad was still brightly lit, and car headlights could be seen streaming out of the city.[116]

As the Dark Eagles moved unseen and unheard above, CNN reporters Bernard Shaw and Peter Arnett were discussing what had happened. As they spoke, Capt. Paul Dolson placed the cross hairs of the targeting system on the fourteen-story Al-Karak telephone and telegraph center. The plane's bomb door opened and a GBU-27 LGB fell free. Millions of people gathered around television sets heard this exchange:

> Shaw: "We have not heard any jet planes yet, Peter."
> Arnett: "Now the sirens are sounding for the first time. The Iraqis have informed us—[static]."[117]

At that instant, the GBU-27 punched through the Al-Karak's roof and destroyed the communications equipment, cutting off CNN. Within five minutes of the 3:00 A.M. H hour, Marcel Kerdavid had destroyed the Al-Kark communications tower, Capt. Mark Lindstrom dropped an LGB through a roof vent on the new Iraqi air force headquarters, while Ralph Getchell struck the National Air Defense Operations Center, and Lee Gustin bombed Saddam Hussein's lakeside palace-command center. As the first bombs exploded, the F-117 pilots saw antiaircraft fire rise above the city.

Major Jerry Leatherman, following one minute behind Dolson, dropped his two GBU-10 LGBs through the hole blasted by the first bomb. Unlike the GBU-27, which was designed for attacking hard targets, the GBU-10 had a thin casing and a greater blast effect. The two bombs gutted the building. As his plane cleared the area, he looked back and beheld the wall of fire he and the other pilots had flown through. He said later, "There were greens, reds, some yellows, and you could see little white flashes all over—the airbursts . . . [The SAMs] move[d] around as they were trying to guide on something, whereas the tracers would just move in a straight line. The 23mm . . . looked like pinwheels the way the Iraqis were using them . . . it looked like they'd just start firing them and spin 'em around."

The F-117As sped away from Baghdad. Some, with both bombs expended, headed home. Others headed for their second target; Kerdavid bombed the deep National Command alternate bunker at the North Taji military complex. Its thirty-feet-thick roof proved too much even for a GBU-27, and it remained intact. More successful were attacks on a communications facility at Ar-Ramadi, the SOCs at Taji and Tallil, and an IOC at Salman Pak.[118]

Between 3:06 and 3:11 A.M., as the F-117As left Baghdad, Tomahawk cruise missiles began striking leadership targets, such as Ba'th party headquarters, the presidential palace, electrical power generation stations, and chemical facilities in and around Baghdad. The Tomahawks directed against the electrical plants shorted out power lines, and all over Baghdad, power went out, not to be restored for the rest of the war.

At 3:30 A.M., the disrupted air-defense network began picking up a huge attack force heading directly toward Baghdad. Comments by air force Chief of Staff Gen. Michael J. Dugan in September 1990 had indicated there would be a raid on Baghdad by nonstealthy aircraft between F-117A strikes. General Dugan was fired for the comments, but the Iraqis still expected such an attack. As the planes neared, the radars came on and the SAMs were prepared to fire.

But they were not airplanes; they were decoy drones. And behind them were navy A-7s and F/A-18s and air force F-4Gs with HARM (high-speed antiradiation) missiles. The HARMs both destroyed radar sites and intimidated Iraqi air defense radar operators to stay off the air.[119]

Even as the Iraqis were attempting to deal with this, the second F-117A wave was closing on Baghdad. It was led by Colonel Whitley. The flight toward Baghdad was a "sobering experience." He later recalled, "At 100 miles plus, you could look out there following the horizon of Baghdad, and it looked like a charcoal grill on the 4th of July." The glow was the continuous firing of nearly four thousand antiaircraft guns.[120]

At 4:00 A.M., the second F-117A wave restruck the air force headquarters and the National Air Defense Operations Center. Other targets hit were the IOCs at Al-Taqaddum Air Base and Ar-Rutba as well as leadership and communications facilities from the Jordan border to Kuwait.[121] In all, the two waves had dropped thirty-three bombs and scored twenty-three hits.

The third wave followed shortly before dawn. Their targets were chemical and biological weapons storage bunkers. The late hour was selected because sunlight would reduce the danger from Anthrax spores. As they approached their targets, a weather front moved into central Iraq, with thin clouds at 5,000 feet. The F-117A's bombing system required a clear view of the target or the LGB would lose its lock. Of the sixteen bombs dropped, only five were hits. These targets were considered less important, but it was a preview of the bad weather that would plague the bombing campaign in the weeks ahead.[122]

As the F-117 pilots turned for home, their mood was somber. They knew they had won a victory, but they were sure the cost had been high. Captain Rob Donaldson said later, "I came out of there on that first night and went 'Whew . . . I survived that one!' But on the way back, I really thought that we had lost some guys due to the heavy volume of bullets and missiles that were thrown up in the air."[123]

At Tonopah East, the ground crews awaited the planes' return. The first wave landed at night, while the second and third came back after sunrise. One by one, the returning planes were counted.

Every one returned.

As the tapes of the strikes were reviewed, it became clear that something remarkable had occurred. In World War II, the RAF had sent huge armadas of bombers on night raids against Germany. Despite years of bombing and the loss of thousands of aircraft and crews, the RAF was never able to knock Germany out of the war, or even win air superiority.

Now, a handful of planes had faced an air-defense network that dwarfed that of Berlin in 1943–44, struck at the heart of the enemy capital, and emerged without a scratch. Each plane's load was a fraction of that carried by a Lancaster bomber, but the results far surpassed all the years of area bombing the RAF had carried out at so heavy a price.[124]

The Iraqi air-defense system died that night; with the headquarters hit and the IOCs and SOCs damaged and out of action, the individual antiaircraft guns and SAM sites were isolated. The operators were unable to operate their tracking and fire control radar, for fear a HARM missile would destroy them. Units in the field had limited communications with each other and with higher command. Electrical power was out in Baghdad. The three F-117A waves, the Tomahawk attacks, and the decoy raid, tightly interrelated in time and space, had left the Iraqis unable to inflict significant losses on Coalition air operations.[125]

The ultimate result of that night of thunder was this: in every war, there comes the time when it becomes clear who will win and who will lose. In the Civil War, it was the Battle of Gettysburg; in the Pacific in World War II, it was the Battle of Midway; in the Gulf War, that was the moment.

THE BLACK JET AT WAR

The ground fire that greeted the F-117As on the second night was described as perhaps the most intense of the Gulf War. Lieutenant Colonel Miles Pound said, "They knew we were at war that second night and they had every gun manned. And they were more than willing to use them." The last ten minutes before the target commanded the pilots' full attention. They could not afford to think about the lethal fireworks outside. Pound explained later, "My own technique was to run the seat down, so I wasn't distracted by what was going on outside. The lower you get in the cockpit, the less you can see outside. I would reduce the amount of distraction to the absolute lowest level and just concentrate on my target."[126]

The targets on the second night included the IOCs and SOCs. Again, the goal was to disrupt Iraqi air defenses. The chemical and biological weapons bunkers not hit on the first night were also targeted. Also destroyed was an Adnan-2 early warning aircraft. In all, two waves were launched. One plane from the first wave had to abort due to system problems. Still a total of nineteen bombs were dropped, with thirteen hits.[127]

On January 19, the weather front stalled over Iraq, and visibility was poor. The F-117As were able to score only six hits. The planes were unable to hit their primary targets in Baghdad, and few alternative targets were open. One of the planned targets was Hawk SAMs in Kuwait. These U.S.-built missiles had been captured during the invasion. The following night, January 20, saw improving weather, and seventeen hits were scored on IOCs and SOCs, ammunition storage sites, an ELINT ground station, bridges, telephone exchanges, and chemical warfare targets.[128]

Although January 19 had been a disappointment for the F-117 pilots, other events that day would have a major impact on their operations. Beginning on the third day, it was planned to send large "packages" of F-16s to Baghdad during daylight. They would attack large targets such as oil refineries and headquarters. The round-the-clock missions—F-117As by night and F-16s by day—would serve notice that Iraq could not defend itself.

The first of these daylight attacks was "Package Q"—seventy F-16s, as well as support aircraft. The strike was directed against the Daura petroleum refinery and the Baghdad nuclear research center. Bad weather, heavy ground fire, and smoke screens resulted in a failure. Almost no damage was inflicted on the reactors, and two F-16s were shot down.[129] It was clear the skies of Baghdad were too dangerous for conventional attacks. After Package Q, plans were dropped for similar raids. All attacks on Baghdad would be made by F-117As or Tomahawks.[130]

This posed a daunting task. The Baghdad nuclear research center was huge, with four reactors and large numbers of support buildings—about one hundred structures in all. Each target would require individual F-117A strikes. The Black Jets would have to destroy Saddam's war machine one aim point at a time.

Eight F-117As attacked the nuclear research center on January 21. Two reactors were destroyed, another was severely damaged, and research buildings were also hit. The Iraqis did not know the attack was under way until the bombs exploded.

The F-117A strike was quite a contrast with Package Q. The eight planes had been supported by just two KC-135 tankers. The Package Q attack on the facility had consisted of thirty-two F-16 bombers, sixteen F-15Cs as air escorts, four EF-111As and eight F-4Gs to suppress the defenses, and fifteen tankers. For the next week, F-117As would make additional attacks on the site.

The second wave of the night was equally successful. The fourteen planes had twenty-six hits and only two misses on a number of Baghdad targets, including a biological weapons plant that was claimed to be a "baby milk factory," with signs conveniently printed in English.[131]

The F-117A strikes were twofold. On the tactical level, the goal was to prevent air defenses from functioning, and chemical or biological weapons from being used against Coalition troops. The strategic goal was long-term.

Attacks on the Baghdad nuclear research center and chemical and biological weapons facilities were intended to prevent Iraq from emerging from the war with the ability to produce such weapons in the future. Bombing such targets as the Ba'th party headquarters and the secret police headquarters was intended to break Saddam's political hold on the country.[132]

As the Black Jet's first week at war ended, there was a shift in activities; airfields were targeted as the F-117As tried to dig the Iraqi air force out of its hiding places.

SHELTER BUSTING

The Iraqi air force showed no more success in countering Coalition air strikes than ground fire. Mirage F1s, MiG 21s, MiG 25s, and MiG 29s fell to U.S. missiles during the first three days.[133] In several cases, Iraqi pilots flew into the ground during dogfights. Rather than face sure destruction, the Iraqi air force tried to ride out the attacks in their hardened aircraft shelters (HAS). These were built to withstand the blast from a nuclear explosion.

Iraq's air force represented both an immediate and future threat. It was feared that the Iraqis might launch an "Air Tet"—a sudden, massive, and suicidal air raid, possibly using chemical weapons, meant for maximum propaganda and destructive effect. Like the Viet Cong's Tet Offensive in January 1968, the goal would be to turn the war against the Coalition, in a symbolic and psychological sense.[134] Such fears were not limited to senior military and political leaders. One U.S. Army reservist recalled having his "head on a swivel" watching for Iraqi planes. His special concern was understandable; he drove a large, *green* fuel truck. (He had dubbed it "Sitting Duck.")[135]

In the long-term, the Iraqi air force represented a potential threat to the whole Mideast. Its destruction would reduce Saddam's future role as a regional troublemaker. In either case, Iraq's air force would have to be sought out and killed.[136]

The F-111Fs began the "shelter busting" campaign on January 21. They were joined the following night by the F-117As, when fourteen Black Jets attacked the Balad Southeast Air Base. One plane was forced to divert to an alternate target, but the others scored twenty-one hits on the HASs. The success was not what it seemed, however. The GBU-10s penetrated the outer layer of concrete but were deflected by the second layer of dirt and rock. The inner concrete shell and the planes inside were unharmed. The Iraqis concluded that the HASs worked, and more planes were hidden away.

On the eighth night of the war, intelligence was received that Air Tet was at hand. It was reported that eight Tu-16 bombers at Al-Taqaddum Air Base were being loaded with chemical weapons for a dawn strike. Several F-117As were switched at the last minute to attack the base. They arrived over the field just before sunrise. The Tu-16 bombers were fully fueled and had support vehicles grouped around them. When the GBU-10s struck three of the Tu-16s, they exploded in huge fireballs. Three of the remaining bombers were destroyed by a conventional strike the following day.[137]

After the first disappointing HAS strike, the F-117A squadrons launched another attack on the night of January 24–25. This time the planes carried GBU-27 penetrator bombs. The night's first wave struck the Qayyarah West, Al Assad, and Kirkuk Air Bases. These targets required the F-117As be refueled by tankers flying inside Iraqi airspace—the best example possible of Coalition air supremacy. A total of twenty hits were made. The GBU-27s penetrated the HAS roofs and exploded inside. In many cases, the armored doors were blown off, and smoke and flame billowed out. It was the funeral pyre of the Iraqi air force.[138]

In the wake of the successful HAS strikes, the Iraqi air force stood down for a day. Then on January 26, Iraqi fighters began fleeing to Iran. In the days ahead, the Iraqi planes ran a gauntlet of U.S. F-15s as they sought to escape. After arriving in Iran, the planes were repainted in Iranian markings.[139]

The escape attempts put more pressure on the shelter-busting effort. The HASs had to be destroyed before all the planes "flew the coop." (A large Iranian air force was no more in the interest of stability than a large Iraqi air force.) To cope with the various demands, additional F-117As were flown to Tonopah East. A total of forty-two planes and nearly all the qualified pilots would see action. This enabled a total of thirty-four sorties per night to be launched, rather than twenty-eight.[140]

The airfield attacks would continue until the end of the war, but the changing situation by the start of the third week of the war had already caused a shift of F-117A activities.

STRATEGIC TARGETS

On January 27, Gen. Norman Schwarzkopf ordered the air force to shift all air activities except F-117As and F-111Fs to Kuwait. The ground war would soon be under way. F-111F crews began spotting tanks at night, and on February 5, the F-111Fs dropped eight GBU-12 LGBs which destroyed four Iraqi tanks and one artillery piece. The following night, a larger F-111F strike was made against Iraqi Republican Guard units. The results were spectacular, and the F-111Fs were immediately shifted out of the strategic campaign. They would concentrate on ground targets, a mission that became known as "tank plunking."[141]

The F-117As would now have to carry the whole weight of the strategic bombing effort. The forty-two Black Jets would cover the wide variety of targets previously assigned to the sixty-six-plane F-111F force.

The main targets for the third and fourth weeks were chemical, biological, and nuclear facilities. In contrast, the HASs accounted for only a small percent. The fourth week showed yet another shift. The number of F-117A attacks against command and control targets increased to nearly equal those directed at chemical targets. Other strikes included leadership targets and military support. The F-117As attacked fixed Scud targets, such as hide sites.[142] The Black Jets also struck SAM sites. On one occasion, sixteen Black Jets took out every SA-2 and SA-3 site from south of Baghdad to Tikrit. Later that night, twenty-four B-52s hit targets in the area without any losses. Without the F-117As to destroy the SAMs, it was probable that several of the B-52s would have been lost.[143]

Whatever the night's target, one factor that remained constant for the F-117 pilots was the ground fire. Leatherman said later, "One thing that surprised me was that they didn't run low on ammunition." At the end of the first week, the minimum altitude of the F-117A strikes was raised to avoid the ground fire. Although the plane could not be detected, a random hit, (called a "Golden B.B.") was still possible. On one night, it seemed to Capt. Rich Treadway there was a "half-price-sale" on 37mm and 57mm ammunition: "You could tell where a bomb went off because the entire sector would be engulfed in tracers."[144] Some of the shots came close; Colonel Whitley recalled hearing the "pop, pop, pop" of rounds exploding nearby and feeling his plane move from the concussion.

The Iraqis tried various techniques to shoot down an F-117A. They began using "barrage fire," directing all the antiaircraft fire into specific parts of the sky, in hopes the Black Jets would fly into it. On one night, they held their fire, watching for the glow of the F-117A's afterburners as the planes sped away from the target area. Once it was spotted, all the guns would fire at the glowing targets. The Iraqis were very confident it would work and even alerted Jordanian reporters in advance. Colonel Whitley and Major Leatherman were over Baghdad when it was tried out. Leatherman recalled "it was eerily quiet—even after our bombs hit." (The F-117A does not have an afterburner, and the exhaust is shielded from the ground.)

One night it seemed an F-117A had taken a hit. A postflight inspection found that RAM on part of the tail was gone. At first, it was thought to be battle damage, but the RAM actually had delaminated. Ironically, there was disappointment—the F-117A had so far escaped damage, and everyone was worried about what would happen if it was hit.[145]

As the fourth week of the Black Jet's war neared its end, the attack plan again shifted. Leadership targets became a high priority, as they had been during the first week. On the night of February 11–12, a total of thirteen strikes were made. The following night, thirty-two strikes were made. The targets included air force headquarters, the Ministry of Defense headquarters, Ba'th party headquarters, several intelligence headquarters, and radio and telephone facilities. One of the targets was the Al Firdos district bunker.

AL FIRDOS

In the early morning hours of February 13, two F-117As dropped a single bomb each on the Al Firdos bunker. Both LGBs penetrated the roof and exploded inside. It was one of twenty-five bunkers that had been built as command posts, but at the start of the Gulf War, it was not in use. On February 5, trucks were observed unloading communications equipment at the bunker. Three black circles were painted on its roof, to simulate bomb hits.[146] Intelligence indicated it was being used as a communications center for one of the Iraqi intelligence agencies bombed out of its original headquarters. Within a week Al Firdos was added to the target list. What no one knew was that the upper floor of the bunker was also a civilian shelter, reserved for the families of the political elite.[147]

When the bombs exploded inside the bunker, a hundred or more women and children were killed. That morning, CNN carried grim film of the bodies being removed from what the Iraqis called "General Shelter 25." They claimed it was an air-raid shelter, not a command post, and that it had been a deliberate attack intended to kill civilians. The Iraqis also showed Western reporters a sign (in English) identifying it as an air-raid shelter. The United States countered the claims, noting that it had been hardened against nuclear attack, was surrounded by a chain-link fence and barbed wire, and had a camouflaged roof. Photos of the "shelter" also showed computer cables in the wreckage. All these were inconsistent with an air-raid shelter.[148]

The dismay and controversy that followed the Al Firdos bunker bombing effectively ended the strategic air campaign against leadership targets in Baghdad. General Schwarzkopf told the air force that they could not hit any targets within Baghdad without his specific permission. For a week after Al Firdos, he was unwilling to give permission to strike any Baghdad target, for fear of civilian casualties.

With leadership targets off-limits, another approach was taken. Rather than the substance of Saddam's regime, three symbols were recommended for attack. These were Ba'th party headquarters, a sixty-feet-tall statue of Saddam, and the huge victory arches commemorating the Iran-Iraq War. The latter were moldings of Saddam's forearms, holding swords that crossed 150 feet above an avenue. Schwarzkopf approved the choices, but the targets ran into a particularly American difficulty—lawyers. The statue and the victory arches, military lawyers said, were "cultural monuments," which could not be bombed under international law. Although they were likened to "Hitler's Nuremberg parade grounds" in the official history, the objection stuck and permission was withdrawn on January 25. It would not be until the eve of the war's end that targets within Baghdad were again approved.[149]

END GAME

With the halt on bombing leadership targets, the F-117As were turning to other targets. Schwarzkopf directed that a seventy-two-hour bombing campaign be planned to destroy nuclear, chemical, and biological targets should a cease-fire be imminent. At the time of the Al Firdos strike, more than a dozen suspected chemical and biological bunkers remained to be hit. Priority was given to research and development sites, however, rather than storage bunkers, to prevent Iraq from having such weapons in the future.

The seventy-two-hour list was headed by the Baghdad nuclear research center. F-117As struck it on February 18, 19, and 23. Weather continued to be a problem; only four of ten aircraft bombed on the eighteenth, while four out of six scored hits the next night. Finally, on the twenty-third, thirteen F-117As bombed the site in good weather, with eighteen out of twenty-six bombs hitting structures in the compound. Other F-117As bombed suspected nuclear facilities across Iraq, including the Al Qaim uranium extraction facility in western Iraq.

United States intelligence did not fully comprehend the scope of the Iraqi nuclear effort. Thus the identity of three nuclear facilities located at rocket development sites was only realized late in the war. The Tarmiya rocket facility was hit by F-117As and B-52s on February 15. By February 23, its nuclear role had been discovered and four F-117As were sent to bomb it. Two of the planes were unable to drop their LGBs due to weather. It became the highest priority target but was not hit again due to weather. The Ash Sharqat facility had been hit by six F-111E raids before four F-117As visited the site on February 16. Its nuclear role was not yet realized, and it was not bombed again. The nuclear site at Musayyib was not discovered until shortly before the war's end; it had been selected to be the final assembly site for the first Iraqi nuclear bomb. A pair of F-117As bombed the site on February 25, and a second group of nine Black Jets finished it off on February 28.[150]

United States intelligence had a much better understanding of Iraqi chemical and biological production capabilities, and all known sites had been bombed by mid-February. As the war neared a close, attacks on chemical and biological weapons storage bunkers increased, as did strikes on the remaining HASs, in an effort to hit as many as possible before the war came to a close.[151]

The F-117As became a kind of flying fire brigade. If a target needed to be attacked on short notice, it was far easier to send a handful of Black Jets, rather than a force of forty to sixty F-16s and support planes. One example of this was the February 15 attack on the fire-trench system in Kuwait. Distribution points, pipe junctions, and pumping stations were hit with 500-pound GBU-12 bombs. The attack was so hurried that the second wave pilots had to use sketch maps, because photos were not available. Despite this, twenty-four out of twenty-seven LGBs found their mark.[152]

Along with the serious business of war, the souvenir business was also booming. The F-117A combat missions were quite popular, and, as the weeks passed, ground personnel, as well as their friends and relatives, wanted a little piece of history. Major Leatherman gave his crew chief one of the bomb arming lanyards from the first mission. F-117 pilots found themselves carrying various items on missions. American flags were most requested, but everything from footballs to Bart Simpson dolls were flown. (Bart was the unit's unofficial mascot.) The pilots would then autograph the souvenirs. One crew chief put a teddy bear on his plane; "Geronimo Bear" came along on every mission that F-117A flew.[153]

VICTORY

It was not until the final nights of the war that the F-117As were again authorized to fly into downtown Baghdad. The target selected was Ba'th party headquarters—a complex covering several city blocks. It had been hit by Tomahawk missiles on the first night and again by F-117As in mid-February. The largest F-117A raid of the war was planned—thirty-two Black Jets on the night of February 25–26 (the night after the start of the ground war). The bad weather that had dogged the air campaign foiled the attack. All F-117A missions were canceled for the night. The following night was no better; only a few targets were hit outside Baghdad.

The mission was finally conducted on the night of February 27–28. The original plan was altered due to the progress of the ground war and the impending cease-fire. The size of the force directed against the Ba'th party headquarters was reduced to release planes for other targets. Still, sixteen of the twenty-plane first wave were directed against it. This became known as "the pick-a-window mission." The complex was left devastated, and a statue in front of the building was also destroyed. The other targets hit by the night's first wave were the biological weapons facility at Salman Pak, and two transport planes at Muthena airfield. It was thought Saddam might use them to flee Baghdad. The second wave struck rocket facilities. The final F-117A attack of the war was on the Al-Athir missile development and production complex.[154]

At 11:30 P.M. the night's third wave was canceled. Then, at 12:15 A.M. on February 28, word was issued that a cease-fire would take effect at 8:00 A.M. that morning.[155]

Just as the Gulf War had begun on live television, so too, did it end. One image was that of Iraqi prisoners. In small groups or huge columns, they willingly surrendered. They had expected the air war to last several days, or a week at most.[156] Instead, it had continued for six weeks. They were helpless before it, without the means to survive or fight back.[157] Another image was General Schwarzkopf's press conference, describing how the 100-hour ground war had been fought. Yet another was the crowds of Kuwaitis welcoming victorious U.S. and Coalition troops.

And there was that final image, which put to rest a ghost from the past. It was a photo of a Blackhawk helicopter, hovering on the roof of the U.S. embassy in Kuwait City. Special Forces troops were climbing from the Blackhawk onto the roof. Sixteen years before, it had been another rooftop—another helicopter, the last helicopter out of Saigon.[158]

In a very real sense, two wars ended that day.

AFTERMATH

The first group of F-117As arrived back at Nellis Air Force Base on April 1. On hand to greet them was a crowd of twenty-five thousand people. To Col. Al Whitley, it was quite a contrast to his returns from two tours in Vietnam. In May, the town of Tonopah held a parade for the F-117 pilots and ground crews. A crowd of some three thousand turned out. A six-foot stone statue commemorating the plane and its crews was unveiled.[159]

In the meantime, the unit was on the move. Even before the Gulf War, the plan was to move the F-117As to Holloman Air Force Base in New Mexico. This would allow the pilots' families to join them and eliminate the need for the Key Airlines shuttle flights. The savings would pay for the move in short order. With the new home came a new name. With the post–Cold War reductions, the air force sought to preserve the names of distinguished units. The 49th TFW, formerly based at Holloman, was one of these units. On July 8, 1992, Colonel Whitley turned over command to Brig. Gen. Lloyd W. "Fig" Newton. The unit was also renamed the 49th Fighter Wing.

Less than a month later, the unit suffered the fourth loss of an F-117A. On August 4, Capt. John Mills took off from Holloman in aircraft number 802. Under the name "Black Magic," it had flown nineteen combat missions in the Gulf. Within moments after taking off, a fire broke out. Captain Mills attempted to return to the field, but the plane began to roll uncontrollably. Mills was forced to eject. He landed with only minor injuries, but the plane was destroyed. The loss was later traced to the improper installation of a bleed air duct just before the flight. This caused the fire and loss of hydraulic and flight-control systems.[160]

Two years after the start of the Gulf War, the F-117As were once more in action against Iraq. A detachment of six to eight F-117As had remained at Tonopah East as part of the U.S. forces in Saudi Arabia. The cease-fire agreement directed that Iraq dismantle all nuclear, chemical, and biological facilities, as well as giving up all Scud missiles and long-range weapons. United Nations inspectors were to enforce compliance. Saddam attempted to interfere with the inspections. There was also an increasing number of border incidents with Kuwait and interference with allied planes policing "no-fly" zones in the north and south of the country. By January of 1993, these demanded a response.

On January 13, six F-117As, each with one LGB, were sent to hit targets in southern Iraq. The Al-Amara IOC was hit, as was an SA-3 SAM site's radar at Ashshuaybah. Yet again, the F-117A's old nemesis, low clouds, caused problems. Bombs dropped against the two radars at Nasiriya lost their locks due to clouds. Weather also prevented one pilot from even finding the Tallil SOC. Bad weather also apparently caused one pilot to bomb a farmhouse a mile from the Basra radar. Still, the results were successful—air defenses again collapsed.[161]

THE FUTURE OF THE F-117

As with other Dark Eagles, the F-117A has undergone improvements to its systems. The planes were sent to the Skunk Works plant at Palmdale for modifications that included color displays, a satellite navigation system, and improvements to the infrared system. One modification was a result of the loss of Mulhare and Stewart in 1986 and 1987. It is called the pilot activated automatic recovery system: at the touch of a button, the plane would return to level flight.[162] Stealth modifications were also made. New coatings were added to reduce radar return. Changes were also made to the exhaust system to reduce both infrared emissions and the visible glow.[163]

In the longer term, in 1992–93 Lockheed proposed a navy version, the F-117N. It would use the basic F-117A fuselage, with a new wing and horizontal stabilizer. The F-117N, like the air force version, would be a "silver bullet" force. It would attack air defenses and other high-priority targets. The navy rejected the proposal, saying the plane was too limited and was not a multimission aircraft. A later proposal called for adding afterburning engines and advanced radar-infrared systems to give the plane both an all-weather air-to-ground and air-to-air capability. Even this later proposal attracted little interest, due to budget problems; there were also questions about the F-117N's carrier suitability and whether it actually was a multimission aircraft.[164]

Major modifications were also proposed for the air force F-117As, to improve payload and range. The major limitation is the plane's inability to carry more than two bombs. Proposals have been made to increase the size of the bomb bay to hold four bombs.

With new engines, the F-117A's payload could be increased to eighteen thousand pounds. Some of the added bomb load could be carried externally, as many as four internal weapons and another four under the wings. To preserve the plane's stealth, the external weapons would be covered with lightweight RAM. Such improvements would have a major impact—if the F-117A's bomb load and radius of action could be doubled, a single squadron could handle a regional conflict such as the Gulf War. Such modifications are problematical, however, given the current budget situation and the reduction in U.S. forces overall.[165]

Nonetheless, with the B-2 force limited to twenty aircraft and the number of F-22 fighters also likely to be smaller than originally planned, the F-117A will continue to be a key strike aircraft. The Black Jet from Groom Lake could still be flying into the second decade of the twenty-first century.

(As the final proofs for this book were being reviewed, two incidents occurred involving F-117A. On April 5, 1995, an F-117A was seriously damaged by a fire after landing. Then, on May 10, an F-117A crashed during a night training flight. Contact was lost at 10:25 P.M. The plane crashed on Red Mesa at the Zuni Indian Reservation, its impact dug a 20-foot deep crater. The pilot, Capt. Kenneth Levens, did not eject and was killed. The Zuni Tribal Police were the first on the scene and secured the area. Due to the remote location, it took the Air Force five hours to reach the crash site. As with the other two fatal F-117A crashes, Captain Levens had only limited flight time in the F-117A (70 hours). News accounts that the crash occurred in an sacred Indian burial ground are not correct.)

The first XP-59A prototype. The plane was both the first U.S. jet aircraft and created the concept of the Black airplane. Bell test pilot Robert Stanley, left, made the XP59A's first flight on Oct. 2, 1942. Colonel Lawrence "Bill" Craigie became the first U.S. military jet pilot when he made the plane's third test flight that same day. *U.S. Air Force*

The second XP-59A prototype. The fake propeller on its nose was used in March and April, 1943, to thwart the curious. *U.S. Air Force*

North Base at Muroc in 1942-43. In background are the hangar, water tower, and "Desert Rat Hotel." In foreground is Bell's portable mission control center during the XP-59A tests. *U.S. Air Force*

The U-2 combined the aerodynamic simplicity of a glider with careful weight control that enabled it to reach altitudes far above those of contemporary Soviet fighters. The U-2 overflew the Soviet Union with impunity for nearly four years. *Lockheed*

The SS-6 test pad at Tyuratam, as photographed in 1959 by a U-2. The pad was used to launch early ICBM tests as well as Sputnik I and Vostok1. *CIA*

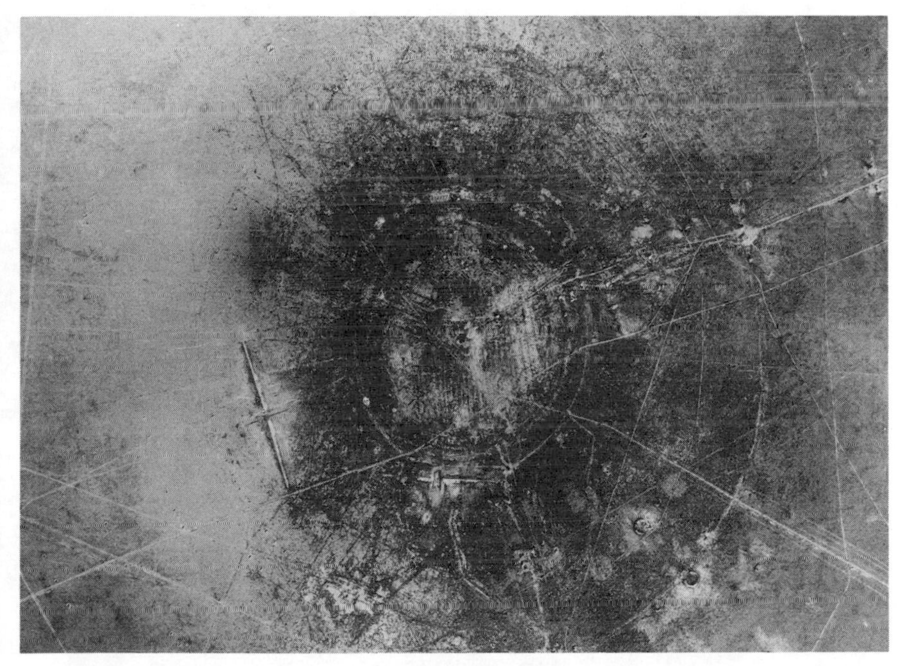

Overflight photo of ground zero of Soviet nuclear test site. *CIA*

U-2 photo of Soviet submarine base. *CIA*

The first A-12 design: the Archangel 1. *Lockheed*

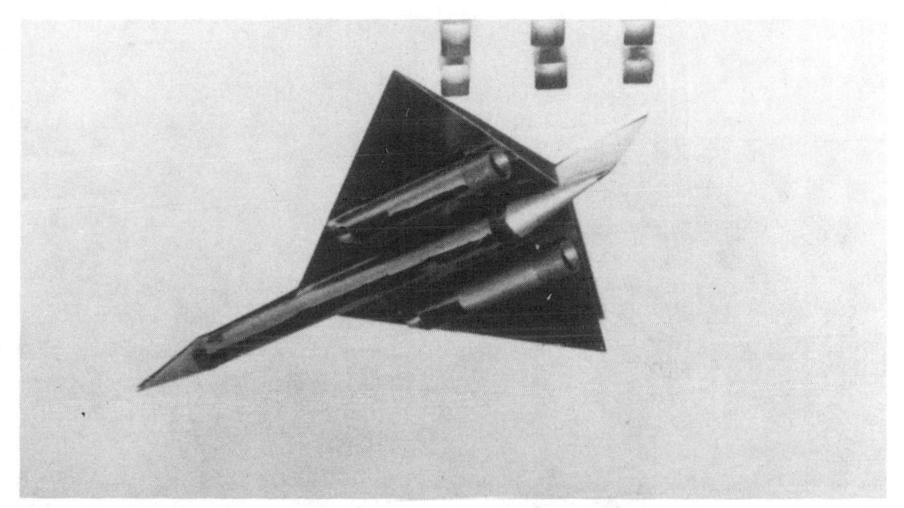

By the time Kelly Johnson and the Skunk Works engineers had worked up to the ninth version of the A-6 proposal, a trend toward a delta with a long forward fuselage had developed. *U.S. Air Force*

The A-11 design with radar absorbing, wedge-shaped chines became the basis for the final A-12. *U.S. Air Force*

Ground personnel around the first YF-12A prototype. Personnel at Groom Lake do not live in the hot, dusty, and isolated area, but are flown to and from the site each week. *U.S. Air Force*

The second YF-12A prototype. This was a modification of the A-12, designed to produce a long-range interceptor. *U.S. Air Force*

A-12 production line at Burbank. Different areas were walled off to prevent individuals working on one project from knowing about the others. The large box in the foreground was used to moves the completed A-12s to Groom Lake for reassembly and flight tests. *U.S. Air Force*

A-12s went through several paint schemes. The first planes were bare metal without any markings. Ultimately, an all-black finish was used (which improved airframe cooling). On missions, the national insignias were removed; the only markings carried a false, five-digit serial number in red paint. *CIA*

The A-12 Article 125 was the fifth Oxcart built and was lost on Jan. 5, 1967, due to a faulty fuel gage. The CIA pilot, Walter L. Ray, was killed. *CIA*

A pre-delivery photo of a Model 147E drone. *Teledyne Ryan Aeronautical photo, courtesy San Diego Aerospace Museum*

Model 147B drone under the wing of its DC-130 launch aircraft. The 147B would then fly its programmed instructions over the target areas and then on to the recovery zone. *U.S. Air Force*

Model 147B drone in flight. The 147G flew missions from October 1965 through August 1967. *U.S. Air Force*

Model 147TE above Edwards Air Force Base. The TE was an ELINT version of the high-altitude photo reconnaissance 147T drone. *U.S. Air Force*

Chinese People's Liberation Army militiamen celebrate the shooting down of a 147H-18. The 147H was the final high-altitude drone to see wide-spread service, with 138 missions flown between 1967 and 1971. *Ryan Aeronautical photo, courtesy San Diego Aerospace Museum*

The explosion of an SA-2 Surface to Air Missile photographed by a Model 147 drone. The SA-2 had a profound impact on post-war Black aircraft development: it put the U2 overflights at risk, which led to the development of the A-12 Oxcart and D-21 Tagboard. *U.S. Air Force*

The Model 147SC "Tom Cat" flew a record-setting 68 missions before being lost on Sept. 25, 1974. (the average for the 147SC drones was 7.3 missions before being lost. *Teledyne Ryan Aeronautical photo courtesy San Diego Aerospace Museum*

Model 154 drone on its handling cart at Edwards Air Force Base. It was designed for high-altitude photo reconnaissance of Communist China, but by the time the Model 154 was operational, these missions were halted, and the drones were scrapped without having ever made an overflight. *U.S. Air Force*

The secret that fell from the sky. The Model 154P-4 Firefly after its landing at Los Alamos in August 1969. The drone was unharmed, but the accident was widely publicized and the Model 154 program was no longer "Black." *Teledyne Ryan Aeronautical*

The D-21 501 on its B-52 launch aircraft. *Lockheed*

D-21/M-21 combination in flight above Nevada. *Lockheed*

The Have Blue 1002—produced under the tightest secrecy since the atomic bomb program of World War II— after final assembly at Burbank. This aircraft was optimized for RCS testing, and had the full range of coatings and radar absorbing materials. *Lockheed*

The Have Blue 1002 above Groom Lake. Unlike its angular upper surface, the plane's underside was arrow shaped. These flat surfaces made it possible to calculate the plane's radar cross-section. *Lockheed*

F-117A 780 as it was rolled out of its Groom Lake hangar before its first flight on June 18, 1981. Activities at Groom Lake, such as the F-117A's first flight, are scheduled around the passes of Soviet reconnaissance satellites. *Lockheed*

The Amber UAV was designed to provide a family of low-cost but effective vehicles capable of undertaking both reconnaissance and strike missions. *General Atomics*

The GNAT 750 was originally designed to be a low-cost, long duration UAV suitable for export. It was used by the CIA to monitor the civil war over Bosnia. *General Atomics*

B-2 Spirit above Edwards Air Force Base. During the late 1980s and early '90s, there were a number of reports about "other" flying wing or triangular Black aircraft. These ranged from Shamu (a sub-scale B-2 test bed) and the TR-3A Black Manta (a reconnaissance aircraft) to an aircraft that, it was claimed, had wingspans of 600—800 feet. *U.S. Air Force*

A strange plane in a strange land. A Soviet-built Yak 23 undergoing flight tests at Wright-Patterson Air Force Base. The plane was "loaned" to the U.S. by an Eastern European intelligence officer, then crated aboard a U.S. C-124 cargo plane. The U.S. markings—thc "X-5"—were used as a cover. The eighth and final test flight was completed on Nov. 4, 1953. The markings were removed; the Yak was crated and reinserted into Eastern Europe without detection. *U.S. Air Force*

The Tacit Blue in flight above Nevada. Long rumored to be a flying wing, it was, in fact, one of the strangest looking aircraft ever built. Its odd shape was to accommodate a specialized radar able to track armor units. The design also provided "all-aspect" stealth. The Tacit Bluc first flew in February 1982, but was not declassified until April 30, 1996.

Artist's conception of a flock of Auroras. From top to bottom, the original "Black Diamond," the "North Sea" Aurora, and finally, the XB-70 Aurora. The stories about the Aurora began to spread in the late 1980s and reached their peak in 1994.

CHAPTER 9

The Return of Black Reconnaissance HALSOL and the GNAT-750

With many calculations, one can win; with few one cannot. . . . By this means I examine the situation and the outcome will be clearly apparent.

Sun Tzu
ca. 400 B.C.

Despite the end of the Model 147 and 154 reconnaissance drone programs, interest remained in what were now called unmanned aerial vehicles (UAV). Most of these were short-range, battlefield-support UAVs. But, just as stealth changed both tactical and strategic aircraft, it also caused a resurgence of interest in strategic reconnaissance UAVs. Two very different systems would emerge during the 1980s and early 1990s. They would also highlight the changing shades of Black.

One advantage of a UAV over a manned aircraft was that of flight duration. This had not been realized when the Model 147 and 154 drones were flown, but was central to the mission of the first of these new Black UAVs. Its planned flight time would not be measured in hours or even days but in months. The technology to build the first of these Dark Eagles did not come from an exotic development program but from a contest.

GOSSAMER DREAMS

In the myth of Icarus, a man had flown by flapping wings attached to his arms. Even with the achievement of heavier-than-air flight, the dream remained: to fly with only the power of one's own body. In 1959, British industrialist Henry Kremer offered a 50,000-pound prize for the first successful man-powered flight. It was to cover a figure-eight course around two markers a half mile apart. The aircraft had to start and finish the course at an altitude over ten feet.

During the 1960s and early 1970s, a number of man-powered aircraft tried for the Kremer Prize, but all were doomed to failure. They were of conventional design, fitted with chain-driven propellers. Principally made of wood, they were too heavy and incapable of turning.

One man realized what was needed. Dr. Paul B. MacCready of AeroVironment Incorporated understood that neither a conventional design nor conventional thinking would work. The Kremer Prize would take an innovative aircraft, designed with no preconceived notions.

The result was the Gossamer Condor. It had a long, tapered wing. Like the Wright Flier, it had a canard and wing warping to provide control. The pilot rode in a gondola suspended below the wing. The Gossamer Condor was designed to fly at extremely low airspeeds. Dozens of times the plane crashed, once as it was approaching the second pylon and victory. Finally, on August 23, 1977, Bryan Allen peddled the Gossamer Condor around the course. Man had flown.[1]

With the Kremer Prize won, other frontiers beckoned. In 1980, AeroVironment was sponsored by DuPont to develop a solar-powered aircraft able to fly from Paris, France, to England. The first test aircraft, the Gossamer Penguin, was so fragile and unstable that it could be flown only in the early morning when the air was calm. Based on the results of these tests, the Solar Challenger was designed. This was a much smaller and more rugged aircraft. It could reach altitudes of over 12,000 feet. Power came from 16,128 solar cells, which covered the entire upper surface of the wing and stabilizer. Despite the large surface area, the array provided a maximum of only 2,600 watts—enough to power two hair driers. The Solar Challenger had a wingspan of 47 feet, but weighed just over 200 pounds. In July 1981, the Solar Challenger successfully completed a 163-mile flight from Paris to RAF Manston, England, averaging a speed of about forty miles per hour at just over 11,000 feet.

THE ETERNAL AIRCRAFT

With the success of the Solar Challenger, AeroVironment began studying the possibility of a high-altitude, solar-powered UAV. If provided with some type of energy storage system, it could fly "eternally" and remain at altitudes above 65,000 feet. Flight duration would ultimately be limited only by such factors as mechanical systems wear.[2]

The design of a solar-powered Gossamer-type aircraft flying at these altitudes faced a number of problems. One peculiarity was engine performance. At takeoff, the Solar Challenger could barely climb; at 11,000 feet, however, it could equal a Cessna light plane. A piston engine loses power as it climbs, due to the thinner air. In contrast, a solar-powered aircraft's climb performance increases as the solar cells "see" a brighter sun, and the colder air improves efficiency. Above 30,000 feet, however, problems start to appear. The air temperature becomes constant, so there is not a continued improvement in efficiency. At the same time, the plane faces an increasing power requirement. Because of the thinner air, the plane must fly faster for its wings to generate the same amount of lift.

Another demand was the need to store power. More than half the current generated by the solar cells would have to be stored in some type of battery to keep the plane aloft during darkness. The "eternal" solar-powered aircraft would need twice the collection area, with no increase in weight, over the Solar Challenger.[3]

The design would have to be extraordinarily light—one-half pound per square foot of wing area, the same weight as foam art board. In contrast, an eagle has a wing loading of four to six pounds per square foot of wing. The eternal aircraft would need a wing loading half that of the Solar Challenger (including the pilot). The Gossamer Penguin's structure was light enough, but was too fragile. Despite these problems, theoretical calculations convinced Ray Morgan, vice president at AeroVironment's Design Development Center, that such an airplane could be built.[4] This also opened Black possibilities.

HALSOL

By 1983, AeroVironment was able to attract government sponsorship from a "classified customer" to build a proof-of-concept test aircraft. (Among the possible "customers" that have been suggested are the National Reconnaissance Office, the CIA, and the Naval Research Laboratory.) It was called "HALSOL," for high altitude solar. Unlike the other Gossamer-type aircraft, this was to be a Black airplane in the classic sense. The HALSOL was developed and flown in secret. Its existence was not to be revealed for another decade.[5]

The HALSOL design was a pure flying wing, with no rudders or canard. It had a span of 98.4 feet. From front to back, the wing was eight feet wide; it was made of five, 20-foot-long segments joined together. The main wing spar was made of thin wall carbon fiber tubes. Attached to this were ribs of Styrofoam reinforced with Kevlar and spruce. The wing was covered with Mylar plastic. Despite what one might think, the wing was far from being weak. It was stressed for a +5/-3g load factor (greater than the U-2). The center segment had two gondolas that enclosed the landing gear.

The aircraft was powered by eight electric motors—two mounted on the center of the wing, two on each inboard wing segment, and one on each outer wing segment. Spreading them out along the full span of the wing distributed the load. The HALSOL propellers had a variable pitch to match the available load on the power source, in order to permit the maximum efficiency. For the test flights, they were powered by silver-zinc batteries. The HALSOL was controlled by radio from the ground. To pitch up or down, an elevator on the center wing section's trailing edge was used. To make a turn, the outermost motor on one side was sped up, while the opposite motors were slowed down.

HALSOL could hardly be called a high-performance aircraft. It flew at twelve knots and had a never-exceed speed at low altitude of twenty-seven or twenty-eight knots. Above this speed, it would go into a nose-down tuck. It was estimated a climb to 70,000 feet would take about six hours. Because it was a test vehicle, the HALSOL was designed to climb, rather than to remain at high altitudes.[6]

Total gross weight of the aircraft was about 410 pounds, with a payload of about 40 pounds.[7] The efforts to control weight led to creative thinking and some unusual solutions. The front wheel assembly on the two gondolas used dual baby-buggy wheels, while the main landing gear assembly had a sixteen-inch bicycle wheel.[8]

The HALSOL made its first flight in June of 1983.[9] Over the next two months, a total of nine flights were completed at Groom Lake. These lasted for thirty to sixty minutes and reached an altitude of 8,000 feet. Although the aircraft was proven to be aerodynamically and structurally sound, studies indicated that 1983-vintage solar cell technology was not efficient enough to permit very long, high-altitude flight. In particular, the solar technology lacked sufficient "power-density," the energy available per pound of the components. The HALSOL program was discontinued, and the aircraft was placed in storage.[10]

PATHFINDER

AeroVironment remained active in solar-energy research. In 1987 (four years after the HALSOL project ended), General Motors (GM) selected AeroVironment to develop the Sun Raycer car, which won the first trans-Australia race for solar-powered vehicles. The following year, GM selected AeroVironment to develop the Impact, a battery-powered commuter car suitable for mass production. Both these electric car projects would have a major effect on the discontinued HALSOL project: by the end of the 1980s, lightweight solar cells, electric motors, and power-storage technology had advanced to the point that the original HALSOL concept became practical.[11]

A mission for such an eternal aircraft had also appeared. An aircraft like the HALSOL could be used to detect missile launches, such as the Scud ballistic missiles Iraq fired against Israel and Saudi Arabia during the Gulf War. This was seen as a preview of future regional conflicts. Scud missiles had been exported and were in production throughout the Third World.

Both Iran and North Korea were active in this area, as well as having ongoing chemical, biological, and nuclear-weapons programs.[12]

With the technology now available and a military need, the HALSOL was taken out of storage in early 1992. Under the direction of the Ballistic Missile Defense Organization (BMDO), the successor to the Strategic Defense Initiative, AeroVironment began a modification program. The basic airframe was retained, with the addition of new systems. One example was the motor-propeller system. The original variable pitch props were replaced with fixed props, with an electronic "peak-power tracker." Removing the original propeller system reduced the number of parts and increased reliability.[13] The original rare-earth DC motors were replaced with brushless AC motors, which also improved reliability and efficiency. The motors also had new custom-designed inverters to improve efficiency. Rows of cooling fins were added behind the propellers to radiate heat. Keeping the motor's temperature within limits while flying at high altitude was a problem due to the thin air. The complete motor and propeller assembly weighed only thirteen pounds.

In addition, the control surfaces were modified. The original HALSOL had only one elevator powered by four servos. In the new version, twenty-six elevators ran the full span of the wing's trailing edge.[14]

The biggest change was in the plane's power source. The original HALSOL was battery powered. (Only a few solar cells had been carried on the HALSOL to test the effect of wing flexing.) The modified aircraft would carry some two hundred square feet of new lighter-weight solar cells. They would cover about one-third of the wing and provide about 3.8 kilowatts of power. This was enough to fly on solar power alone after 9:30 A.M. In practice, however, the aircraft would fly on dual solar-battery power.[15]

The modification work was completed in the late summer of 1993—a full decade after the HALSOL's last flight. Because of the extensive modifications, a new name was given to the aircraft. It was now called "Pathfinder."

OUT OF THE BLACK

It is important to note that, up to this time, there had been no hint that there was a Black Gossamer-type UAV or that it had been flown a decade before. The HALSOL-Pathfinder finally came out of the Black with an October 1, 1993, air force press release, which announced the aircraft's existence and described its history. It stated that the Pathfinder would make a series of low-altitude test flights at Edwards Air Force Base in October–November 1993. It was explained that the project had been declassified to allow use of commercially available technology and open discussion of technical ideas. At the same time, a photo of the original HALSOL was released.[16]

The Pathfinder's first flight was made on October 20, 1993. It lasted forty-one minutes and involved six trips around a 1.2-mile racetrack course on the Edwards lake bed. The top speed was sixteen knots, and the plane's altitude was limited to 200 feet. The Pathfinder was a majestic sight as it flew slowly above the tan lake bed like some huge transparent bird. As it flew, the wing tips arched up, forming a U-shape. The propellers made a humming noise as they spun. At least 60 percent of the Pathfinder's power was provided by the solar cells. The Pathfinder was controlled from the ground by Ray Morgan, who originally conceived of the project.[17]

The Pathfinder's public debut came on November 23, 1993, at the Edwards Air Force Base Air Show. Six weeks before, even its existence had been a secret; now, more than 200,000 people saw it on display. Never before had a Black airplane's first public showing come so soon after its existence had been revealed. The Pathfinder was in a roped-off area of the main hangar, and there were armed guards nearby. This was not due to any secrecy about the plane—people were kept back because the light in the hangar was enough to start the motors, and a spinning propeller might injure a spectator. The guards were to protect the other plane in the enclosure, a B-2 Spirit.[18]

The October–November flight-test series pushed the Pathfinder to the edge of its low-altitude flight envelope and measured the stability and performance of the modified aircraft. The data collected was used to develop autopilot software to allow the aircraft to operate independently of constant ground control.[19]

Following the first flight tests, the Pathfinder was returned to AeroVironment for modifications that would enable it to undertake high-altitude flights. Control of the program was also transferred from BMDO to NASA due to budget reasons. Rather than detecting Scud launches, it would be used for atmospheric research.[20]

While at AeroVironment, about 70 percent of the wing surface was covered with solar cells. (For aerodynamic reasons, the wing's leading edge cannot be covered.) These cells were lighter than those used for the first flights.

The high-altitude flights were seen as critical to demonstrate the feasibility of the Pathfinder concept. It was planned that the tests would run from late August through mid-October 1994 at Edwards.[21] The addition of the solar cells and the structural modifications took longer than expected, however, and by the time the Pathfinder was ready, it was late in the flight season. With the sun lower in the sky, it was felt it would be better to wait for more favorable conditions.[22]

The high-altitude tests were rescheduled for April–July 1995. As the Pathfinder lacks an energy-storage system, the flights would be made during daylight hours. The plane would take off at dawn and climb all day. Peak altitude would be between 60,000 and 65,000 feet. The Pathfinder's flight characteristics at high altitudes are very different than the heavier and faster-flying U-2s. At altitude, the U-2s' minimum and maximum speeds, limited by wing flutter, provide only a small margin for flight. With the lighter and slower Pathfinder, both speeds are considerably slower and therefore much less susceptible to structural failure. This is important, as tight limits would be very demanding for the automatic control system. The estimated cruising speed at altitude would be around 100 knots.[23] At sunset, electrical power would be lost from the solar cells, and the Pathfinder would start down. Due to the plane's high glide ratio, it would not land until about 2:00 A.M.

Following the high-altitude tests, long-duration flights could be attempted. The Pathfinder could take off from Alaska and fly over the North Pole for weeks at a time to measure ozone levels. During the summer, there is nearly continuous sunlight over the pole, so the Pathfinder would not need any energy storage system. The continuous flight time could be 2,000 to 3,000 hours, or 80 to 120 days aloft.[24] The Pathfinder's solar array would be equipped with double-sided cells. These would use the light reflected from clouds, the ice cap, and the atmosphere to produce power during the evening hours, when the sun is low. The plane would also have a set of small computers, gyroscopes, and a four-antenna satellite navigation system to allow automatic flight.[25]

HELIOS

The HALSOL-Pathfinder was designed to act as a technology demonstrator for the eternal aircraft, which would be a much larger and heavier aircraft. Currently called "Helios" or solar rechargeable aircraft (SRA), it would have a wingspan of 200 feet, made up of five, 40-foot-long segments. There would be four landing gear gondolas. The solar array would produce 21.6 kilowatts, powering the eight motors—the same motors as were used on Pathfinder. The Helios-SRA would weigh about 1,100 to 1,200 pounds, and have a payload of 150 to 200 pounds. Its operational ceiling is planned to be about 60,000 to 65,000 feet; top speed would be around 100 knots. Unlike the HALSOL-Pathfinder, the Helios-SRA would be optimized for high-altitude cruise, rather than climb.[26]

To allow around-the-clock flights, the Helios-SRA required an energy-storage system—a proton-exchange membrane fuel cell. Excess electrical current from the solar array would electrolyze water into hydrogen and oxygen gas. The gases would then be compressed and stored separately in the wing spar. At night, the hydrogen and oxygen would be allowed to recombine into water. The energy released would be converted back into electrical power for the motors. These fuel cells would be carried in the wing leading edge. The system has less than half the weight of rechargeable batteries.[27] Low-level research is currently under way on the fuel cells. This is a two- to three-year program. Although the technology is understood, the problem is putting it into a lightweight package. The Helios-SRA could be ready in 1997 or 1998.

INTO THE WHITE

In the post–Cold War political environment, attention is shifting to dual-use technology—military systems that can also have civilian applications. Pathfinder and Helios-SRA are prime examples. The ability to reach high altitudes, then stay there for months at a time, opens numerous possibilities in the area of atmospheric research. Sporadic, one-shot measurements of atmospheric conditions cannot give the needed baseline data to detect changes. To distinguish normal atmospheric variations from changes caused by natural and man-made sources, it is necessary to take measurements throughout the day-night cycle.

The Pathfinder, for example, could be flown continuously over the Yukon Valley in Alaska during the summer. It could monitor the flow of air pollution from Russia as well as the effect of local volcanoes. The data would be provided at very low cost, compared to satellites. Another advantage is that the Pathfinder's electric motors give off no exhaust emissions. It could carry sensitive chemical detectors without their measurements being contaminated by the plane itself.

Other research projects would utilize the Helios and its ability to fly continuously through the night. One possibility would be storm tracking. Major storms develop in the subtropical areas of the Atlantic and Pacific, then move toward North America over several weeks. A Helios could be "parked" over the storm, monitoring its development. Continuous data on atmospheric conditions and water temperatures would result in better predictions of storm strength and movements.[28]

Another application is high-altitude astronomical observations. The Helios can fly above most of the atmosphere, which would allow instruments to observe in ultraviolet and infrared wavelengths. These observations cannot be made from ground-based telescopes because the atmosphere and water vapor absorb these wavelengths.

A Helios could also serve as a "pseudosatellite" for communications. If communications links are disrupted by a natural disaster, such as an earthquake or severe storm, cellular phone equipment could be carried by a

Helios, which would act as a long-distance relay station. As MacCready observed, "mouths water when they hear about this."[29]

The final, indirect result of a Helios would be improved electronic and solar technology. The payload limitations would encourage development of ultralightweight electronics and sensors, much as satellites did in the 1950s and 1960s.[30] Just as the solar-powered car efforts made Pathfinder practical, a Helios development effort could improve solar technology.[31]

It has been estimated that a Helios-SRA would cost between $2 and $3 million, assuming a minimum production run of twenty aircraft.[32] Although the possible uses of such an eternal aircraft are many, it still faces many technological problems. In particular, the development of flyable fuel cells are the key to day-night flights. As a White project, Pathfinder-Helios now faces a political environment that views new and innovative ideas with disinterest or hostility.

AMBER

The HALSOL was not the only Black UAV program. In the early 1980s, DARPA began a Black study of long-endurance UAVs under the code name "Teal Rain." This looked at both short- and long-term possibilities.[33]

In December of 1984, DARPA issued a $40 million development contract to Leading Systems of Irvine, California, to build a medium-range, low-cost tactical UAV. Leading Systems, in the best tradition of high-tech companies, had started in a garage in 1980. The control system for the UAV was built in the living room of one of the founders.[34] The navy, army, and Marine Corps soon became involved with the project. The navy was given control. As with earlier Black airplanes, the goal was the rapid building of the prototypes. The program office had few government personnel, while the number of support contractors was kept small. This kept the amount of program reviews and paperwork to a minimum.[35]

Called "Amber," it could carry either a warhead or reconnaissance or ELINT equipment. The inverted "V-tail" stabilizers hung down from the rear fuselage. The long wing was mounted on a small pylon atop the fuselage. On the strike version, the wing would be separated from the pylon; the UAV would then fall to the target. The piston engine was located at the rear. It used technology originally developed for Indy car engines. The Amber took off and landed on a set of long and sticklike retractable landing gear. Although it was as large as a light airplane, the Amber was more akin to the simple battlefield support UAVs, rather than the complex strategic drones used earlier.

Once the contract was issued, Leading Systems began work on six prototypes called the Basic Amber—three A45s, with the pointed nose of the strike version, and three B45s, with the bulged nose section of the reconnaissance version. Flight tests began in November 1986 at the Leading Systems test facility at the El Mirage Airport, in the Mojave Desert, near Edwards Air Force Base. On June 16, 1987, a 20.25-hour flight was successfully made. The initial test program was completed in June, and the Basic Amber was judged to have performed "extremely well."[36]

During September and October of 1987, the Basic Ambers underwent a second series of tests at the army's Dugway Proving Grounds in Utah. These proved out the low-drag aerodynamics, lightweight structure, control system, and engine. The third Amber prototype was lost during the tests. When the UAV was placed in a high-angle-of-attack or high-g condition, it would start to oscillate.[37] This was corrected, and by 1988 flight durations of thirty hours at 17,000 feet, and thirty-five hours at 5,000 feet had been demonstrated. The maximum altitude reached was 27,800 feet.[38]

Up to this point, the Amber program had been a secret. During the fall of 1987, information began to be released on the design, possible payloads, and future activities. However, many details such as the engine technology and performance specifications, remained "highly classified."[39] The first public appearance by the Amber was at the 1988 San Diego Air Show.[40]

Another highly publicized display of Amber took place in June 1988. For the Fifteenth Annual Association of Unmanned Vehicle Systems Technical Symposium and Exhibit, Leading Systems decided to attempt to break its UAV world endurance record. The Amber took off from El Mirage at 7:48 A.M. on June 6. As the flight progressed, the elapsed time was posted in Leading Systems' booth. At 10:10 P.M. on June 7, the Amber was brought to a landing. The total elapsed time was thirty-eight hours and twenty-two minutes, a new record.[41]

During this same time, Congress was becoming dissatisfied with the large number of UAV programs under way. In 1987, it ordered a consolidation of the programs and froze UAV funding pending submission of a master plan for its approval. In June 1988, the Joint Project Office (JPO) for Unmanned Aerial Vehicles was established, under the U.S. Navy's Naval Air Systems Command. UAV funding was now provided at the Office of the Secretary of Defense level. The idea was that with control centered in a single, high-level group, duplication between the UAV programs would be eliminated.[42]

With these bureaucratic changes accomplished, work began on the production Amber I UAVs. It had a length of 14.8 feet and a wingspan of 29.5 feet. Payload was a television camera or a FLIR system weighing 110 to 140 pounds. These were mounted under the nose and were covered with a plastic bubble.[43] The Amber I's maximum altitude was 25,000 feet, and it had a range of 1,200 nautical miles. The UAV's endurance was thirty-eight hours at 5,000 feet while flying at 85 to 110 knots.[44] The airframe was made of composites for stealth. The Amber I was controlled by an autopilot and a command data link. The operator at the ground station flew the Amber I using a nose-mounted television camera. The powerplant was a sixty-five-horsepower, four-cylinder, liquid-cooled engine.[45]

The second series of Amber I test flights was made during October 1989 at Fort Huachuca, Arizona, (headquarters of U.S. Army Intelligence, underlining the tactical mission of Amber). These were called the "maturation" tests and were to last for 500 hours of flight time. They would test the Amber design's ability to meet service specifications for reliability, availability, and maintainability. The payload, data links, and ground control were also integrated. Ten Ambers (three B45s and seven Amber Is) were used in the tests. Some expected that all ten would be lost during the tests, but in fact, there were no mishaps.[46] The Amber Is were turned over to the government between December 1989 and January 1990.[47] A total of thirteen Amber UAVs had been built—three A45s, three B45s, and seven Amber Is.[48] The Amber project was subjected to repeated budget cuts, however, and after the Amber Is were delivered, they were put into storage and the program was canceled.[49]

The end of the Amber I program was one sign of a growing trend. Airborne reconnaissance was being reduced to a second-class mission. In 1990, the SR-71 was retired due to the high cost of operations. The move was widely regarded as an act of folly because it created a gap in overhead reconnaissance. At the low end, the air force and navy were dependent on RF-4C reconnaissance planes and F-14s carrying camera pods. At the high end, photo reconnaissance satellites would provide worldwide coverage. Their resolution was as good as six inches under ideal conditions, but they could provide coverage *only* when they orbited over a target. There was nothing to provide deep coverage on a continuing basis. The events of January and February 1991 would make this shortcoming clear.

UAVs Over the Gulf

Throughout the Gulf War, intelligence was a problem. Both army General Schwarzkopf and air force Gen. Charles Horner expressed dissatisfaction about the quality of intelligence reports they were given. Schwarzkopf said later that the reports were outdated, as well as "caveated, disagreed with, footnoted, and watered down."

There were major disagreements over the bomb damage assessments (BDA) of Iraqi forces. The BDA controversy was the result of the differing data and "platforms" being used, as well as the mindsets of the analysts. The air force intelligence officers in the field were using the videotapes from the strike missions. Back in Washington, the CIA and DIA were using satellite photos. Based on this, the CIA-DIA consistently reported Iraqi forces had greater strength and Coalition air strikes had lesser effectiveness than the air force estimated.[50]

An example of this was the tank-plinking effort. When an Iraqi T72 tank was hit, the ammunition and fuel would ignite. A jet of flame would erupt from the hatch, and the tank's interior and crew would be incinerated. All this was clearly visible on the strike video. But once the fire burned out, the tank would have little visible external damage—perhaps only a small entry hole. When a satellite photographed the area hours or days later, the tank would be listed as intact. In one case, a tank was declared operational, until it was pointed out that the turret had been blown a foot out of position. Schwarzkopf complained that the guidelines for assessing damage were so stringent that a tank had to be on its back "like a dead cockroach" before it would be counted as destroyed. In other cases, the T72 would undergo such a violent secondary explosion that it would be blown apart. Then it would be claimed that the revetment had been empty when it was bombed.

The result was widely differing figures. As the ground war was about to begin in late February 1991, the air force estimated that 40 percent of the Iraqi tanks and artillery had been destroyed. The CIA-DIA estimate was between 20 to 30 percent. Some estimates were as low as 15 percent. Not surprisingly, the air force was accused of inflating its damage estimates. In fact, 60 percent of the Iraqi tanks and artillery had been destroyed during the bombing campaign.

In the final assessment, the strike videos, which recorded events as they happened, proved more accurate than "National Technical Means" such as satellites.[51] Clearly, tactical intelligence required continuous real-time data, which the strategic systems could not provide.

This was reinforced by the experience of UAVs in the Gulf War. Three different systems were used by U.S. forces—the Pointer, Pioneer, and Exdrone UAVs. The Pointer, built by AeroVironment and used by the marines, was simplicity itself. It was a radio-controlled model airplane, similar to the ones built by hobbyists. It was hand-launched and powered by an electric motor. The battery provided about an hour of operation. The Pointer carried a television camera that transmitted its photos back to the operator. The Pointer was used for real-time BDA, artillery adjustment, and reconnaissance-early warning.

The Pioneer was based on an Israeli design. It was powered by a piston engine and had a twin-boom, high-wing design. The Pioneer had a flight time of five hours and a range of 100 nautical miles. In the Gulf, one Pioneer unit was aboard each of the battleships *Missouri* and *Wisconsin,* three units were assigned to the marines, and one to the army. (Each unit consisted of eight Pioneers and support equipment.) The Pioneers flew 533 sorties; at least one was aloft at all times during the war. They suffered heavy losses—

of the forty UAVs, twelve were lost and another fourteen or sixteen were damaged. Several were sent on one-way missions—the UAV was kept over a target until the fuel ran out, in order to produce the maximum amount of information.

The Pioneers undertook a range of missions. The army would fly them along the routes to be taken by Apache helicopters; the pilots would watch the live video, then climb into their helicopters and take off. The navy used them to spot targets for the 16-inch battleship guns, then correct their fire. During one such mission, forty Iraqi soldiers were seen coming out of a bunker and waving white cloths at a Pioneer in an attempt to surrender to it. This was the first time that humans had surrendered to a machine.

About fifty-five to sixty Exdrone UAVs were also used by the marines in the Gulf. These were television-equipped expendable drones. The marines used them to observe minefields and barriers in southern Kuwait City. When their video showed that the Iraqis had abandoned their positions, the marines moved forward a full day and a half earlier than originally planned.[52]

In each of these cases, the UAVs were able to provide real-time intelligence directly to the units that needed it. Following the Gulf War, interest within the air force grew in the development of long-range UAV systems that could keep watch on a specific area day and night. This could do much to clear up the shortcomings in reconnaissance that the war had made apparent. In the years following the Gulf War, a new trouble spot appeared that would lead to operation of such a UAV. A descendant of the Amber, it was called the GNAT-750.

GNAT-750

Work on the GNAT-750 project had actually begun at Leading Systems in 1988. It was designed for export to friendly countries. The GNAT-750 eliminated several features that were not needed by non-U.S. users and had a simplified structure and lower cost than the Amber.[53]

The fuselage was 16.4 feet long and had sloping sides for a low RCS. The GNAT-750's long and narrow straight wings spanned 35.3 feet. Like Amber, the GNAT-750 had an inverted V-shaped tail, the same sixty-five-horsepower engine, and took off and landed on a spindly tricycle landing gear. Mounted under the nose was a "skyball"; a movable turret that could carry a stabilized forward-looking infrared system, a daylight television camera, and a low-light-level television camera. In all, 132 pounds of payload could be carried in the nose and 330 pounds under the wing. The GNAT-750 could fly for forty-eight hours continuously, giving it a maximum radius of 1,512 nautical miles. It could also fly out to a target area 1,080 nautical miles away, then remain on station for twelve hours before having to return.[54]

Compared to the Amber, the GNAT-750 was larger, lighter, had a heavier payload, and a ten-hour greater endurance. Despite this, when one sees the GNAT-750, the first impression is of a model airplane. (Admittedly, a *big* model airplane.) Perhaps it is the wooden prop at the rear of the plane. As with Amber, the goal is a low-cost, long-duration UAV.

The prototype GNAT-750 made its first flight in the summer of 1989, which began a series of successful test flights. The first eight GNAT-750s were scheduled to start production in December 1989.[55] Despite this success, the confusion and technical problems with the UAV program continued. The JPO seemed unable to bring order to the situation. Leading Systems was also on the verge of bankruptcy. In 1990, its assets were bought out by General Atomics; this included the Amber and GNAT-750 projects.

General Atomics continued development of the GNAT-750. In 1992, a prototype made a continuous flight of over forty hours at El Mirage, reaching altitudes of over 25,000 feet. General Atomics then won a contract from the Turkish government to supply it with GNAT-750s; deliveries began by the end of 1993.[56]

In the summer of 1993, world events again spurred interest in UAVs. The Joint Chiefs of Staff requested immediate development of a UAV for use over Bosnia and Serbia. Following the end of Communist rule in Eastern Europe, Yugoslavia had disintegrated into a barbaric civil war between Serbs and Moslems. It was marked by ethnic cleansing—wanton killing, torture, and starvation. The UAV was to keep track of the warring factions' troop movements, artillery emplacements, and antiaircraft weapons, in support of UN peacekeeping forces.

Pentagon acquisition chief John Deutch endorsed the recommendation and called for the development of a UAV that could fly 500 nautical miles, then remain on station for twenty-four hours or more; carry a 400- to 500-pound payload; fly at 15,000 to 25,000 feet; and be equipped with a combination electro-optical, infrared and/or radar system.[57]

It was decided to use the GNAT-750. It would be rapidly modified for the mission, then equipped with an off-the-shelf sensor package. The GNAT-750 was available for export; if one was lost, no sensitive technology would be compromised.[58] And it was designed, as the sales brochure put it, with "sensitive" and "high-risk missions" in mind.[59] The images would be transmitted to local UN commanders via a relay aircraft. Two GNAT-750s would be built for the program. The effort would cost $5 million.

Because the situation in Bosnia was highly fluid, the GNAT-750s needed to be operational by October 1993 at the latest. A development effort with so short a time frame could only be accomplished, however, if freed of the funding and acquisition constraints of a military program. There had been a similar situation some forty years before. The solution was the same; the CIA was given control of the program.[60]

The GNAT-750 had now become a Black airplane.

A PALER SHADE OF BLACK

By the late summer of 1993, the CIA GNAT-750 effort was under way. The CIA acquired two GNAT-750s from General Atomics, which were then modified with the relay antennae. A large teardrop-shaped dome was added to the top of the GNAT-750 to house the data link. The flight tests were done at the El Mirage site. The CIA GNAT-750s were to be ready for operations by October 1, 1993; once operational, they were to be based in Italy.

In spite of its new Black status, the project did not go smoothly. The CIA GNAT-750 program had to face the political realities of spying in the 1990s. Not even congressmen had been told of the U-2 overflights. In 1993, however, the autonomy of CIA Director R. James Woolsey Jr. was far less than that enjoyed by Allen Dulles in 1956, when the U-2 program was about to begin. Now it was necessary for the CIA to consult with the Defense Department and Congress on the GNAT-750 effort.

The overall UAV program itself was caught in a tug-of-war between the various factions who were soon faulting the CIA's handling of the GNAT-750 program. They objected to the CIA's unwillingness to tell others what was going on with the project. One congressional staffer complained (rather incongruously), "They tried to turn it into a secret program."[61]

Not surprisingly, word of the CIA GNAT-750 program leaked. The early details were fragmentary, with one report referring to the GNAT-750 as "an unmanned SR-71 follow-on." The image that phrase brings to mind is far from a plane with a wooden propeller.[62] But more detailed accounts were soon appearing.

Clearly, the GNAT-750 was a very different shade of Black than the U-2. The concept of Black itself had also changed. HALSOL was classic Black—secret from start to finish. Since then, the concept had become blurred. Teal Rain was Black, yet the end result, the GNAT-750, was White. It was even sold to an Allied country. Then, suddenly, the White GNAT-750 became Black with the start of the CIA program.

The net result was that the CIA GNAT-750 effort was a Black project that was almost open. It was a very strange picture; yet it would become stranger still.

The political infighting over the CIA GNAT-750 was aggravated by technical problems. It was reported that the GNAT-750 was suffering from computer and software problems, which resulted in cost overruns, schedule delays, and "annoyed congressional staffers." A "Pentagon official" complained that the CIA lacked experience with systems integration or developing tactical communications links. Defense officials spoke of a "hobby shop approach" to development and said that the CIA was hurrying the project too fast.[63]

By late October, the GNAT-750 was judged ready for its final systems integration test flight. On the night before the flight, the CIA contract personnel modified the software program. The new program shut down the GNAT-750's motor and data link when its speed fell below 40 mph. It was assumed that if the UAV was at this low a speed, it would be on the ground.

During the flight the next day, the GNAT-750 was hit by a gust of wind. This produced a low-speed indication, and the software shut down the motor and data link. The GNAT-750 rolled over on its back, went into a flat spin, and crashed. The surviving GNAT-750 was grounded pending an investigation. In all, the technical difficulties and the crash caused a three-month delay, as well as a $1 million cost overrun.[64]

The program regrouped from the mishap. On December 2, 1993, a GNAT-750 flying over the El Mirage test site successfully transmitted imagery of moving targets, such as a tank, to a relay aircraft, which then passed it on to the Pentagon. Tests also indicated the GNAT-750 was quite stealthy. The radar at Edwards Air Force Base did not pick it up. It was believed that this was due to several factors—the shaping of the fuselage, its heavy use of carbon epoxy materials, and, ironically, its slow speed of 90 knots. The radar's own software filters out such weak and slow-moving targets as returns from birds or weather.[65]

BOSNIAN OVERFLIGHTS

Under the revised schedule, the remaining GNAT-750 was to be operational by February 1, 1994. CIA Director Woolsey reportedly demanded that it be operational before his next appearance at congressional budget hearings. There was, however, a new political problem. The Italian government was having second thoughts about playing host to the CIA operation.[66] By late January 1994, they formally turned down the U.S. request, citing the possible danger to civil aircraft from the GNAT-750.[67]

The UAVs had lost their base of operation. But the world had changed since the U-2 had taken off on that first overflight. There were other possibilities, ones that would have been unthinkable in the mid-1950s. The CIA GNAT-750 would be based in Albania.

When the Italians refused permission, the U.S. government made a direct appeal to senior Albanian officials. They approved the operation. The CIA unit would be flown in from Germany in a C-130. The single GNAT-750 and a satellite transmission station would be set up at an Albanian military base on the Adriatic coast. The three prime candidates were Scutari, Durres, and Tiranë, the capital.

Albania's location had advantages for the GNAT-750 overflights. They would have direct access to Montenegro, Kosovo, and Macedonia without having to overfly a third country. It was in these regions that U.S. ground troops were assigned as part of the UN peacekeeping force. Bosnia Herzegovina was also well within range of the GNAT-750, as was the Serbian capital, Belgrade.[68]

It was a stunning example of how the world had changed in just a few years. Between 1949 and 1953, the CIA and the British Secret Intelligence Service had tried to establish networks of agents within Albania in an attempt to overthrow its Communist government. The effort ended in disaster—the agents were caught and killed. Throughout the 1970s and 1980s, Albania remained a closed, Stalinist country. It was the last Communist country in Eastern Europe to undergo the Revolution of 1989.[69]

The CIA unit was in place by early February. Its deployment in Albania was marked by a major article in *Aviation Week and Space Technology* magazine. The results of the GNAT-750's Bosnian overflights were mixed. It was reported that of the thirty attempted overflights, only twelve were successful, due to continuing bad weather, maintenance problems, and difficulties with the GNAT-750's data link.[70] One of the UAV's missions was to follow UN convoys. Other flights were targeted against entrenchments; from 6,000 feet they were able to identify decoy artillery and SAM sites. Airfields, troop and artillery movements, supply dumps, and tank locations were also monitored. Best resolution was eighteen inches. The GNAT-750s were quite stealthy—at no time did people on the ground realize the UAVs were overhead.

The overflights did reveal several problems. The relay aircraft was a two-man Schweitzer RG-8 powered glider. Because of Albania's location, it could only spend about two hours on station. Another six hours was spent flying to and from the area. Having to coordinate the GNAT-750s with the manned RG-8 meant the UAVs could only be flown in specific areas, and at certain times and altitudes. Real-time changes in the route, to more closely examine side roads and buildings, also disrupted the GNAT-750 activities.

By the summer of 1994, the U.S. European Command requested the renewal of GNAT-750 overflights. The CIA preferred to fly from an Italian base, but this was again refused. The unit was based in Croatia. This allowed better use of the RG-8 relay plane. The GNAT-750s were also equipped with an improved IR scanner.[71]

The CIA also wanted to expand its GNAT-750 unit with three more UAVs and an additional ground station. Two of the UAVs would be used for reconnaissance, while two others would be used as relay aircraft. The reconnaissance GNAT-750s were also expected to be fitted with ELINT receivers that could pick up both radar signals and transmissions from walkie-talkies.

This would be done by buying GNAT-750s originally ordered by Turkey. The Turkish government was in a budget crunch and lacked the money to pay for the three GNAT-750s already delivered, much less the four still at the General Atomics factory. Each GNAT-750 cost $800,000, while the ground station cost $1.2 million. The CIA asked Congress for authority to reprogram funds to buy the UAVs.[72]

THE CIRCLE CLOSES

The HALSOL and GNAT-750 represent the future of airborne reconnaissance. The smaller post–Cold War air force is unlikely to have the billions of dollars needed to develop and fly a new manned reconnaissance aircraft. In contrast, a unit of UAVs can be built and operated for less than one-tenth the cost of a single F-15E or F-117A. The GNAT-750's similarity to a big model airplane is an advantage, rather than a shortcoming. This allows it to combine long-duration flight times with the simplicity of the small tactical UAVs.

The CIA GNAT-750 also showed the continued value of Black projects. Despite all the problems, it was still done faster and at a lower cost than conventional methods could have accomplished. The criticism was based as much on jealousy as on the problems. At the same time, the GNAT-750 was a very different Black than the U-2, A-12, or even the Have Blue (which also started out White and then went Black). The involvement of congressional staffers meant that the lines of authority were less clear and direct.

The GNAT-750 effort also marked the return to Black aviation by the CIA after two decades. The Nationalist Chinese U-2 overflight program had ended in 1968. The CIA A-12s were retired the same year. Between 1969 and 1971, the CIA-sponsored D-21 Tagboard made their few disappointing flights. By the mid-1970s the CIA U-2 operation had been closed down, and the surviving planes had been transferred to the air force and NASA.

With this, the GNAT-750 project brings the story of the U.S. Black airplanes full circle.

The A-12, D-21, and HALSOL had all been kept secret for a decade or more before their existence was revealed. Up to that time, there had been no hint of these Dark Eagles. Such secrecy, along with the large number of stories about the stealth fighter, inevitably gave rise to speculation about other Black airplanes, ones that were still secret. Reports and sightings of these secret airplanes were soon being whispered about. From time to time, the stories would be published.

They were tales of darkness and shadows.

Chapter 10

The MiGs of Red Square
Have Doughnut and Have Drill

Therefore I say, "know the enemy and know yourself; in a hundred battles, you will never be in peril."

Sun Tzu
ca. 400 B.C.

The longest continuing U.S. Black airplane program is the secret test flying of MiGs and other Soviet aircraft. This effort's tentative beginnings were in the mid-1950s, before the U-2. It began in earnest a decade later, contemporarily with the A-12 and D-21, and has continued to the present day. Unlike the other Black airplane programs, such as the Have Blue, F-117A, or HALSOL, MiG operations still remain Black. The program can not even be acknowledged.

It is not known exactly the actual number or types of aircraft involved, where they came from, or the complete history of the program. There are only a few, limited accounts, and it is probable that many of these are, at best, incomplete, and at worst, wrong. In one case, a published MiG tale proved spectacularly wrong.

It is known that these Dark Eagles brought about a fundamental change in air-combat tactics. They revitalized the art of dogfighting at a time when, seemingly, it had nearly been forgotten. The knowledge gained from these planes was reflected in the success of U.S. Navy air operations over North Vietnam in the final year of the war, as well as in the founding of the Navy's Top Gun school.

THE BLACK YAK

The program started with a C-124 Globemaster landing at Wright-Patterson Air Force Base in the mid-1950s. Its payload was a Soviet-built Yak 23 Flora. This was a small single-seat, single-engine jet fighter. The aircraft had first been exported to the Polish, Czechoslovakian, Bulgarian, and Romanian air forces in 1951. Although quickly superseded by the more advanced MiG 15, it was the first jet operated by these Eastern European air forces. For the U.S. Air Force, it introduced the secret testing of Soviet aircraft.

According to one account, an Eastern European intelligence officer contacted an American intelligence officer and offered the "loan" of the plane. The deal was made and the Yak 23 was packed in shipping crates and left aboard a railroad car. After its arrival, the parts were photographed, to insure they could be repacked correctly.

The project pilot for the Yak 23 flights was Capt. Tom Collins. In September 1953, he had become the first U.S. pilot to fly a MiG 15. It had been flown to South Korea by a defecting pilot. Unlike that effort, the tests Collins conducted with the Yak 23 were top secret. The plane carried U.S. Air Force insignias, while its buzz number of "FU-599," and the "0599" serial number actually belonged to an F-86E. If the curious asked what kind of plane it was, they were told it was an "X-5."

The flight tests of the Yak 23 lasted about a month. The plane was then disassembled, loaded aboard a C-124, and flown away. It would be another forty years before the existence of a U.S. Air Force Yak 23 would be revealed. In December 1994, four photos of the plane were published. Where the plane had come from, and where it went after leaving Wright-Patterson, are secret still.[1]

In the years following the Yak 23's brief stay at Wright-Patterson, a new generation of fighter aircraft, such as the F-4 Phantom II, was developed. The F-4 was the first fighter designed from the start with only air-to-air missiles—the radar-guided Sparrow and the shorter-range Sidewinder infrared-guided missile. With the new missiles came the new attitude that dogfighting was obsolete. The air-to-air training given to new navy F-4 crews was extremely limited. It involved about ten flights and provided little useful information. By 1964, few in the navy were left to carry on the tradition of classic dogfighting.[2] Then came Vietnam.

THE AIR WAR OVER NORTH VIETNAM

The early years of the air war over North Vietnam showed the faith placed in missiles was terribly in error. Between 1965 and the bombing halt in 1968, the U.S. Air Force had a 2.15 to 1 kill ratio. The navy was doing slightly better with a 2.75 to 1 rate. For roughly every two North Vietnamese MiG 17 Frescos or MiG 21 Fishbeds shot down, an American F-4, F-105, or F-8 would be lost. This was far worse than the 10-plus to 1 kill rate in Korea.[3]

More serious, the percentage of U.S. fighters being lost in air-to-air combat was growing. During 1966, only 3 percent of U.S. aircraft losses were due to MiGs. This rose to 8 percent in 1967, then climbed to 22 percent for the first three months of 1968.[4]

In 1968, navy Capt. Frank Ault was assigned to learn the reasons for this poor showing against the MiGs. The "Ault Report" was issued on January 1, 1969. It found 242 problems that ranged from hardware to crew training. The Sidewinder and Sparrow missiles showed very poor reliability. A full 25 percent of Sparrows failed because their rocket engines never fired. The Sidewinder and Sparrow were both limited to 2- to 2.5-g maneuvers, as they had been designed for use against nonmaneuvering bombers rather than fighters.[5] It took a full 5.2 seconds to fire a Sparrow; yet, the average time the F-4 crews had to fire was 2.2 seconds. To hit the target, the F-4's radar beam had to be kept on the MiG. This was extremely difficult in a turning dogfight.

Far more important was the training of the crews. Few F-4 crewmen knew the firing parameters for the missiles. These changed with altitude, and whether it was a tail, head-on, or side attack. The crews lacked the knowledge to judge the ever-changing parameters in the midst of the fight. The result: of some six hundred missiles fired between 1965 and 1968, only one out of ten or eleven had any chance to hit its target.[6]

Finally, the emphasis on interception meant the F-4 crews had only the sketchiest knowledge of dogfighting. The design of the F-4 made it ill-suited for a tight-turning dogfight. In contrast to the MiG 17, the F-4 was large and heavy. When a tight turn was made, the F-4 would lose speed. The MiG 17's superior turning capability then allowed it to close to gun range. All too often, hits from the MiG 17's "outmoded" cannons would then destroy the F-4.

The key to survival in the skies of North Vietnam, as it had been in every air war, was to make the enemy pilot fight on your terms. This meant knowing his weaknesses, while using your plane's strengths to maneuver into position to destroy the enemy airplane. Of course, the problem was acquiring the knowledge.

Out in the Nevada desert, a MiG 21 awaited.

HAVE DOUGHNUT

In 1967, the Defense Intelligence Agency secretly acquired a single MiG 21. The country the MiG 21 came from, and the means by which it came to the United States, remain secret to this day. Because U.S. possession of the MiG 21 was, itself, secret, it was tested at Groom Lake. A joint air force–navy team was assembled for a series of dogfight tests. The project was code-named "Have Doughnut."[7]

Comparisons between the F-4 and the MiG 21 indicated that, on the surface, they were evenly matched. At a speed of Mach 0.9 at 15,000 feet the instantaneous turn rates of the two planes were nearly identical, at 13.5 degrees per second. At Mach 0.5, the MiG 21 held the edge at 11.1 degrees versus 7.8 degrees for the F-4.[8] But air combat was not just about degrees per second of turn rate. In the final analysis, it was the skill of the man in the cockpit. The Have Doughnut tests showed this most strongly.

When the air force pilots flew the MiG 21, the results were a draw—the F-4 would win some fights, the MiG 21 would win others. There were no clear advantages. The problem was not with the planes, but with the pilots flying them. The air force pilots would not fly either plane to its limits. To avoid accidents, restrictions had been placed by the air force on air combat maneuvers.

One of the navy pilots was Marland W. "Doc" Townsend, then commander of VF-121, the F-4 training squadron at NAS Miramar. He was an engineer and a Korean War veteran and had flown almost every navy aircraft. When he flew against the MiG 21, he would outmaneuver it every time. The air force pilots would not go vertical in the MiG 21. Townsend would make his pass, then pull up into a vertical climb, do a roll as he came over the top, spot the MiG 21, then line up on its tail. He recalled years later, "It was a piece of cake. . . . Easiest plane I've ever fought in my life."

The Have Doughnut project officer was Tom Cassidy, a pilot with VX-4, the navy's Air Development Squadron at Point Mugu. He had been watching as Townsend "waxed" the air force MiG 21 pilots. Cassidy climbed into the MiG 21 and went up against Townsend's F-4. This time the result was far different. Cassidy was willing to fight in the vertical, flying the plane to the point where it was buffeting, just above the stall. Cassidy was able to get on the F-4's tail. After the flight, they realized the MiG 21 turned better than the F-4 at lower speeds. The key was for the F-4 to keep its speed up.

On the third day, Townsend and Cassidy met for a final dogfight. The fight started with the F-4 and MiG 21 coming toward each other. When Townsend spotted the MiG, he lowered the F-4's nose and pulled into a high-g turn, maintaining a speed of 450 knots. The MiG 21 could not follow and lost speed. Townsend then pulled the F-4 into the vertical. The MiG 21 lacked the energy to follow, and Cassidy dove away. Townsend rolled over the top and pulled behind the MiG 21. Nothing Cassidy did could shake the F-4. Finally, the dogfight was called off when the MiG 21 ran low on fuel.[9]

What had happened in the blue sky above Groom Lake was remarkable. An F-4 had defeated the MiG 21; the weakness of the Soviet plane had been found. The means to reverse the 2 to 1 kill ratio was at hand. It was also clear that the MiG 21 was a formidable enemy. United States pilots would have to fly much better than they had been to beat it. This would require a special school to teach advanced air combat techniques.

And it would require more MiGs.

007 AND THE WAYWARD PILOTS

The Iraqi MiG 21 pilot saw the two Israeli Mirage III fighters closing in. The MiG pilot reduced his speed, wiggled his wings, and lowered his landing gear. When one of the Israeli planes pulled alongside, the MiG pilot signaled he wanted to land. The Israeli pilot indicated the MiG was to follow him. It was August 16, 1966, and the first MiG 21F-13 fighter had reached Israeli hands.[10]

At a press conference that evening, the Iraqi pilot, Capt. Monir Radfa, explained that he had defected due to his revulsion against attacks on Kurds in northern Iraq. He was also a Roman Catholic, one of only five or six in the Iraqi air force, and he felt discriminated against in the predominately Moslem country. He had requested a transfer to MiG 21s and spent four months training in the Soviet Union. Once he decided to defect, he had sent his family out of Iraq. He also sent a letter to Israel announcing his intention to defect. Dated August 3, it said in part,

> I am a MiG 21 pilot of the Iraqi air force and I have decided to come to your country. This decision I have arrived at after very hard thinking and for important reasons that I shall explain to you personally. . . .
>
> I plan to carry out this decision within three or four weeks from now. Please tell your pilots not to shoot me down because I come for peaceful feelings. I should come in high altitude between 0700 and 1100 hours G.M.T. flying east to west and over the Jordan River near the Dead Sea. With the help of God I hope to land at one of your bases.

It was signed, "Yours faithfully, a MiG 21 pilot."

Monir Radfa took off from Rashid Air Base outside Baghdad on a navigation flight, then headed for Israel. As he crossed Jordan, two Hawker Hunter jet fighters climbed toward his plane, but the MiG was flying too high and too fast for them to catch it. He landed at Hatzor Air Base, to the astonishment of the pilots and ground crews.[11]

In fact, the defection was not a spontaneous action. For three years, the Mossad had been attempting to acquire a MiG.[12] The actual planning had taken nearly six months. Reportedly, an Israeli air force pilot had even taken a check ride with him, to ensure he could fly and navigate well enough to make the flight.[13]

The MiG 21 was repainted with the Israeli Star of David insignia, yellow recognition stripes, and the number "007" (a subtle reflection of the secret means behind its acquisition). For the next several months, it was subjected to a series of flights to learn its strengths and weaknesses. These were made by Lt. Col. Dani Shapira, the Israeli air force's chief test pilot. He recalled later, "We found out, for example, that at high speed it had trouble maneuvering as well as the Mirage, which meant we had to try to get it into tight turns at high speeds." (This was the same conclusion drawn in the Have Doughnut tests the following year.) At low speeds, the MiG 21 would tend to spin out in tight, low-altitude turns.

The MiG 21 was then used to train Israeli Mirage pilots. Some 100 hours were spent flying in mock combat with Israeli Mirage III fighters. By the end of the year, every Mirage squadron had been familiarized with the MiG 21.[14]

It was unique information, but it does not appear it was shared. The newspaper *Lamerhav* said two days after the defection that "supreme political interests" and defense responsibilities required that Israel not give any information to other countries. "The Government would do well if it left no room even for unfounded suspicions," the paper said. The *Ma'ariv* newspaper said, "Israel should not allow foreign experts to examine the new Soviet plane. The pilot landed his plane in Israel and only Israel should be able to glean benefit from this important event." The reason given was the need to avoid worsening relations with the Soviet Union.[15]

The information gleaned from 007 proved valuable to the Israelis. On June 5, 1967, Israel launched a preemptive strike on Egypt, Syria, Jordan, and Iraq. By the end of the first day, the Arab air forces had been destroyed on the ground and Israel had complete air superiority. During the Six Day War, thirty-seven MiG 21s were shot down (out of seventy-two air-to-air kills). Two Israeli aircraft (out of forty-eight lost) were shot down in air-to air-combat—a 36 to 1 kill rate.[16]

It was during the Six Day War that, reportedly, a group of MiG 21s fell into Israeli hands. Six Algerian MiG 21 pilots flew into El Arish Air Base in the Sinai Desert. The Egyptian and Syrian governments had concealed the magnitude of their losses, so the Algerian pilots did not know El Arish had already been captured by the Israelis.[17]

This was not the only example of wayward pilots delivering MiGs to Israeli control. On August 12, 1968, two Syrian air force lieutenants, Walid Adham and Radfan Rifai, took off in a pair of MiG 17s on a training mission. They lost their way and, believing they were over Lebanon, landed at the Beset Landing Field in northern Israel. (One version has it that they were led astray by an Arabic-speaking Israeli.)

The first plane overran the end of the strip, plowed across a field, and stopped just short of a stand of cypress trees. The second MiG 17 circled again and made a successful landing. Neither plane was damaged, nor were the pilots injured. Yossi Yitzhak, who had witnessed the landings while working in a nearby hay field, told them they were in Israel. He was quoted later as saying, "They both went white with shock." Both pilots were taken away as POWs.[18]

The two MiG 17s were repainted with Israeli insignia and red recognition markings, then were test flown from Hatzor Air Base. As with 007, Dani Shapira flew the missions.[19]

The Israeli MiGs had, by this time, acquired an importance far beyond their intelligence value. In the wake of the Six Day War, France cut off military supplies to Israel. Since the 1950s, France had provided most of Israel's aircraft (as well as secret support for its nuclear program). Israel turned to the United States, seeking A-4 attack aircraft and F-4E fighters. The U.S. government was reluctant, but Israel could offer in exchange captured Soviet tanks, SA-2 SAMs, and the MiGs. A deal was struck, and 007 and the two MiG 17s were shipped to the United States in 1969.

The U.S. MiG operations were extremely secret—personnel connected with the project were told that if they leaked any information, they would "disappear." Therefore, the item in the "Industry Observer" column in the February 17, 1969, issue of *Aviation Week and Space Technology* magazine came as a shock. It read:

> Soviet MiG 21 fighter was secretly brought to the U.S. last spring and flight tested by USAF pilots to learn first-hand its capabilities and design characteristics. The aircraft, which engaged in simulated combat against U.S. fighters, was highly regarded by the pilots who flew it. The MiG 21 was particularly impressive at altitudes over 25,000 ft. The evaluation was part of a broad effort by USAF to detail the threat of Soviet air power in planning new aircraft, such as the F-15 fighter.[20]

Out at Groom Lake, the MiG program accelerated. Again, it was the U.S. Navy that took the lead.

HAVE DRILL

In April 1969, Jim Foster was named commander of VX-4. He had been an F-8 pilot during the early years of the Vietnam War and had seen the problems with the F-4. He quickly set about to "acquire" the new MiGs for the navy.

Soon after taking command, he selected Foster S. "Tooter" Teague to head the navy exploitation effort. Teague was a fighter pilot who had completed two combat tours in Vietnam. Teague did not know about the MiGs even after his arrival at VX-4. Initially, he was only told he was being sent "into the desert" for two hundred days. After being told about the MiGs, Teague and his personnel quickly wrote up a test plan for the aircraft. The report was typed and bound in time for the first project meeting. (The air force personnel, in contrast, were only getting started.) The project was code-named "Have Drill."[21]

The Have Drill tests were conducted by a small group of air force and navy pilots under Teague and VX-4 test pilot Ronald "Mugs" McKeown. As with the earlier Black airplanes, the pilots' introduction to the MiGs was a strange and wonderful experience. Foster recalled being taken to the "hidden location" (i.e., Groom Lake) and entering a huge well-lighted hangar where the gleaming silver MiGs were kept. He was amazed by the "secret site." The MiGs' instruments were in Russian, so the pilots had to be briefed extensively. Their systems were very different than those on American aircraft. This made even such basics as taking off and landing a learning experience.

Taxiing was an "unnatural" experience for American pilots, as the MiGs used air brakes activated by a lever on the stick. (United States aircraft used toe brakes on the rudder pedals.) The American MiG pilot tended to wander around as he taxied out to the Groom Lake runway. An F-4 chase plane followed along behind, giving instructions. Before taking off, one of the officers would climb up on the MiG's wing and give the pilot final words of advice.

To lower the landing gear on an F-4, the pilot threw a switch. On the MiG 17, the pilot had to lift a toggle switch, turn on a pump, wait until the pressure built up, and then put the gear handle down. Once the gear was locked, the pilot turned off the pump and closed the toggle switch.

Flying the MiG 17 reminded Foster of the T-33 trainer in the way it handled. Teague called it the best turning aircraft he had ever flown. To demonstrate, Teague would let an F-4 get into firing position on the MiG 17's tail. Within seconds, the F-4 would be out in front and the MiG 17 in firing position. There was simply no way an F-4 could win a close-in turning fight with a MiG 17.

Very soon, the MiG 17's shortcomings became clear. It had an extremely simple, even crude, control system which lacked the power-boosted controls of American aircraft. Although tight turning at low speeds, the control system would lock up at high speeds. At 425 knots, the plane began to roll to the left as the wing began to warp. Above 450 knots, the plane was uncontrollable—it was actually possible to bend the stick without any control response at all.[22]

The solution became obvious—keep the F-4's speed above 500 knots and take advantage of the MiG 17's poor roll rate. One technique for an F-4 to evade a MiG 17 on its tail was to fake a roll one way, then make a hard turn the other direction. When the MiG tried to follow, the F-4's better roll rate and acceleration would open the distance between them, so the F-4 could turn around and reengage. Another maneuver was to make a turn at a right angle to the MiG's flight path, then accelerate away from the MiG.

These maneuvers also made use of the F-4's advantage in acceleration. The F-4's twin engines were so powerful it could accelerate out of range of the MiG 17's guns in thirty seconds. It was important for the F-4 to keep its distance from the MiG 17. As long as the F-4 was one and a half miles from the MiG 17, it was outside the reach of the Soviet fighter's guns, but the MiG was within reach of the F-4's missiles.

To turn with the MiG 17, a technique called "lag pursuit" was developed. The F-4 would follow behind and outside the MiG 17's flight path. Rather than trying to turn inside the MiG (which was impossible), the F-4 would stay to the outside. It would use its higher speed and turn rate to stay behind the MiG 17 and close to missile range.[23]

Tests of the MiG 21 indicated a different set of techniques was needed. The MiG 21 was almost as fast as the F-4, so running was not an option. The MiG 21's afterburner put out only a few hundred pounds of thrust. Foster noted, "I hit the burner and I didn't feel any thrust." The delta-wing design of the MiG 21 also lost energy in turns faster than the F-4 did. The technique therefore was to force the MiG 21 to make vertical turns until its air speed fell below a critical value. With the MiG slowed down, the F-4 was in the superior position.

In a turning battle, as first discovered in the Have Doughnut tests, the F-4 had to keep its speed up, to insure the turn rate was even. The MiG 21 was also optimized for high altitudes—above 30,000 feet. At lower altitudes, the F-4's higher-thrust engines gave it an advantage.[24]

The data from the Have Doughnut and Have Drill tests were provided to the newly formed Top Gun school at NASA Miramar. The Top Gun instructors had been studying the Ault Report, accounts of dogfights, and intelligence reports on North Vietnamese tactics for six months. Now they were shown a film of the Have Doughnut tests; then they were given clearances to see and fly the MiGs.

The selected instructors would fly in pairs to Nellis Air Force Base to be briefed by Teague or his assistants. They would be told to fly to "a certain spot" and wait for the MiGs. The event was eagerly anticipated; the instructors had known for weeks they would be going up against the MiGs in simulated combat. One instructor later recalled his heart racing as the MiGs drew closer on radar. Finally, at about three and a half miles, they were spotted visually.

The first step was for the MiGs to pull alongside the F-4s so the instructors could get a good look at them. The MiGs were not much to look at; there were bumps and rivets that would never have been on an American plane.

The F-4 would then follow along behind as the MiG 17's good low-speed and poor high-speed maneuverability was demonstrated. The tendency of the MiG 21 to lose speed in turns was also shown. The flights were very helpful in a number of areas. Ironically, although the instructors were combat veterans, few had ever seen a MiG. A close-up view of the MiGs lessened the surprise. The radar intercept officers (RIOs—the rear seat crewmen who operated the F-4's radar) also got the chance to study the MiG's radar return. Finally, the instructors were able to actually try out the techniques they had been developing with the "real thing."

By 1970, the Have Drill program was expanded; a few selected fleet F-4 crews were given the chance to fight the MiGs. Eventually, for the vast majority who could not go to Top Gun or participate in Have Drill, a thirty-minute film called *Throw a Nickel on the Grass* (the name of a fighter pilot's song) was produced. It was shown to every navy squadron. In the film, Teague noted, "The most important result of Project Have Drill is that no Navy pilot who flew in the project defeated the [MiG 17] Fresco in the first engagement."[25]

The Have Drill dogfights were by invitation only. The other pilots based at Nellis Air Force Base were not to know about the U.S.-operated MiGs. To prevent any sightings, the airspace above the Groom Lake range was closed. On aeronautical maps, the exercise area was marked in red ink. The forbidden zone became known as "Red Square."[26]

Although much of the MiG operations were connected with such training, tests were also being conducted. Several MiG 17 flights were made to test fire the plane's twin 23mm and single 37mm cannons. (When the two MiG 17s landed in Israel, their cannons were fully loaded.) A special dogfight test was conducted with navy and air force F-4 crews. The navy RIOs were specially trained for their task; in contrast, the air force used pilots for the position. The tests showed that the navy's pilot-RIO team worked better than using two pilots. The navy crews were able to detect the target at longer ranges and go over to the offense faster. The air force crews scored fewer simulated kills. Soon after, the air force started specialized training.[27]

The MiGs were also used for familiarization flights by high-ranking naval officers. At the 1969 Tailhook Convention in Las Vegas, Foster convinced several admirals to go with him to Groom Lake to fly the MiGs.

Once special permission had been granted, Foster loaded them into a transport and flew them out to Groom Lake. The admirals were carefully briefed, then strapped into the MiGs. The Have Drill personnel were worried about the high-ranking pilots flying the irreplaceable MiGs. At one point, Teague, who was flying chase, thought that an admiral was flying a MiG 17 too hard. He radioed, "Goddamnit, Admiral, put it on the ground."[28]

Another senior officer to fly the MiG 17 was Marine Corps Gen. Marion Carl. He had become the first marine ace over Guadalcanal in the dark early days of World War II. On August 25, 1947, he set a world speed record of 650.6 mph (Mach 0.82) in the Douglas D-558-I Skystreak. On August 21, 1953, he set a world altitude record of 83,235 feet in the Douglas D-552-II Skyrocket. He had also served as commander of the first marine jet fighter squadron and flew secret reconnaissance missions over Communist China. He was a living link between the days of propeller fighters and the missile-armed, supersonic fighters of the 1960s.

General Carl was flown to "a secret desert site" in a T-33 trainer. He flew two missions in the MiG 17, totaling 1.7 hours of flight time. (He was not allowed to log the time, of course.) The first was a simulated dogfight with several F-8s, while the second was with A-4s. Carl was very impressed with the MiG 17's maneuverability. It could be stalled flying straight up and allowed to fall off but would be under full control within 1,500 to 2,000 feet. The only drawback he noted was the cockpit size—pilots over six feet tall were cramped.

While at Groom Lake, General Carl ran into one of the security precautions surrounding the MiGs. About midday, the ground crews began pulling the MiGs into the hangars. Carl asked about this and was told a Soviet reconnaissance satellite would soon pass overhead. It was a clear day, so the MiGs were being hidden to prevent any photos being taken of them by the satellite.[29]

In spite of the valuable lessons learned from the MiG program and from Top Gun, the future of both programs was by no means assured. High-level support was needed. Air-combat training was dangerous, and several aircraft had been lost in accidents. In one simulated dogfight with a MiG 17, a brand-new F-4 was lost when it went into a flat spin. The crew ejected at the last moment. The Vietnam War was also winding down—there had been only one MiG 21 shot down since the bombing halt in 1968.

Top Gun and Have Drill had produced a small group of navy pilots skilled in the art of air combat, but they had yet to prove the value of their training. Just after midnight on March 30, 1972, the North Vietnamese Army launched the Easter Offensive. On May 8, President Richard Nixon ordered the mining of North Vietnam's ports and the start of the Linebacker I bombing campaign.

Two days later would come the greatest dogfight in the history of jet aviation.

THE CUNNINGHAM-TOON DOGFIGHT

On May 10, 1972, the F-4J crew of Lt. Randall H. "Duke" Cunningham and Lt. (jg) William R. "Willie" Driscoll was assigned to an air strike on the railyards at Hai Duong, located halfway between Hanoi and Haiphong. Cunningham and Driscoll had already shot down two MiGs—a MiG 21 on January 19 and a MiG 17 on May 8. As the strike force neared the target, twenty-two MiG 17s, MiG 19s, and MiG 21s rose to challenge the navy planes.[30]

Cunningham and his wingman, Lt. Brad Grant, dropped their cluster bombs on a storage building. As Cunningham looked to his right at the building, two MiG 17s were coming in from the left. Grant saw them and radioed a warning, "Duke, you have MiG 17s at your seven o'clock, shooting." Driscoll also saw the MiGs as two black dots approaching fast and yelled, "Break port." Cunningham could see the muzzle flash from the lead MiG. The MiG 17s were moving fast, so he knew the controls would be hard to move. He turned toward the lead MiG, and it overshot the F-4. Cunningham recalled later, "The MiG driver just didn't have the muscle to move that stick." While Grant chased off the second MiG 17, Cunningham reversed and fired a Sidewinder. The MiG 17 was within the missile's minimum range at launch, but due to the MiG's high speed, the distance had opened to 2,500 feet when the Sidewinder hit. The MiG 17 was blown apart. The whole engagement lasted only fifteen seconds.[31]

Cunningham evaded a MiG that attempted to follow him by using his afterburner to accelerate to 600 knots. Cunningham and his wingman then climbed to 15,000 feet. Below them were eight MiG 17s in a defensive circle with three F-4s mixed in. Cunningham later wrote, "The scene below was straight out of *The Dawn Patrol.*" The F-4s had slowed to 350 knots and were vulnerable. Cunningham told Grant to cover him, then dove toward the circle.

As they did, one of the F-4s broke from the circle and nearly collided with Cunningham's plane. It was flown by Comdr. Dwight Timm, the second in command of VF-96. Following behind his plane were a MiG 17 about 2,000 feet away; a MiG 21, 3,000 feet behind; and, unseen by Timm or his RIO, Jim Fox, a MiG 17 in his blind spot below and to the outside of the plane.

As Cunningham closed on the three MiGs, Driscoll warned him, "Duke, we have four MiG 17s at our seven o'clock." They had broken out of the circle and were pursuing his F-4. For the moment, they were out of range, but Timm's arcing turn was allowing them to close the distance. Then

Driscoll gave another warning, "Duke, look at two o'clock high." Cunningham looked up and saw two bright flashes—two MiG 19s beginning a diving firing pass on the F-4. Cunningham reversed and the two MiG 19s flew past.

Cunningham was in a lag pursuit with the MiG 17 on Timm's wing, but one of the MiG 17s was now within firing range of Cunningham's F-4. Flying at 550 knots, he could outrun the MiG, but he had to turn to stay behind Timm. This allowed the MiG 17 chasing Cunningham to close the range with his plane. Cunningham told Driscoll to watch the MiG 17 following them. When it opened fire, Cunningham straightened out to open the range. He then turned back toward Timm's F-4 and yelled for him to break right. Timm, thinking he had evaded the two MiGs (and still not aware of the third MiG), turned right. The fast-flying MiG 17 was not able to follow, and Cunningham had a clear shot.

Cunningham fired a Sidewinder, which homed in on the MiG 17's hot exhaust and destroyed it. Just as the missile hit, Timm and Fox finally saw the MiG 17. It turned into a fireball, and the North Vietnamese pilot ejected. So far, only about two minutes had passed. It was Cunningham and Driscoll's fourth kill; they were now tied with air force Col. Robin Olds as top-scoring American aviators of the Vietnam War. But the day's events were not yet over.[32]

Now, two or three of the MiG 17s pursuing Cunningham had closed in. Four more MiG 21s were also bearing down on the F-4. It did not seem possible they could escape. Driscoll yelled, "Break! Break! Give me all you've got." Cunningham turned toward the MiG 21s. Moments later, they were alone in the sky. Cunningham had become separated from his wingman, and it was time to leave. Cunningham turned toward the coast.[33]

As they headed east, Driscoll kept watch behind them for any MiGs that might try to catch them, so it was Cunningham who spotted the next MiG 17, ahead and slightly below them. Cunningham altered course to make a close, head-on pass. This prevented the MiG pilot from using the lateral separation to make a quick turn. Cunningham could then outrun the MiG.

As they closed the distance, the nose of the MiG 17 lit up as the pilot fired his guns. Cunningham hauled back on the stick and went into a vertical climb. He expected the MiG 17 would continue on. When Cunningham looked back over his ejection seat, it was with considerable shock that he saw the MiG 17, also in a vertical climb. Cunningham could see the North Vietnamese pilot's leather helmet, goggles, scarf, and even the expression on his face. Cunningham recalled, "There was no fear in this guy's eyes."[34]

The MiG 17 began to fall behind, and Cunningham lit the F-4's afterburner. As he rolled over the top, the MiG 17, on the verge of a stall, nosed over and fired its guns. Normally, North Vietnamese pilots would fire a continuous stream. This pilot fired only a brief burst, to conserve his ammunition. Driscoll recalled, "It was like, who is this guy? . . . He knew exactly what he was doing."[35]

Both planes began to descend and started a vertical rolling scissors. In this maneuver, the planes would go into a weaving climb whenever one flew lower than the other. This forced the opposing plane to overshoot. Whenever the F-4 pulled out in front, the MiG pilot would fire a short burst. Each time Cunningham made a maneuver, the MiG 17 would counter him. Cunningham was holding his own, but the F-4 was losing speed faster than the MiG 17. It was now down to 200 knots, and Cunningham decided it was time to "bug out." He lowered half flaps, then made a looping turn at a right angle to the MiG's flight path. The F-4 dove away with full afterburner. Before the MiG could follow, Cunningham had accelerated to 500 knots and was out of range.

He then turned back toward the MiG 17. The two planes met in a head-on pass, then went into a vertical climb. Again, as they descended, the planes began a rolling scissors. Neither plane could get enough of an advantage for a killing shot. Again, the F-4 slowed to 200 knots. When Cunningham saw the MiG's nose turn slightly, he extended his own turn to the outside of the MiG's flight path. Cunningham then pulled into a vertical loop. Remembering the previous escape maneuver, the MiG pilot also pulled up into a loop to counter. The two planes passed each other going in opposite directions. Cunningham continued through the loop, extended out under the thrust of the afterburners, and was quickly out of range. With the F-4's speed back up, he turned again, into the fight. Neither pilot attempted to run; one plane would go home, the other would be shot down.

As the two planes headed toward each other in their final pass, Cunningham pulled up into a climb. The MiG 17 did the same, then turned toward the F-4. As it did, Cunningham pulled the throttles to idle, lowered the flaps, and popped the speed brakes. The F-4 went from 550 to 150 knots, and the MiG 17 overshot. The MiG 17 was now 500 to 1,000 feet ahead of the F-4, too close for a Sidewinder. Both planes were on the verge of a stall—at low speed and nose high. Cunningham had to use full afterburner. Using only the F-4's rudder to avoid stalling, he rolled in behind the MiG. The MiG pilot saw the F-4 cross into his plane's six o'clock position, then lost sight of the plane. He made a maximum turn to the right to keep the F-4 in sight, but the MiG 17 stalled. Its nose went down, and the pilot dove away to regain flying speed.

Cunningham, his own plane near a stall, pulled up the flaps and air brakes, dove, and swung to the right to increase separation. The MiG continued to dive away, attempting to run, and Cunningham pulled in behind the MiG. Now 3,000 feet behind the MiG, Cunningham fired a Sidewinder. It homed in on the MiG's afterburner. There was a small flash, then a burst of flame and black smoke. The MiG 17 flew into the ground. The pilot did not eject.[36]

Cunningham and Driscoll were now the first American aces of the Vietnam War. But they were still far from home. As they headed toward the coast, an SA-2 missile damaged their plane. Both hydraulic systems failed, and Cunningham had to use the rudder, throttle, and speed brakes to control the plane. For twenty miles, he fought to control the F-4. Just as it crossed the coast, an explosion shook the plane. Cunningham and Driscoll ejected and landed in the mouth of the Red River. As American planes chased off patrol boats, junks, and a freighter, they were picked up by marine helicopters. They were soon returned to the USS *Constellation,* to be greeted by the entire crew.[37]

On May 10, 1972, eleven North Vietnamese MiGs were shot down—eight by the navy and three by air force pilots. Two air force F-4s were lost to MiG 19s. During the remainder of the war, the navy kill ratio climbed to 8.33 to 1. In contrast, the air force rate improved only slightly to 2.83 to 1.[38]

The reason for this difference was Top Gun. The navy had revitalized its air combat training, while the air force had stayed stagnant. Most of the navy MiG kills were by Top Gun graduates, Cunningham among them. Top Gun was soon made a separate command, ending the need to beg, borrow, or steal aircraft, fuel, or supplies. The air force also saw the need and, starting in 1975, began the Red Flag exercises. This went far beyond Top Gun, in that Red Flag was a war game involving not only fighters but attack aircraft, bombers, and transport in simulated combat against units that mimicked Soviet forces.[39]

Yet behind the success of Top Gun, there was Have Drill. The Cunningham dogfight is an example. The maneuvers he used in that epic fight were all taught at Top Gun. But they had been originally developed as part of Have Drill. These included the use of high speed to counter the MiG 17's maneuverability, the close head-on pass, turning away at a right angle to the MiG's flight path, and the pilot-RIO teamwork. Cunningham and Driscoll had used all these to survive and triumph, even when outnumbered by seven or eight to one. And when it was one on one.

"IT WAS LIKE, WHO IS THIS GUY?"

There was one loose end, however. Cunningham called the pilot he had fought in the fifth kill the best he had ever faced—to quote Driscoll, "Who is this guy?" Clearly, he was not a typical North Vietnamese pilot. Cunningham was later told, based on intercepts of North Vietnamese radio traffic, that he was "Colonel Toon" (also spelled "Tomb"). He was described as being the North Vietnamese air force's leading ace with thirteen kills. A MiG 21 with thirteen kill markings was thought to be his.[40]

An examination of North Vietnamese propaganda, however, revealed no mention of a pilot named Toon or Tomb.[41] The North Vietnamese were quite willing to describe the exploits of their pilots, even while the war was going on. Toon and Tomb are also not standard Vietnamese names. In the two decades since the dogfight, there has been considerable speculation about the identity of Colonel Toon.

The day after the fight, the North Vietnamese announced the death of Col. Nguyen Van Coc, a MiG 17 pilot with nine claims (including the only F-102 shot down during the war). It was not stated that he had been shot down, however. Cunningham doubted that Coc was Toon. It has also been speculated that he was a Soviet, North Korean, or even Warsaw Pact pilot.

In the early 1990s, former Soviet pilots stationed in North Vietnam indicated that Toon was actually the nickname given by the Soviets to a North Vietnamese pilot named Pham Tuan. Tuan, however, survived the war. In 1980, he was selected by the Soviets to make an eight-day flight to the *Salyut 6* space station. He subsequently became a senior air force officer.[42]

From these fragments, it is possible to reconstruct the events of May 10, 1972. The MiG 17 pilot Cunningham and Driscoll shot down was, in fact, Col. Nguyen Van Coc. Despite Cunningham's doubts, several factors point in this direction. The rank, timing of his death, type of aircraft he flew, and skill level all match. More important, Colonel Coc was the North Vietnamese air force's highest scoring ace with nine kills. This would match the claim that Toon was also the top ace.[43]

The story of a North Vietnamese pilot with thirteen kills was apparently based on the photo of a MiG 21 with thirteen stars painted on its nose. The thirteen kill markings were the total for the *plane,* however, and were claimed by several pilots.[44] It seems likely that the story of a hotshot pilot named Toon and the thirteen-kill MiG 21 were connected.

It is possible that Tuan was aloft during May 10, 1972, and, with the confusion of the large numbers of North Vietnamese MiGs in the air, was mistaken for the pilot Cunningham shot down. Radio intercepts are often difficult to interpret. One need only recall the decades of confusion from the intercepts made during the shooting down of Powers's U-2.

In the end, all that matters is that the two best pilots met in the sky that day. They flew their aircraft to their limits, and Cunningham won—thanks to a Black program.

U.S. MiG OPERATIONS IN THE 1970s

With the end of the Vietnam War, U.S. MiG operations seem to have undergone some changes. The navy's Have Drill program was ended; the air force now had sole control of the MiGs.[45] Top Gun and Red Flag had their own "Aggressor units," which used Soviet-style tactics. Top Gun used A-4s to simulate MiG 17s, while F-5s stood in for MiG 21s. They did not need the special security of the MiGs, and the flight characteristics of the aircraft were close enough for training.

The MiGs still had a role, however. During the early 1970s, the prototypes of the F-14, F-15, and F-16 fighters made their first flights. This new generation of U.S. fighters was designed with the lessons of Vietnam in mind. They were much more agile than the F-4; they also packed a 20mm cannon—something the original F-4 lacked. The MiGs might have been used in tests of these new fighters. This would allow development of tactics for the new fighters, much as the Have Drill tests were used to develop F-4 tactics. The orientation flights for senior officers also continued.

It appeared that the U.S. MiG operation grew to squadron size. During the Vietnam War period, it appears that there were two MiG 17s and two MiG 21s (the two captured Syrian MiG 17s, the original MiG 21, and 007). At some point, according to reports, four of the captured Algerian MiG 21s were also sent by Israel to the United States.[46] If true, this would raise the total to eight MiGs (two MiG 17s and six MiG 21s). The MiG operations were based at the former A-12 test hangars at the north end of the Groom Lake complex.

With a squadron-sized force, there was also need for a formal unit to operate them. This was the 4477th Test and Evaluation Squadron, also known as the "Red Eagles." Its unit patch featured a red-eagle design with a white or black star.[47] The MiGs had less flamboyant markings—the MiG 21s were in bare metal with only the U.S. star and bar insignia on the nose.[48]

Part of the attraction of working on a Black airplane project is the challenge. The U-2 and A-12 pushed the limits of aviation technology. The MiGs, in contrast, were ordinary operational fighters. Still, they posed a challenge—the challenge of keeping them flying. The simple and cheap design of the MiGs helped in this regard. It was reported that the MiG 17s were flown 255 times without a "down."[49]

Still, high-performance jet aircraft require regular maintenance. This means a supply of spare parts. Soviet jet engines were known for having a short operating lifetime. Tires were also a problem—they were good for only a few landings before replacement. It was reported that one of the reasons U.S. MiG operations were kept so secret was the problem of clandestine parts support (especially engines). The work done on the MiGs was described as "first rate."

The quality of the U.S. MiG pilots was also high. During the spring of 1977, a MiG 21 reportedly suffered a landing gear problem. The air force pilot was able to keep the skid under control and fairly straight. This limited major damage to the landing gear and the engine. The MiG was restored to operational status within a few days. Such high-quality flying and maintenance were critical. The MiGs were irreplaceable assets.

As during the Have Drill period, control of the MiGs was a source of conflict. During 1977, the air force and CIA fought over the planes. It was reported that the air force was barely able to retain custody of the planes and responsibility for their operation. It may have been that the value of the MiGs for training and evaluation outweighed any use the CIA may have had for the planes. It is worth noting that the Have Blue project was nearing its first flight. The Have Blue was tested against Soviet radar; it would be probable that the MiGs would also be used.

In 1977 there was a second leak about the existence of the U.S.-operated MiGs. The September 1977 issue of *Armed Forces Journal* carried a two-page article titled, "Soviet Jets in USAF Use." It began, "The United States has been flying a squadron of stolen or captured Soviet-built fighters for years. Some of the planes are very recent models; others date back to the Korean War. By one informal estimate, close to 20 MiGs are now in USAF service. Another source says that the U.S. has managed to obtain more than 25 planes, but has been able to keep only about five in flying condition on a regular basis." The article reported that the MiGs were being flown from "at least one base" in the United States to perfect air-to-air tactics. (The name Groom Lake had not yet been published.)

The magazine said it had been about to publish the MiG article several months before but had held off because of an "informal" request by a "senior official." *Armed Forces Journal* subsequently learned that the existence of the MiGs, as well as some test data from joint air force-navy tests, had inadvertently been compromised to a potentially unfriendly nation.

The article noted that estimates varied on the total number of MiGs the United States had. One source said it was about twenty, while another said it was about half that, "say eight, ten, or maybe twelve." Still another source, who said he had checked with "someone who knows," said that twenty "is high by a factor of four or five" (i.e., there were only four or five MiGs). In retrospect, the lower end of the range was more accurate.

The article also estimated that about one hundred U.S. pilots had flown the MiGs, while three or four hundred pilots were given the opportunity to fly against them. The article noted that this included pilots from the navy and Marine Corps. Again, these numbers seem high. Assuming four to eight MiGs operational at a given time, six to ten pilots would be involved (one and a half pilots to one plane). The Have Drill flights were limited to Top Gun instructors, a few selected fleet pilots, and senior officers like Gen. Marion Carl. A more probable guess would be around seventy-five pilots involved with the MiG operations in one way or another by 1977.

On a more whimsical note, the article asked, "Does USAF's secret air force explain a lot of verified, but unexplained UFO (Unidentified Flying Object) sightings?"[50]

THE EGYPTIAN MiGS

As the 1970s neared their tumultuous end, U.S. MiG operations were undergoing another change. In the late 1960s, the MiG 17 and MiG 21F were still frontline aircraft. A decade later, they had been superseded by later-model MiG 21s and new aircraft, such as the MiG 23. Fortunately, a new source of supply was available—Egypt.

In the mid-1970s, relations between Egypt and the Soviet Union had become strained, and Soviet advisers were ordered out. The Soviets had provided the Egyptian air force with MiGs since the mid-1950s. Now, with their traditional source out of the picture, the Egyptians began looking west. They turned to U.S. companies for parts to support their late-model MiG 21s and MiG 23s. Very soon, a deal was made.

According to one account, two MiG 23 fighter bombers were given to the United States by Egyptian president Anwar Sadat. The planes were disassembled and shipped from Egypt to Edwards Air Force Base. They were then transferred to Groom Lake for reassembly and study. [51]

The MiG 23 Flogger had a very different design philosophy than previous Soviet jet fighters. The MiG 17 and MiG 21 were small, short-range, tight-turning, point-defense fighters. The MiG 23 was more akin to multirole Western fighters. For one thing, it used variable sweep wings. They could be pivoted forward for takeoff and landing, then swept back for high-speed flight. The aircraft was larger and heavier than earlier MiGs. It was also faster, had a longer range (some three times the MiG 17's range and half again that of the MiG 21), and a heavier weapons load (four air-to-air missiles versus two for the MiG 21). The MiG 23 could function both as a fighter and a bomber.[52]

Unlike earlier Soviet fighters, the MiG 23 had a high-wing loading and thus poor maneuverability. In fact, the F-4 was more maneuverable than the MiG 23. The U.S. fighter closest in performance was the F-104. (Both planes went like a bullet and turned like one.) Like the F-104, the MiG 23 was a demanding aircraft; it had a high accident rate in Soviet and Warsaw

Pact service. On the plus side, the MiG 23 had a radar comparable to that in the F-4E. (Radars had been a weak spot in earlier MiGs.)[53]

According to some reports, the U.S. MiGs also received special designations. There was the practical problem of what to call the aircraft. This was solved by giving them numbers in the century series. The MiG 21s were called the "YF-110" (the original designation for the air force F-4C), while the MiG 23s were called the "YF-113."[54]

With the MiG 23s, operations expanded. Up to 1978, North Base at Edwards Air Force Base was largely inactive. The hangars were used for storage by NASA. Soon after, new security arrangements were put into place. It now appears they were in support of the MiG operations. It also appears that the MiG squadron was reorganized. The 4477th TES was replaced by the 413th Test Squadron (Special Operations) at some point in the late 1970s–early 1980s. The unit patch showed a Russian bear wearing a red hat, six red stars, and the slogans "Red Hats" and "More With Less."[55]

U.S. MiG 23 TRAINING

The would-be U.S. MiG 23 pilot underwent a three-phase training program. Phase I involved both ground training and six flights. Ground training covered review of the flight manual (general aircraft information, engine fuel system, electrical system, auxiliary equipment, operating limitations, flight characteristics, stall-spin characteristics, system operations, normal and emergency procedures, and performance data). The pilot then underwent ejection seat training, three hours of cockpit training, and briefings on the local area traffic patterns, restricted areas, and navigation aids. The U.S. MiG 23 pilot would then undergo written proficiency and emergency exams and an oral emergency exam.

In preparation for flying the MiG 23, the pilot would make a supervised engine start and then a high-speed taxi. The six flights were broken down into familiarization (TR-1, 2, and 3), MiG systems (TR-4, and 5), and a qualification check (TR-6).

Phase II was mission qualification training. The pilot would have to demonstrate his knowledge of flight test techniques and effective aircraft-systems handling. This demonstration could be made during or after the TR-6 flight.

Phase III was continuation training, in which the pilot was required to demonstrate mission qualification events, approach, and a normal landing. This phase underlined the low flight rates of the U.S.-operated MiGs. Only one flight every forty-five days was required. Clearly, the MiG program was more akin to the flight rates of the X-planes, rather than the day-in, day-out operations of a regular fighter unit. If a pilot did not make the minimum of one flight in forty-five days, he would have to be recertified. Depending on the time elapsed, this would vary from reviewing the exams and procedures, then making a flight, up to undergoing nearly the complete training cycle,

A U.S. MiG 23 pilot could also be upgraded to instructor following a single flight to demonstrate mission qualifications, instructional capability, and situational awareness. All the pilots had to complete open- and closed-book exams and make an annual qualification check flight.[56]

LEAKS

Coinciding with the arrival of the Egyptian MiGs, there were several leaks about both the MiGs and Groom Lake. In September 1978, a man named John Lear took a panoramic photograph of the Groom Lake facility from public land at the north end of the lake bed. When enlarged, it showed a MiG 21 on the parking apron. It appears to be one of the later model MiG 21s, with a broader tail and larger dorsal hump than the MiG 21F. (The photo was not published until 1991.)[57]

The following year came the notorious *Las Vegas Review-Journal* article on Groom Lake activities. In addition to naming Groom Lake as the location of the test site and revealing the existence of the stealth test aircraft (while getting the details wrong), it also stated that "three unrelated sources" had said two MiG 23s had been provided by Egypt.[58] During the early 1980s, several different photos were published of U.S.-operated MiG 21Fs.[59]

At the same time, the U.S. MiG squadron grew considerably. According to one account, there were about twenty MiG 21s and four MiG 23s in service by the mid-1980s. They operated not only from Groom Lake and North Base at Edwards, but also from the Tonopah Test Range,[60] during this same time as the first F-117As were becoming operational at TTR. Over all, U.S. Black airplane activities in the early 1980s rivaled those of the 1960s. Not only the MiGs, but also the F-117A, HALSOL, and possibly another Black airplane were undergoing flight tests.

The MiGs operated by the United States during this time were "new" aircraft. The original group of MiGs had been retired. The two MiG 17s and 007 had been returned to Israel in the mid-1970s. One of the MiG 17s and 007 were placed on display at the Israeli Air Force Museum. The other U.S.-operated MiG 21s were apparently retired, stored, stripped of useful parts, or scrapped. One of the MiG 21s was later used by the navy as an RCS test article.[61]

Despite the leaks, only minimal attention had been drawn to U.S. MiG operations. Then, in the spring of 1984, a crash would make it front-page news.

THE DEATH OF GENERAL BOND

At 10:18 A.M. on April 26, 1984, a plane crashed on the Nellis Air Force Base range. Witnesses at a cafe in nearby Lathrop Wells reported hearing an explosion and seeing smoke in the area of Little Skull Mountain.[62] A few hours later, the Air Force Systems Command at Andrews Air Force Base outside Washington, D.C., issued a brief statement: "Lt. Gen. Robert M. Bond, vice commander, Air Force Systems Command, was killed today in an accident while flying in an Air Force specially modified test aircraft."[63] General Bond was a thirty-three-year veteran of the air force, had flown in both the Korean and Vietnam Wars, and had more than five thousand hours of flight time. He was alone in the aircraft when it crashed.

Three-star generals do not generally fly test missions, so Bond's death attracted press interest. The fact that the air force also refused to identify the type of plane also raised questions. Early reports claimed he had been flying "a super-secret Stealth fighter prototype."

Within a week, stories were published that it actually was a MiG 23. It was said that the MiG 23 was used to test the Stealth fighter in simulated combat and that Bond might have been involved in such tests at the time of the crash.[64] Still later, it would be claimed that the crash was caused by a loss of control at high speed.[65]

The flight was the second of two orientation flights for General Bond. The mission was planned to include a high-speed run, followed by a systems-radar familiarization. Bond's aircraft was accompanied by a T-38 trainer as chase plane. Engine start, taxi, and takeoff were normal. The two planes climbed to 40,000 feet while ground control reported they had about fifteen minutes of airspace time. Ground control gave them the distance to the turn point. Reaching it, they turned right. The T-38 pilot told Bond to check his fuel and calibration. They had descended slightly to about 37,000 feet, and Bond climbed back to 40,000 feet.

Bond increased the throttle and began the speed run. Bond then reported to ground control that he had reached the planned speed without problems. The much slower T-38 was now some distance behind. Ground control radioed Bond that he was four miles from the next turn. At 10:17:50 A.M. Bond asked, "How far to the turn?" Ground control responded at 10:17:53 A.M. with "Turn now, right 020." Bond responded with two clicks of the radio.

At 10:18:02 A.M. Bond radioed, "I'm out of control. I'm out of . . ." Ground control informed the T-38 pilot that Bond was twenty-two miles away. At 10:18:23 A.M. Bond radioed, "I've got to get out, I'm out of control." Ground control warned the T-38 pilot that Bond's plane was nearly at the edge of the airspace. Soon after, radar contact was lost.

Bond had ejected, but he was killed.

An investigation was started, even as speculation about the accident grew. The plane had hit the ground in a high-speed, 60-degree dive and was destroyed. Three major sections were examined—the tail, the engine compressor and turbine blades, and the engine inlet. Examination of the debris showed that all the damage was due to ground impact. The engine was running normally, at a throttle setting of about 80 to 90 percent. There was no evidence of fire, an "overtemp," heat distress, or an engine stall. Checks of the fuel, hydraulic fluid, and lube oil showed no contamination or abnormal wear. The accident report concluded that the plane had crashed due to loss of control during high-altitude, supersonic flight.[66]

Although U.S. MiG operations continued after the death of Bond, its days were numbered. The TTR operations were closed down in the mid-1980s, apparently due to the growth of F-117A activities. It was reported that in late 1988 or early 1989, the MiGs were grounded. This was caused by the problem of getting spare parts. Most of the planes were placed in storage at North Base at Edwards Air Force Base. Others were described as being on display at Groom Lake.[67] It seemed that MiG operations had ended.

So it seemed.

A MiG OF ONE'S OWN

Ironically, as the U.S. MiG operations were ended, MiG 15s and MiG 17s began arriving on the civilian market. Communist China and the Eastern European countries began selling the old MiGs to anyone with hard currency. By the early 1990s, supersonic MiG 19s and MiG 21s were for sale. A MiG 15 could be bought for $175,000 (a fraction the cost of a flyable P-51 Mustang) and operated for about $10,000 per year. This made it practical to fly the planes for air shows and movie work.[68] For anyone who grew up during the Cold War, it was a strange and delightful experience to see a former Communist-bloc MiG sitting on the ramp of a U.S. airport. As one aviation magazine put it, "Watching an American citizen strap into a MiG 15 is a lot like watching Captain Kirk flying a Klingon battle cruiser."[69]

Among those who took advantage of the privately owned MiGs was the U.S. Air Force Test Pilot School. In 1992, students at the school were given the opportunity to fly in a MiG 21U Mongol (the two-seat trainer). One student found it responsive, given its age. He judged the engine very good. The flights were short, on the order of twenty minutes in length (understandable, given it was designed as a "manned SAM" for point defense of targets).[70] Two years later, a MiG 15UTI Midget trainer was used at the school for student flights.[71]

There was also an attempt to organize a White version of the MiG operations. In 1988, Combat Core Certification Professionals imported from Poland four MiG 15s, a MiG 15UTI, and six MiG 17s for the Defense Test and Support Evaluation Agency (DTESA). They were intended to be used in air-combat training for both U.S. and Allied forces. One such exercise was reportedly held at Kirtland Air Force Base, New Mexico, in September 1988. The program ran into a legal problem, however—the MiGs had been acquired through "sole source procurement." This was a violation of Department of Defense policy. As rules were judged more important than results, the MiGs were put into storage at Kirtland until the legal problems could be sorted out. Soon after, the U.S. MiG program ended. [72]

FADE TO BLACK

The most significant information to be revealed about U.S. MiG operations was published in the 1990 book *Scream of Eagles.* Although primarily concerned with U.S. Navy air combat over North Vietnam and the founding of the Top Gun School, it did include material on the MiGs. It revealed the acquisition of the MiG 21 in 1967, the Have Doughnut tests, the Have Drill activities, and the film "Throw a Nickel on the Grass." The account was based on interviews with the navy pilots who had taken part in the tests.[73]

The following year, a few of the MiG 21s came out of storage. The unveiling was as clandestine as the planes' original acquisition. In March 1991, the Strategic Air Museum at Offutt Air Force Base received a call from the Air Force Museum at Wright-Patterson Air Force Base. They were told that a new airplane would be arriving soon. The caller did not identify the type of plane nor any details. The next morning, when the SAC Museum employees arrived at work, they discovered four large crates had been dropped off by their back fence. It literally had been done in the middle of the night. When they opened the crates, they found a MiG 21F.

When it was put on display a year later, questions arose about its source. A civilian employee of air force intelligence said that the U.S. government had agreed never to divulge the source of such planes, even years after the deal was made. The director of the museum, Jim Bert said, "Officially, the Air Force neither confirms nor denies the existence of that aircraft."[74]

Another MiG 21F was given to the National Air and Space Museum. It was explained that it had been used in a classified display of Soviet weapons, and that, with the end of the Cold War, the display was dismantled.[75] Still another MiG 21F was put on display at the USAF Armament Museum at Eglin Air Force Base. A close examination showed the plane had Chinese characters inside some of the access panels. It is actually a J-7, a Chinese-built version of the MiG 21F.[76] The final example was a MiG 21U trainer on display at Wright-Patterson. It had been used to train intelligence officers in determining the capabilities, performance, and technology of enemy aircraft.[77] The DTESA MiG 15s and MiG 17s were also loaned to museums in 1992. The Pima Air Museum received a MiG 15, MiG 15UTI, and a MiG 17. None of the later-model MiG 21s (understood to be in storage at North Base) nor any MiG 23s were released.[78]

Although some of the U.S.-operated MiGs were now on public display, they were still Black airplanes, in the darkest shades. When the author filed a Freedom of Information Act (FOIA) Request for the unit history of the 4477th Test and Evaluation Squadron, the air force responded: "It has been determined that the fact of the existence or nonexistence of records which would reveal a connection or interest in the matters related to those set forth in your request is classified. . . . By this statement, the United States Air Force neither confirms nor denies that such records may or may not exist."[79]

It is clear that the air force had created a security "firebreak" around the MiGs. The fact the air force had some MiGs was not considered sensitive. How the MiGs came into U.S. possession, or any records connected in any way with this, were out of bounds. In some fifteen years of filing FOIA requests, this was the first time the author had run across information considered this sensitive.

Another reason the air force was so protective of a project that seemingly had ended became apparent in March 1994. In an article on Groom Lake in *Popular Science,* a photo was published of an Su 22 Fitter in flight. The plane was painted in a green and tan finish. The Su 22 is a swing-wing, light-attack aircraft. It is currently in frontline Russian air force service and has been exported widely to Eastern European and Third-World countries. The article also said that MiG 23s had been seen flying above Groom Lake.[80]

U.S. Air Force MiG operations had resumed in 1993, when Germany exported nine MiG 23s and two Su 22s to the United States. With East and West Germany now unified, there was an ample supply of both Soviet-built planes and the spare parts needed to support them.

There were also suggestions that operations were not limited to MiG 23s and Su 22s. In October 1994, *Aerospace Daily* reported that "reliable observers" had sighted an Su 27 Flanker on two occasions. The Su 27 is the Russian's most advanced interceptor. It is in operation with both the Russian and Communist Chinese air forces. The first sighting took place in late August near Lake Tahoe, while the second occurred near Yosemite National Park in September. In both cases, an F-117A had been seen a few minutes before the Su 27. The F-117A was flying in the opposite direction at about the same altitude as the Su 27.[81] It has also been suggested that the sightings were actually of F-15s, which resemble the Su 27.

While U.S. Air Force operation of Soviet-built aircraft is still sensitive, the U.S. Army use of ex-Soviet helicopters is not. These include Mi 2 Hoplites, Mi 8/17 Hips, Mi 24 Hinds, Mi 14 Haze, and a Ka 32 Helix. The helicopters are flown out of Fort Bliss and Fort Polk and are used for training. The pilots were particularly impressed with the Ka 32, due to its lifting power. Although their existence is not a secret, some things remain classified. A photographer was not allowed to go inside the helicopters, as this would indicate the country of origin.[82]

According to one account, MiG operations are conducted under the name "Special Evaluation Program." During the early 1990s, this was budgeted at only a few million dollars per year. In 1993, this jumped to $336 million. The budget in 1993 for "Foreign Material Acquisition" was $500 million. This, however, covered all such activities, not just purchasing the MiGs.[83]

THE MiG CRASH AT RACHEL—A CAUTIONARY TALE

In reconstructing the history of Black airplane programs that are still secret, such as the MiG operations, where reports are many, but the confirmed facts are few, one must take care not to be led astray. One observer noted, "Those who need to know, know; the trouble is that those who don't know are the ones doing the talking."[84] One need only recall the confusion over the D-21 and F-19 to understand the truth of this statement.

One such cautionary tale is the report of a MiG crash that occurred near the town of Rachel, Nevada. In 1994, an article on Groom Lake was published in the *Las Vegas Review-Journal.* It claimed that three MiGs had crashed, "including one that landed in a woman's backyard in Rachel."[85]

The possibility that debris from a crashed MiG was in someone's backyard attracted attention. An individual went to Rachel and was able to track down the crash site. It was located next to a dirt road and near a trailer. When he asked the owner how she knew what type of plane it was, she said that "the word around town was that it was a MiG." On this note, he looked at the impact point. It had been cleaned up very well; the only visible debris was tiny bits of aluminum. He got out his rake and started "farming aluminum." The rake was soon pulling up compressor blades, hose clamps, and a data plate. These had part numbers, contract numbers, and inspection stamps. The way the debris was spread out was also consistent with a plane crash.

All this proved conclusively that the "MiG" had been built at General Dynamics Fort Worth. It was an F-16!

Later, he found a person in Rachel who had kept a scrapbook of newspaper articles about the town. The crash had occurred on July 10, 1986, (the day before the loss of Maj. Ross E. Mulhare's F-117A). A Royal Norwegian

Air Force F-16 at a Red Flag exercise clipped another plane, and the pilot was forced to eject. Several people in town saw the collision. The F-16 impacted about seventy-five yards from two different trailers and three hundred yards from the town's gas station. The Norwegian pilot landed safely by parachute and was picked up by one of the townspeople.

The local sheriff's office was called and told of the accident and that the pilot was safe. (The wingman's plane was not badly damaged, and he was able to return to Nellis.) No property on the ground was damaged, and the fire was allowed to burn itself out. The sheriff, his deputies, the paramedics from Alamo, and the Lincoln County SWAT team all showed up. Helicopters brought in an air force recovery crew. Because of the possibility of toxic fumes from the burning metal, it was decided that people living in the immediate area should go to the Stage Stop Saloon.

The next day, the locals set up a picnic for the air force recovery crew. A Major Flynn talked to the townspeople about the crash. Channel 8 from Las Vegas came out to film the crash site, and the *Las Vegas Review-Journal* called several people for interviews. On July 12, a truck came and carried away the remains of the F-16. Letters of appreciation were sent by the Nellis base commander and the Norwegian pilot.[86]

Over the subsequent years, the story had gotten "better." The facts had been forgotten, and the myth started to grow. The F-16 was transformed into a "MiG." Then the myth was printed, and the myth became "fact."

Beware.

Despite the secrecy that still surrounds MiG operations, there is no doubt that the United States has owned and flown MiGs. The public displays and photos of the MiGs in flight are proof enough. But stories are also told of another Black airplane. There are no photos of this Dark Eagle, no confirmation of its existence. There are only vague stories of a plane whose shape seems to change with every telling.

It is called "Shamu."

Chapter 11

Still Black
The Enigmatic Shamu

In such a case I must be deep and subtle. Then I can assess the truth or falsity of the . . . statements and discriminate between what is substantial and what is not.

Sun Tzu
ca. 400 B.C.

By the mid-1980s, with stories of the stealth fighter increasing in frequency, suspicions began to grow that there were other Black airplanes. A small technology demonstrator like the Have Blue could be built at low cost and in a short time. Speculation grew that there was a kind of Black X-plane program, where different stealth configurations could be tested—one- or two-of-a-kind prototypes or aircraft with a very limited production run. Reports, stories, and sightings of still-secret Dark Eagles began to appear. Many of these dealt with the enigmatic "Shamu."

SHAMU

The story of Shamu began with the June 1, 1981, issue of *Aviation Week and Space Technology*. The magazine carried a report that a fighter-sized stealth aircraft built by Northrop was to make its first flight "soon."[1]

Northrop had lost the competition to build the Have Blue, but interest in stealth aircraft was not limited to attack aircraft. A stealth strategic bomber could render the huge Soviet air-defense network useless. As support for a "stealth bomber" grew in 1980–81, Northrop soon emerged as the leading contender. In the 1940s, John K. Northrop had designed the XB-35 and YB-49 flying wings. Perhaps the most graceful bombers ever flown, they were beset with a number of problems that prevented them from entering service. The XB-35's propeller system was redesigned several times but never proved reliable. With the emergence of jets, a propeller-driven bomber was considered obsolete. The YB-49 was an all-jet conversion, but it suffered from short range and instability. The second prototype crashed during a stall check, killing the crew (including Capt. Glen Edwards, for whom Edwards Air Force Base is named).

Before the program ended, the YB-49 made several flights against a coastal radar site at Half Moon Bay, California. The aircraft proved hard to detect.[2] Even given the primitive state of radar technology in the late 1940s, it was clear that a large flying wing had good stealth properties. With RAM and the proper shape, a flying-wing stealth bomber could be as hard to detect as the Have Blue.

Such an aircraft would use a different design philosophy. Both the Lockheed and Northrop Have Blue designs used faceting, in order to make RCS calculations easier. The Northrop stealth bomber would use "smooth" stealth. Given the limitations of computers in the late 1970s, this would be much harder to design.

It was therefore *assumed* the "Northrop" aircraft described in the *Aviation Week* report was a Have Blue–like technology demonstrator, built to test smooth stealth. It could also insure the stability system would overcome the control and stability problems that had doomed the YB-49.

SIGHTINGS AND REPORTS

Several sightings were made in the 1980s of triangular aircraft from areas near Groom Lake. People who worked at various sites in Nevada said that such a triangular shape was quite familiar and "has been around a long time."[3] It was assumed these were sightings of the Northrop test aircraft.

Another sighting reportedly occurred in 1986. Despite the Groom Mountain land seizure, Greenpeace continued efforts to infiltrate the nuclear test site. During one such attempt, a group from Greenpeace was trespassing near the Groom Lake area. They reported seeing a black, triangular-shaped aircraft flying slowly overhead. The "group's hair stood on end" at the sight of the plane. Years later, several of the group said they thought it was a subscale flying prototype of the B-2.[4]

In 1988, an article claimed that Northrop had built three of the scaled-down prototypes for its stealth bomber. They had been flying since 1981.[5]

It was not until 1990 that the Northrop aircraft received its name. In a book on the F-117A, it was stated that the project was under way in the early 1980s, the same time as the FSD tests. The plane was described as "a Northrop Stealth prototype" which was "a demonstrator for the 'seamless' design philosophy." Lockheed engineers nicknamed the plane Shamu, because it resembled the killer whale at Sea World.

According to one story, the Lockheed engineers were initially not cleared to see the Northrop aircraft, while the Northrop personnel were not to see the Lockheed F-117A FSD aircraft. The result was that each group had to remain indoors whenever the other plane was in the air. This disrupted operations for both programs, and each group was eventually cleared to see the other's airplane.[6]

Again, the report implied Shamu was a flying-wing design. Seen from the side, the B-2 fuselage does look whalelike. According to some stories, Shamu was 60 percent the size of the B-2 (the same scale as the Have Blue). This would indicate a wingspan of 103 feet and a length of 41 feet. It was also said that Shamu could fit sideways inside a C-5 transport. The Northrop aircraft is understood to have operated from a hangar at the south end of Groom Lake, separate from those used by Lockheed. (It also housed the T-38 chase planes.)[7]

Yet, there were inconsistencies. If Shamu really had a wingspan of 100-plus feet, it was rather large for a subscale test aircraft, and far larger than the "fighter-sized" plane *Aviation Week and Space Technology* had described. If it was this large, the claim it could fit inside a C-5 is in error—the interior volume of a C-5 is 121 feet long but only 19 feet wide.[8] Additionally, Northrop personnel and the air force have repeatedly denied that a subscale test aircraft for the B-2 was built.[9]

"TR-3A BLACK MANTA"

It was believed by some that Shamu had given rise to a family of flying-wing Black airplanes. Over the years, a number of different reports circulated about triangular or manta-ray-shaped Black airplanes. The smaller of these had a wingspan of about 60 feet. It was described as having a rounded nose and wingtips, with a trailing wing edge that was slightly curved, rather than the W shape of the F-117A.

The sightings began in mid-1989, when daylight operations of the F-117A started. The normal pattern was for several F-117As to fly the same route, separated by eight to ten minutes. Observers would report that three or four F-117As would pass, then the triangular plane, then three or four more F-117As. The plane was described as both larger and quieter than the F-117As. It showed a similar lighting pattern to the F-117As—amber lights at the wingtips and a red light near the nose.

There were also sightings of the triangular plane on moonlit nights. Several such sightings occurred on the night of May 3, 1990. Five different observers reported seeing it over Mojave, Lancaster, Palmdale, and Tehachapi, California, (near Edwards Air Force Base) during a four-hour period.[10]

The triangular plane was soon given the name "TR-3A Black Manta." As the "history" of the program was reconstructed, it is claimed that Northrop was awarded a contract in late 1978 to develop a stealth test aircraft. It was a flying-wing design called the tactical high altitude penetrator (THAP).

The story continues that the test aircraft made its first flight in 1981 (a reference to the 1981 *Aviation Week* report and the Shamu story). In 1982, Northrop is described as receiving a contract to build a production aircraft, based on the test aircraft. This, the stories continued, became the TR-3A.[11]

The TR-3A is described as a tactical reconnaissance aircraft. It was speculated its reconnaissance data and target information were relayed directly to the F-117As in near real time. It was also claimed the TR-3A could designate targets for the F-117As, that is, illuminate them with a laser so the F-117A's LGB could strike. Reports said it was likely that the plane was operational and had been used in the Gulf War.[12]

THE BIG TRIANGLES

The TR-3A was not the only triangular aircraft reported. Months before the B-2 made its first flight, sightings were made of a large flying-wing-shaped aircraft. Later, sightings were made of this aircraft on moonlit nights while the B-2 was grounded. The aircraft was described as having a wingspan of 150 feet. (This is similar to the descriptions of the large Shamu.) It is also described as being highly maneuverable. In one case, it was claimed that the aircraft turned 90 degrees on its wingtip. It was speculated that this was a prototype B-2, a technology demonstrator for the B-2, or a one-of-a-kind experimental prototype.[13]

Even more remarkable are claims of a huge black boomerang-shaped flying wing that moves silently at speeds as low as 20 knots. It is described as being 600 to 800 feet in wingspan. Several Antelope Valley residents have reported seeing it slowly moving through the night sky. One said it was moving so slowly that he could jog along with it. The big triangle is also described as making such "unlikely maneuvers" as coming to a full stop, tipping upright, and hovering in this vertical position. It is also claimed that a pattern of small white lights on the triangle's black underside provide "constellation camouflage" against the starry night sky.

Although noting this was "a craft that simply strains credulity," and that "such sightings encourage those who link the military with unearthly technology," some suggest that it could be a lighter-than-air craft propelled by slowly turning propellers. The vertical hovering suggests to some that it acts as a reflector for a bistatic radar system. In this system, a transmitter emits a "fence" of radar signals, which is reflected back to a receiver. If a plane crosses the fence, it would be detected. Another possibility put forth is that the aircraft is used for troop transport or covert surveillance.[14]

By late 1993, two new triangular Black airplanes began to be talked about. One was the Northrop "F-121 Sentinel." It is described as being an equilateral triangle with the tips flattened. Its three sloping sides met at a point, giving it the appearance of a pyramid. At the center of the F-121's underside was a depression and a dome-shaped nozzle. It is claimed that a squadron of the planes are based at the Northrop RCS facility at Tehachapi. The site had no runway, but the planes take off and land vertically from the underground bunkers.[15]

The other new Black airplane is called the "Artichoke." The plane is described as similar to the F-117A in shape; but at least 20 percent bigger, with a two-man crew and a larger bomb load. The rear of the plane has several spikes, giving it the appearance of an artichoke. It is further claimed that the design was tested with a subscale flying model about the size of the Have Blue. The Artichoke is claimed to be based at TTR. Despite the move of the F-117As to Holloman Air Force Base, security has been beefed up at TTR. It is suggested the Artichoke is the F–117A's replacement. The F–117As would, in turn, be modified for Wild Weasel SAM–hunting missions.[16]

Despite the sightings, these stories of large triangular Black airplanes received, at best, a limited following. Shamu, on the other hand, seemed more likely to be real. There were the eyewitness accounts of both the Lockheed engineers and Greenpeace protestors. It seemed likely there was a flying-wing test bed, which had pioneered the technology for the B-2 and the "TR-3A."

Then, suddenly, Shamu changed shape.

THE "NEW SHAMU"

In early 1994, all the assumptions about Shamu were challenged. Rather than a flying-wing design, the "new" version of the Shamu story held that it was a Northrop design for a one-of-a-kind electronic warfare aircraft. The aircraft was large and has been described as "conventional"—one source likened it to the Boeing Stratocruiser airliner of the 1950s.

Shamu's fuselage was now described as rounded and bulbous. Like the Stratocruiser, the cockpit windows wrapped around the front of the fuselage. The side of the fuselage was described as flat and covered with antennae for the jamming equipment. It also had a very small vertical fin. This gave it the whalelike appearance. The wings were said to be slightly swept and were set low on the fuselage. The plane's four jet engines were buried within the wings (similar to the Comet airliner). According to another (allegedly) "reliable source," Shamu *looked funny.* People actually laughed when they saw it for the first time. It was likened to an unnamed cartoon character. (Security prevented saying which one.) The timescale of the project also changed. It now was reported to have been under way in 1977–79. This was the same time as the Have Blue flight tests.[17]

TACIT BLUE—THE TALE OF THE LONELY WHALE

In the months that followed the hardcover publication of this book, "Shamu" was still believed to be a flying wing. By the spring of 1996, there was even a description—a small flying wing with a transparent windshield in the leading edge for the pilot and two inwardly curved fins, code-named "Tacit Blue." By late April, reports began to circulate that the air force would soon unveil a "fighter-sized" Black airplane. On April 30, Tacit Blue was finally unveiled. To the surprise of all but the author, it was "funny looking."

Tacit Blue began in 1978, when DARPA asked Northrop to develop a stealth reconnaissance aircraft. It would have to operate behind front lines and carry radar able to detect armored units. The data would be relayed in real time to a ground command post.

The plane's designers, John Cashen (who later designed the B-2) and Steve Smith faced a difficult task. The F-117A was designed to have a minimum radar return from the front and back. This was because it flew towards and away from a target. The Tacit Blue would fly in circles, and thus be exposed to radar signals from all directions. This required "all-aspect" stealth. They also had to fit a large radar inside an airframe that was about the same size as an F-15.

The design that emerged was a slope-sided box with straight wings, two angled square fins, and a shovel-like chine on the nose. The pilot's windshield wrapped around the front like that on the Stratocruiser airliner. (The unnamed cartoon character it resembled was Howard the Duck.) The plane was 55.8 feet long, had a wingspan of 48.2 feet, and weighed 30,000 pounds. As with the Have Blue, the Tacit Blue made use of existing components—F-5E landing gear, an F-15 ejector seat, and two Garrett ATF3-6 high-bypass turbofan jet engines from the Falcon 20 business jet. The engines were supplied with air from a flush inlet atop the fuselage, while a slot exhaust was between the fins.

As with the other stealth aircraft, the Tacit Blue was highly unstable. If allowed to "weathervane" in a wind tunnel, a model would actually fly backwards. A nose up or down attitude would cause it to roll over. A General Electric digital quadruple-redundant fly-by-wire control system was needed to make Tacit Blue controllable.

Northrop test pilot Dick Thomas was selected to make the first flight. Thomas, Cashen, and the rest of the team were keyed up the night before. When two beers didn't help to relax them, Thomas and Cashen played a one-on-one basketball game until they both dropped. The next day, February 5, 1982, Thomas made a successful flight.[18]

The Tacit Blue proved to have limited performance—an operating altitude of 25,000 to 33,000 feet, a speed of 250 knots, and an approach speed of 120 knots. The Hughes radar performed well. At first glance, use of radar violated a basic tenet of stealth—remain silent. It was designed, however, to use low power and techniques which made the signals seem to be only background noise. Flight test showed the inlet worked well, but crosswinds on the ground caused stalls during engine startups. The cockpit was also too wide for a single-seat plane—the pilot had to lean in different directions to see.[19]

Despite the success, it was soon clear the Tacit Blue would never enter production. It would have had to operate over the lines. Although invisible to radar, it could still be seen during daylight. In all, 135 flights were made; the last on February 14, 1985. Five pilots flew the plane—Dick Thomas, Lieutenant Colonels Ken Dyson, Russ Easter, and Don Cornell, and Maj. Dan Vanderhorst. The primary accomplishment of the program was to develop all-aspect stealth, later used on the B-2 and F-22/23 programs.[20]

As with other Black airplanes, Tacit Blue had its secret symbols. The walls at Northrop were adorned with pictures of whales, reflecting the secret plane's secret nickname of "The Whale."

While "Shamu" was real, the other triangular planes were not. The "TR-3A" sightings were of F-117As (which are delta-shaped). The "F-121" was described as taking off vertically with an anti-gravity engine.

The unveiling of Tacit Blue was a source of personal satisfaction. On Saturday, May 4, 1996, I was told that an individual found it suspicious that I had become interested in Black airplanes and UFOs, that my previous book, *Watch the Skies!*, had been published by Smithsonian Institution Press, that Dr. Carl Sagan had liked the book, and that I had been right about Tacit Blue. He concluded I was spreading government disinformation. It was my birthday, and this was the best gift I had ever received.[21]

At the Tacit Blue press conference, the name of another Black airplane came up. It is rumored to be the most remarkable flying machine ever built.

It is called "Aurora."

Chapter 12

Tales of Darkness and Shadows
The Illusive Aurora

. . . we listen carefully for distant sounds and screw up our eyes to see clearly.

Sun Tzu
ca. 400 B.C.

For the past several years, strange sounds have been heard coming from the skies above the western United States. These sounds are described as a "rumbling," akin to a small earthquake or like the sky is being ripped open. The source of these sounds is claimed to be the "Aurora"—a Mach 6 Black aircraft developed and flown in secrecy. The Aurora's speed is such that it could, according to the stories, fly from Washington, D.C., to Baghdad in ninety minutes. With its claimed top speed of 4,500 mph (six times the speed of sound, or 1.25 miles per *second*), the Aurora is seen as the epitome of a century of aerospace development. It is described as embodying an otherworldly technology.

These tales of darkness and shadows have been told in the technical press, weekly news magazines, and television news and entertainment programs. The air force has repeatedly denied the existence of the Aurora, yet the stories have continued to spread and grow. Ever more details have been added to the stories: what it looks like, what it smells like, and how its propulsion system works. The stories have also spread worldwide—sightings and "hearings" have been reported from a remote field in Scotland and on a small island in the Pacific. Aurora has also been used to raise questions about the development of such Black aircraft, and the role and necessity of such advanced weapons in a post–Cold War, one-superpower world.

There is just one little problem . . .

PRESENT AT THE CREATION

In August of 1976, the author was attending a convention of the Western Amateur Astronomers at Palomar Junior College. While there, a Caltech graduate student told the author that he had heard that Lockheed was developing a high-speed, high-altitude replacement for the SR-71 reconnaissance aircraft. That was the start.

In 1979, *Aviation Week and Space Technology* magazine published an article describing high-speed aircraft studies undertaken at the Skunk Works. By the 1990s, the article speculated, a Mach 4, 200,000-foot aircraft could be developed. The aircraft would relay its image and radar intelligence data via satellite in real time. It would be powered by an advanced turbo-ramjet engine—a J58 with a ramjet surrounding it. The airframe would be an aluminum-beryllium alloy. The aircraft would be delta-shaped, similar to the X-24B lifting body. The main problem was cost—the engine alone would cost $1 billion to develop. By the year 2000, Lockheed estimated it would be possible to build a Mach 7, 250,000-foot aircraft which used supersonic-cruise ramjet (scramjet) propulsion.[1]

At the same time, the air force was looking at advanced bomber and cruise missile designs. One area of study was penetration altitude. Since the mid-1960s, when the SA-2 SAM had chased them out of the stratosphere, bombers had gone in at low altitude. The A-12/SR-71 showed a high-speed and high-altitude profile could work. Some wanted to use a very high, very fast attack profile, harkening back to the B-12 studies. One possibility looked at was a hypersonic cruise missile. It would be capable of Mach 6 at altitudes of 150,000 to 200,000 feet.[2]

At this time, it was assumed that any future Black airplane would rely on ever higher speeds and altitudes to survive. It was not commonly understood that stealth and high speeds were incompatible.

Rumors of such high-speed projects continued to circulate during the 1970s and early 1980s. The only published account was a brief report that Lockheed had, by 1982, already flown a Mach 6 research aircraft.[3] This was also the time the Have Blue was flown, the existence of stealth was unveiled, and the stories about the stealth fighter grew.

In February 1985, the project gained a name. The Department of Defense issued a budget document, a declassified version of the P-1 weapons procurement document for fiscal year 1986. Under "Strategic Reconnaissance" there was an entry titled "Aurora." The project would receive $80.1 million in fiscal year 1986, and $2.272 billion in fiscal year 1987. This was an unusual amount—$80.1 million was a very small amount for an aircraft development program, while $2.272 billion was very large. Such a sudden growth was also remarkable. It was speculated that the Aurora entry should have been removed before the document was declassified. It was also speculated that Aurora was the B-2 or F-117A program. Soon, another possibility began to be discussed.[4]

On January 28, 1986, space shuttle *Challenger* was destroyed during launch, killing the seven crewmembers. In the wake of the tragedy, there was discussion of possible future shuttle replacements. These centered on single-stage-to-orbit vehicles. They would take off from a runway, then accelerate to Mach 25 speed and go into orbit. The technology necessary would be tested in the X-30 National Aero Space Plane program. Some remembered cases where White projects had Black counterparts. Examples included the White SR-71 and Black A-12 Oxcart. Another was the Hubble Space Telescope. This scientific instrument was similar in design to the Lockheed Big Bird photo reconnaissance satellite. If the X-30 was a White project, might the rumored superfast aircraft be its Black counterpart? Such a Black project, it was suspected, could have already cleared up some of the X-30's technological unknowns. It was also noted that the air force wished to retire the SR-71 fleet due to cost. It was presumed that an SR-71 replacement was in the wings.

AURORA

In early 1988, there were several articles on the alleged Aurora. On January 10, the *New York Times* reported that the air force was developing an aircraft able to reach Mach 5 (3,800 mph) and altitudes of over 100,000 feet. The aircraft would incorporate stealth technology to evade detection. The article quoted one official as saying, "With the SR-71, they know we're there but they can't touch us. With the new technology, they won't even know we're there." The progress of the development effort, when it would become operational, and the specifications could not, the article said, be determined. Nor could the contractor be determined, although it noted Lockheed was reportedly building the F-19 stealth fighter.[5]

That same month, *Armed Forces Journal* published a financial analysis that indicated Lockheed's income far exceeded that which could be explained by the C-5B, C-141, or C-130 transport aircraft, TR-1 reconnaissance aircraft, P-3 antisubmarine aircraft, F-117A, or YF-22 stealth aircraft. The article speculated this hidden income was from "something *very big*—perhaps a *very* black program *within* a black program." It was suggested that Lockheed was building a "super-stealth" replacement for the SR-71 and that flight tests had begun in 1987.[6]

It is worth noting that neither of these articles used the term Aurora as the name for the super-high-speed aircraft. That was done by the third article, which was published in the February 1988 *Gung-Ho* magazine. It described Aurora as being fueled by methane and capable of Mach 7 (5,000 mph) and 250,000-foot altitudes. The aircraft was described as having a three-man crew and as being operational since the mid-1980s. It was also claimed that inflight refueling was done by special KC-135Qs. Among other things, the article quoted one official as saying, "We are flight-testing vehicles that defy description. To compare them conceptually to the SR-71 would be like comparing Leonardo da Vinci's parachute design to the Space Shuttle." A retired colonel was quoted as saying, "We have things that are so far beyond the comprehension of the average aviation authority as to be really alien to our way of thinking." Finally, a retired Lockheed engineer was reported as saying, "Let's put it this way. We have things flying in the Nevada desert that would make George Lucas drool."[7]

In 1989, an eyewitness came forward to claim, among other things, that he had seen Aurora close up while working at Groom Lake. He described it as fueled by liquid methane and requiring the entire three-mile runway at Groom Lake to take off. During takeoff, he said, "it sounds like a continuous explosion." Aurora was described as able to reach speeds of Mach 10 and altitudes of 250,000 feet.[8] Aurora was later described "as an X-15 on steroids—fat and chunky with short stubby wings."[9]

In 1989 and 1990, a number of reports began to appear that described a very loud, deep rumbling engine noise, sometimes punctuated by a one-hertz "pulsing" sound. The aircraft also left a "sausage-link-shaped" contrail. The first reported sighting occurred in July 1989, at about 3:00 A.M., near Edwards Air Force Base. The "pulser," as it became known, was reported to be flying at medium altitude and visible as a "white glow."

On October 18, 1989, another "hearing" occurred during the early evening hours. The sound seemed to take off from North Base at Edwards. It was described as "extremely loud, with a deep, throaty rumble" which shook houses sixteen miles away. People came into the streets trying to locate the source of the sound. No light or glow was seen, but the roar continued for about five minutes; it seemed to be heading north and climbing into the sky. One witness said, "Your eyes tended to follow the noise; something was climbing at a very steep angle." Local residents said the sound "was like the sky ripping" and was unlike anything heard at Edwards for years. One witness compared it to the *Saturn 5* rocket engine tests of the 1960s.

A double sighting occurred on June 19, 1990, near Mojave, California, (in the Edwards area). The first occurred at 3:44 A.M., while the second was at 4:50 A.M. Both were headed to the northeast, and it was not clear if the sighting was of one aircraft twice or two different vehicles. There were eight separate reports of the "pulser" from Mojave. All occurred between midnight and 5:00 A.M., and all the objects were headed northeast.

Similar sightings were also reported from central Nevada during 1989–90. Again, it was a rumbling noise, with a one to two-hertz pulse rate and heard in the early morning hours. One Nevada sighting occurred on August 6, 1990. The aircraft was reported to have left the Groom Lake range and overflown a small town. A witness said it was "the loudest thing I've ever heard. It wasn't breaking the sound barrier, but it was rattling the windows!"[10]

FIRST ARTICLES

The sighting reports sparked the interest of the technical press, and *Aviation Week and Space Technology* began a series of articles on the subject. The first appeared in the December 18, 1989, issue, against the background of the fall of the Soviet Empire in Eastern Europe. Much of the article was a discussion of the role of Black development in the post–Cold War world. It did note the sighting reports and quoted "officials close to the program" as saying, "Aurora is so black, you won't see anything about it [in public] for 10 or 15 years."[11]

A pair of articles followed in the October 1, 1990, issue. These gave more details about the sightings and included artist conceptions of possible designs. They also noted that many in government were "extremely skeptical" of the reports. One official said he was confident that there was no such thing as a family of high-speed aircraft. The articles argued, however, that very few officials would be told of such a project.[12] The *Aviation Week and Space Technology* articles sparked a flurry of newspaper articles on Aurora during November and early December 1990.[13]

In its December 24, 1990, issue *Aviation Week and Space Technology* published an extremely detailed account of a "theoretical" hypersonic aircraft "which could be cruising the skies tonight." The vehicle was shaped like a flattened diamond or football—110 feet long and 60 feet wide. The fuselage's edges were rounded, and the aircraft's contours were likened to a skipping stone. All surfaces were covered with black ceramic tiles. The article said, "They have a scorched, heat streaked appearance, and seem to be coated with a crystalline patina indicative of sustained exposure to high temperature. A burnt-carbon odor emanates from the surface. The aft body tiles are distinctly more pockmarked and degraded than those on the forward half of the aircraft, as if they had experienced the most heat."

The aircraft was powered up to supersonic speeds by jet engines. The "external burning mechanism" then took over. The jet engines were shut down, and the inlet and exhaust ports were closed. Misted fuel was sprayed from ports at the midsection of the fuselage and ignited. The shape of the aft fuselage and the shock wave formed at the midpoint acted as a "nozzle." This external burning was used to propel it to Mach 6 to 8.

In the fuselage's underside, the article said, were ports for 121 nuclear weapons. The ports were covered with heat-tile-covered caps. The weapons were cone-shaped, like an ICBM's warhead, and sat in the ports nose down. The airplane must slow to subsonic speeds to drop the weapons. The outer cap was discarded, the weapon was ejected, and then a second tile-covered cap moved into position. The aircraft was unmanned—it was preprogrammed, but could also be controlled via a satellite or ground station. Such Black aircraft were given as "the reason the Iron Curtain fell."

All these articles were written against the background of Iraq's invasion of Kuwait, and the buildup of Allied forces. The *Aviation Week and Space Technology* article asked some pointed and angry questions:

> As Persian Gulf tensions continue into 1991, one must question whether the U.S. commander in chief and his defense secretary are fully aware of super-black weapon systems' potential. . . .
>
> Hard as it may be to fathom, there is reason to wonder whether complete knowledge of the most exotic aircraft may reach "The Top," all for super-security.
>
> One would like to think America's staggering black-world expenditures have yielded weaponry that could neutralize Iraq President Saddam Hussein's most valued military and political assets quickly. Some say that capability is in hand and could be used . . . if the right people choose to do so. If they do not, why not?
>
> If so, why are almost 400,000 U.S. and allied troops dug into the sand in Saudi Arabia, prepared to slug it out in a bloody ground war?
>
> . . . the lives of those troops are worthy of wider consideration.[14]

Following the flurry of reports in late 1990, there was a drop-off in articles during 1991. What was being published was little more than a rehash of the earlier articles. For its part, the air force denied that Aurora existed. Most statements were "no comment," but some were more direct. An air force public affairs officer stated, "We have no aircraft matching these descriptions."[15] A senior government official privately told the author that when the reports were published, they sparked a flap inside the U.S. government. An investigation was launched at a high level, which, despite diligent effort, could not discover such an aircraft. He officially stated that there was no such aircraft, as described in the articles.[16] Believers in Aurora were quick to point out the air force had also denied the existence of the F-19 stealth fighter.

RUMBLES IN THE MORNING

Early 1992 saw a flood of Aurora stories. These were sparked by four "brief rumbles" in the Los Angeles area over a ten-month period. All occurred between 6:30 and 7:30 A.M. on Thursdays. The specific dates were June 27, October 31, and November 21, 1991, and January 30, 1992. Believing they were earthquakes, radio station KFWB called Caltech's Seismo Lab for the location and magnitude of the quakes. Caltech seismologists looked at computer records from a network of sensors and determined no earthquakes had occurred.

Jim Mori, a research seismologist at Caltech, concluded that the rumble was actually a sonic boom. The seismic record of a sonic boom is different than a small earthquake, even though they have the same magnitude. As Mori explained, "Sonic boom records look like a short pop; earthquakes have a longer decay period." At first, Mori thought it might be from a meteor, but the regular occurrence, always between 6:30 and 7:30 A.M. and always on a Thursday, ruled this out. That left an airplane. Since there are no commercial supersonic aircraft, KFWB called local air force and navy bases. They denied any of their aircraft were flying at the time. This seemed to point toward Aurora.

In 1989, Mori had worked on a team studying seismograms recorded during space shuttle landings at Edwards Air Force Base. Based on this, some information about the mystery plane could be determined. It was possible to calculate how fast and how high the plane was flying, based on the pattern of the arrival times of the sonic boom across the seismic network. By looking at the entire network, it was possible to determine the flight path and direction. Using the experience of twelve shuttle landings and one SR-71 flight, Mori concluded:

> On three of the four days, the records showed two events, which means that there were two planes. They flew about one minute apart as they traveled across the seismic net from south to north. The planes traveled at about two to four times the speed of sound, at an altitude of 10,000 to 40,000 feet. From the frequency of the event on the seismogram, we think that the aircraft is smaller than the Shuttle. It appears that they did not land at Edwards Air Force Base, but kept going north to southern Nevada. The last time the planes flew, on January 30, they passed over the Los Angeles area about one-half hour before the Space Shuttle was due.[17]

Because the top end of the speed range was above that of conventional aircraft and because its flight path would take it toward Groom Lake, many assumed the mystery plane was Aurora. A fifth "rumble" occurred on April 16, 1992, (again a Thursday) at about 7:00 A.M. A public affairs director for the Tournament of Roses described it as "kind of a rumble—very short." A restaurant manager said, "It rattled the glass door. And there was a roar, a slight roar." By November 1993, a total of eight such rumbles had been heard.[18]

On April 20, 1992, the *NBC Nightly News* carried a feature on Aurora. This included details on the rumbles and on the aircrafts' flight path. There was a brief film clip taken from a site near Groom Lake of a light hovering in the night sky. This light was described as being an extremely maneuverable Black airplane.[19]

As Mori noted, "If they were really trying to keep this secret they wouldn't fly it over downtown L.A., over and over again." The seeming public display of Aurora indicated to some that a disclosure was imminent. The believers detected a "subtle language shift" in the air force denials. One spokesman said, "I have nothing for you on that." Steven Aftergood, an analyst with the Federation of American Scientists—which had long opposed new U.S. weapons programs—said about the air force statement, "it's a non-denial denial. Coming from the air force, it amounts to a confirmation." Aftergood continued, "I would say the triggering of earthquake sensors is a leak from the air force. This is a form of signaling to the people who watch these things that (they've) arrived." Just when the "unveiling" of Aurora would come was unclear to observers. Some thought it would occur in the spring of 1992, recalling Lyndon Johnson's A-11 and SR-71 announcements during the 1964 presidential campaign. (1992 was also an election year.) Others thought it would be within a year. (The F-117A was unveiled shortly after the 1988 election.) Others thought it still might be several years away.[20]

Another reason for the belief in an imminent disclosure was reports of an Aurora sighting in Scotland. It was claimed that in November 1991, a Royal Air Force air traffic controller had picked up a radar target leaving a NATO-RAF base at Machrihanish, Scotland. He tracked it at a speed of Mach 3. When he telephoned the base to ask what the plane was, he was told to forget what he had seen. Another witness reported hearing an extremely loud roar at the same time as the radar sighting.[21]

A number of English newspapers and magazines carried the story. It was claimed that Aurora took off from Groom Lake and headed west to a landing at Kwajalein Atoll in the Pacific. It then continued west and overflew Iraq, using "high-powered cameras and infrared radar" before continuing on to Machrihanish. After taking off, Aurora was refueled with liquid methane by tankers based in England. It then flew back to the United States. To hide the landings, an F-111 flew in close formation to confuse civilian radar. As the reports spread, Defence Minister Archie Hamilton told Parliament that the existence and operation of Aurora was a "matter for the American authorities." When the air force was asked, they did not confirm or deny its existence.[22]

The story crossed the Atlantic and the *Antelope Valley Press* carried a story on March 6, 1992. It included an estimate that the development cost ran into the tens of billions, with each Aurora in a twenty-plane fleet costing $1 billion apiece.[23] An editorial published on March 12 expressed the hope that the Aurora would produce high-paying jobs in an area hard hit by defense cutbacks.[24] The May 25, 1992, issue of *Time* picked up the Machrihanish story and carried a brief report in its "Grapevine" column. (Another item was a poll on the question of which presidential candidate had the best haircut.) It said the code name for Aurora was "Senior Citizen."[25]

Nor were sightings limited to the wilds of Scotland. On February 25, 26, and 27, 1992, there were nighttime sightings of an unknown aircraft with a "diamond-pattern" of lights at Beale Air Force Base (the former SR-71 base). On the first two nights, a KC-135Q took off at about 6:15 P.M., followed by the aircraft. It had a red light near the nose, two "whitish" lights at the wingtips, and an amber light at the tail. The aircraft had a distinctive engine noise, described as "a very, very low rumble, like air rushing through a big tube." The plane then joined the tankers in a close formation and extinguished its lights.

On one of these nights, at 9:30 P.M., two T-38 trainers took off with an unknown aircraft between them. The third aircraft did not turn on its lights until it was about three miles from the runway. It also showed a diamond pattern.

Finally, on February 27, a formation of a KC-135Q, two F-117As, and the unknown aircraft took off. The wing lights were described as about twice as far apart as those on the F-117s, and the length was about 50 percent longer than the F-117s.[26]

What was thought by some to be a ground test of the Aurora's engine was also reported. Late on the night of February 26 (the second night of the sightings) a series of "booms" was heard coming from the base. These occurred every two to three seconds and continued for around thirty minutes. They were described as "like artillery fire," and "deep bass notes, not like sonic booms." It was thought these might be "light-off" tests of the engine. It was speculated the plane used a pulse detonation wave engine (PDWE). The noise and low frequency would, it was said, be consistent with a PDWE. The "light-off" was thought to be the most difficult phase to control, and the sound may have been from technicians "trimming" an engine.[27]

CONTRAILS AND RADIO INTERCEPTS

Three months later, photos were published of a "doughnut-on-a-rope" contrail seen over Amarillo, Texas. It had long been reported that Aurora left such contrails, but this was the first time it had been photographed. The sighting took place at 8:30 A.M. on March 23, 1992. The person who took the photographs stated that he had heard a "strange, loud, pulsating roar . . . unique . . . a deep pulsating rumble that vibrated the house and made the windows vibrate." He added that the sound was "similar to rocket engine noise, but deeper, with evenly timed pulses."[28] It has been reported that the pulse rate was three hertz.[29] The photographer later said he talked with an engineer at Convair Fort Worth, who said the contrails were formed only when the PWDE was operating outside its design parameters. Another theory was that they became visible only when the aircraft descended out of the thin air of high altitudes and into the thicker air of low altitudes.[30] This powerplant was also referred to as an "impulse motor."[31]

There was also unusual radio traffic monitored on April 5 and 22, 1992, at about 6:00 A.M., between the Edwards Air Force Base radar-control facility (call sign "Joshua Control") and an unknown high-altitude aircraft with the call sign "Gaspipe." Two advisories were recorded—"You're at 67,000, eighty-one miles out," and "Seventy miles out, 36,000. Above glide slope." Fighter aircraft, such as F-15s and F-16s, do not fly above 50,000 feet, as they lack pressure suits. When *Aviation Week and Space Technology* contacted Edwards Air Force Base, they said Joshua Control had no record of an aircraft with the call sign Gaspipe on those dates. No U-2Rs or the NASA-operated SR-71s were in flight at the time.[32]

A year later, another unusual contrail was observed. On April 15, 1993, weather satellite photos received at the University of Leicester in England showed a spiral-shaped contrail. When they were published, one letter writer was struck by the similarity with the "doughnuts-on-a-rope."[33]

THE XB-70 AURORA

The original 1990 *Aviation Week and Space Technology* articles were illustrated with artist conceptions of a rounded-delta design. This was followed by the flattened-diamond aircraft. After the Scottish reports, drawings were published of a pure-delta Aurora. In the summer of 1992, still another design was publicized. It was similar in shape to the XB-70 bomber flown during the 1960s.

The first sightings were made in the late summer of 1990. On September 13 and October 3, 1990, sightings were made at Mojave (near Edwards) in the late evening. Another sighting was made north of Edwards in April 1991 at about 11:00 A.M. On May 10, 1992, a writer with CNN saw the plane flying near Atlanta, Georgia, at about 5:00 P.M. The final sighting occurred on July 12 at 11:45 P.M. at the Helendale Airfield, near Barstow, California. This field is located next to a Lockheed radar cross section test range. Lockheed aircraft land at the field to bring in workers to the test range. The witness said the aircraft turned on its landing lights while quite high, then descended quickly in an S-pattern. There was bright moonlight, which allowed the witness a good look at the plane as it landed. Although the weather was clear at Helendale, there were severe thunderstorms in Las Vegas and the Groom Lake area. The implication was that the sighting was an emergency divert. On January 6, 1992, there had also been a sighting of a shape being loaded on a C-5 cargo plane at the Skunk Works facility in Burbank. It was described as looking like the forward part of an SR-71 fuselage, except the chines were rounded. It was about 65 to 75 feet long and 10 feet high. The C-5 was cleared to Boeing Field in Seattle.

The aircraft was described as having a large delta wing and a long forward fuselage. The wingtips were upturned to form fins. The edges of the wing and fins had a black tile covering, while the rest of the fuselage was white. The rear fuselage had a raised area with a black line extending down it. Some witnesses reported seeing a long-span canard near the nose. Because some did not recall seeing the canard, it was thought to be retractable. (A large delta wing, long-forward fuselage, and canards were prominent features on the XB-70.) It was described as being about 200 feet long; witnesses said it "dwarfed" an F-16 chase plane. There were two rectangular engine exhausts, and it produced a "very loud, low-pitched roar" with a rhythmic beat to it.[34]

It was speculated that the XB-70–like aircraft was the first stage of a two-stage satellite launcher developed following the loss of space shuttle *Challenger* in 1986. The aircraft would reach a speed of Mach 6 to 8, then the second stage, attached to the raised section of the rear fuselage, would fire to put a small satellite into orbit. Such a procedure would be ideal for a quick-response launch of a reconnaissance satellite in a crisis.[35]

It was noted that the Groom Lake facility had recently undergone an expansion, which believers pointed to as support facilities for the Aurora. The old housing area, built for the A-12 personnel, was demolished and replaced by 180 new units. An indoor recreation facility and a new commissary were also built. Four water tanks were built on the hillside behind the base for fire-fighting purposes. There was also an extensive runway upgrade program, which included the addition of a second runway. Another improvment was construction of a new fuel tank farm at the south end of the base. This was believed to store the liquid methane that fueled Aurora.[36] About midway down the Groom Lake flight line, a large hangar was built. It had a high roof. Believers thought this was the hangar used to load Aurora's upper stage.[37]

In a separate incident, a United Airlines 747 crew reported a near miss with an unknown aircraft. It occurred at 1:45 P.M. on August 5, 1992, as the airliner was headed east out of Los Angeles International Airport. The crew reported the plane was headed directly toward them and passed five hundred to a thousand feet below them. The crew thought the plane was supersonic as the closure rate was two to three times normal. They described the plane as having a lifting-body configuration, much like the forward fuselage of an SR-71 with some type of tail, and was the size of an F-16. It was speculated the plane was a drone that had "escaped." The sighting took place near the Edwards test range. The FAA and Edwards radar records were examined, but no target was recorded when the crew said the near miss occurred.[38]

DENIALS MADE AND DENIALS REJECTED

The flood of reports on Aurora generated a number of denials by Air Force Secretary Donald Rice. In a letter to the *Washington Post,* he said,

> Let me reiterate what I have said publicly for months. The Air Force has no such program either known as "Aurora" or by any other name. And if such a program existed elsewhere, I'd know about it—and I don't. Furthermore, the Air Force has neither created nor released cover stories to protect any program like "Aurora." I can't be more unambiguous than that. When the latest spate of "Aurora" stories appeared, I once again had my staff look into each alleged "sighting" to see what could be fueling the fire. Some reported "sightings" will probably never be explained simply because there isn't enough information to investigate. Other accounts, such as of sonic booms over California, the near collision with a commercial airliner and strange shapes loaded into Air Force aircraft are easily explained and we have done so numerous times on the record. I have never hedged a denial over any issue related to the so-called "Aurora." The Air Force has no aircraft or aircraft program remotely similar to the capabilities being attributed to the "Aurora." While I know this letter will not stop the speculation, I feel that I must set the record straight.[39]

On July 23, 1992, Rice told reporters, "I can tell you that there is no airplane that exists remotely like that which has been described in some articles."[40] On another occasion, Rice called Aurora "fantasy." On October 30, 1992, Rice said, "The system that has been described in those articles does not exist. We have no aircraft program that flies at six times the speed of sound or anything up close to that."

Such explicit denials did not stop believers from quickly trying to cast doubt on Rice's honesty. It was suggested that "We" meant the air force, while Aurora was operated by the National Reconnaissance Office (NRO), which controlled spy satellites. "And maybe," one believer continued, "Mach 5 or Mach 8 is 'not anything up close to' Mach 6 by Secretary Rice's reckoning." The believer concluded, "This particular exchange of question and answer typifies the Air Force's practice of avoiding any direct and unambiguous denial."[41]

The suggestion that the NRO operated Aurora was later denied by the NRO's director, Martin C. Faga:

> If there ever was a follow-on to the SR-71, that aircraft could be assigned to the NRO . . . in the case of a hypersonic vehicle, as has been widely speculated—by which I mean a vehicle faster than an SR-71—Mach 3, 4, 5, 6, 8—we at NRO have no such vehicle [flying] or under development. I'm not aware of any such activity, and the Air Force has said the same. I don't know what the Navy, Army, NASA or anybody else is doing. I'm just saying that NRO doesn't have an "Aurora" or anything else like it. It's a fascinating mystery.[42]

The air force also examined the "skyquakes." Massachusetts Institute of Technology's Lincoln Laboratory was hired to analyze a seismic recording from Catalina Island. Lincoln Lab concluded it was from an F-14 on a test flight off the California coast. The believers were quick to attack the F-14 explanation; one said, "This explanation doesn't hold water" and called it "an attempt to discredit the 'skyquake' evidence." They also quoted Edwards Air Force Base as saying a sonic boom from a jet at 50,000 feet only extended about twenty-five miles.[43]

Aurora also became a political issue during 1992. The October–November 1992 issue of *Air and Space* magazine published an essay on Black aircraft. Written by Steven Aftergood and John E. Pike of the Federation of American Scientists, it claimed: "In fact, it appears that Black aircraft programs are designed only to penetrate Congressional airspace. That is, wasteful, dangerous, or highly speculative programs will have a much better chance of being funded by Congress if they are highly classified."

The secrecy surrounding Black airplanes was described as exceeding all reasonable justification, not being effective, blocking technological development, and "promoting fraud and abuse."[44] Aftergood had earlier said, "It inhibits the oversight process and it puts these programs outside the sphere of democratic activity."[45]

The editor of a private newsletter echoed these remarks, saying,

* * *

> The real reasons behind the secrecy is becoming very clear to the American taxpayer. The cost of these programs must be enormous and the Pentagon is afraid that Congress might suffer an attack of sticker shock when they find out how much money the military is vacuuming out of the nation's treasury. It could also be that buried in the budgets of black programs could be evidence of monetary fraud, kickbacks and wasteful expenditures. It seems that many of these programs are designed not to evade radar but to evade accountability to Congress and the American taxpayer.[46]

Following Bill Clinton's election as president, some saw Aurora as a test of his "pledge for a more open government—and as a means of gauging his appetite for tackling the secretive, deep-rooted and conservative intelligence community."[47]

THE NORTH SEA SIGHTING

At the end of the year came word of the most important Aurora sighting yet. The December 12, 1992, issue of *Jane's Defence Weekly* carried an account of a sighting by a North Sea oil-drilling engineer. The witness said the sighting took place in August 1989 from the drilling rig Galveston Key. The day was bright, with a hazy layer at high altitude. The sky above the rig was part of the Air-To-Air Refueling Area 6A. The witness reported seeing a KC-135 tanker, two F-111s, and a triangular aircraft a little larger than the F-111s flying north. The unknown aircraft appeared to be refueling from the KC-135. It was black and had a 75-degree sweep angle. The witness was also a twelve-year veteran of the Royal Observer Corps and had been a member of its international aircraft recognition team. The witness made a sketch but did not send it into *Jane's* until the fall of 1992, after the publicity following the "skyquakes." (In 1989, he was also still in the Royal Observer Corps and subject to the Official Secrets Act.)[48]

The sweep angle was nearly identical with that of McDonnell Douglas hypersonic designs studied between the late 1960s and the 1980s. The aircraft was estimated to be 90 feet long, with a 45- to 50-foot wingspan. As with most of the speculated Aurora designs, it was described as burning liquid methane.[49] The new Aurora shape was widely publicized, with a major feature story in *Popular Science,* and shorter articles in other magazines.[50] In mid-1993, a short book on Aurora was also published.

Reports also claimed sightings had been made at Kwajalein Atoll and that a loud sonic boom, which had caused damage in Holland, was from Aurora. Other sonic booms were heard near the White Sands Missile Range. It was said that they came from the Aurora. It was also said that Royal

Australian Air Force aircraft had tracked and chased aircraft flying above Mach 6.[51]

As the reports continued, believers became more strident in their attacks on the air force's denials. The Aurora book's first chapter was titled, "Would Your Government Lie to You?" One article began with the statement that "the Pentagon would like you to believe" Aurora does not exist, but people following the story "know differently." It went on to say that "the Pentagon continues to deny the existence of the Aurora. In an attempt to protect its black projects the USAF has gone so far as to tell the world's leading aerospace experts that they are seeing things that aren't there, similar to the way they handled UFO sightings."[52]

THE TESTOR'S AURORA MODELS

As with the F-19, the Testor Corporation released two models of the Aurora in November 1993. The "SR-75 Penetrator" was their version of the XB-70–like Aurora. Launched from its back was the "XR-7 Thunderdart." This Aurora was loosely based on the North Sea sighting.[53]

And as with the F-19 kit, the idea of a Black airplane at your local hobby store attracted press attention. The *CBS Evening News* carried a report on November 11, 1993. Dan Rather introduced the spot by saying, "Does the United States military have a new Top Secret—or at least used-to-be Top Secret—mystery plane . . . there have been several sightings, but not where you'd expect." The story began: "Presenting Aurora, the Pentagon's secret weapon . . . For years it was only whispered about. Now you can see it with your own eyes, right next to the '57 Chevy at this year's model and hobby show." After showing a copy of the Rice letter, the reporter asked, "So who are you going to believe: the Secretary of the Air Force or the toy maker?"[54] When the air force was asked about the kits, they responded, "We're not saying no comment—we're saying such a plane does not exist."[55]

THE END OF AURORA

By late 1993 and early 1994, numerous articles about Aurora had been published. There was a qualitative difference with the stories about the stealth fighter published during the 1980s, however. The nightly flight activities at TTR clearly indicated the stealth fighter was operational years before the public announcement. These were rich in details, many of which, in retrospect, proved correct. The Aurora stories, in contrast, were fragmentary and stood in isolation. The Aurora had a distinctive sound and left a distinctive contrail, both of which could be heard and seen for many miles. Yet, months would pass between sighting reports. Among some believers, the suspicion grew that all was not well with the Aurora project.

After claiming that secrecy "has often been a cover" to hide problems, one journalist suggested Aurora had performance shortcomings, such as range, had suffered cost overruns, or had been designed for an obsolete nuclear war fighting mission.[56]

John Pike of the Federation of American Scientists said in December 1993, "My current theory is that they spent $10 billion to $15 billion on a very fast, very high-flying airplane." Only "one or two prototypes" of the Aurora were built, however. Plans to build several dozen production Auroras were abandoned by the NRO in 1989–90. Pike continued, "The main secret they have on the Aurora is not that it exists, but that they spent $15 billion and don't have anything to show for it."[57]

Aurora was depicted as yet another Black project that had ended in an expensive failure. Even as Aurora faded away, attacks on Black projects became more strident. The editor in chief of *Popular Science* wrote in the March 1994 issue that the $14.3 billion Black budget was "beyond the scrutiny of even the most powerful congressional oversight."[58] The same issue quoted "a congressional source," described as having "the highest level of security clearance," as believing that "a mysterious technology development effort" had been under way for several years at Groom Lake. He added, "This is not a part of the official program of the U.S. government. I think this is some sort of intelligence operation, or there could be foreign money involved. . . . It's expensive, and is immune to the oversight process. This defrauds the American government and people. You go to jail for that."[59]

The believers in Aurora express confidence the cover-up will end. "Sooner or later," one said, the "story will come out." He concluded, "Nothing thrives in the dark, except ghosts, mushrooms, and bad decisions."[60]

This is the story of Aurora, as told by some of the foremost magazines, newspapers, public interest groups, and aerospace writers. Reading through the inch-thick stack of material, one might think the case for Aurora was proven beyond a reasonable doubt—that the United States had, in fact, built a remarkable aircraft, unlike any the world has ever seen, then systematically lied about it. So one might think.

But perhaps the reader can think of another case where sightings were also made by "reliable witnesses" of remarkable vehicles, capable of speeds and maneuvers beyond those of conventional aircraft. These include 1,700-mph speeds, at a time when manned supersonic flight had not yet been achieved, and right-angle turns. These objects also showed a wide range of shapes and sizes. In that earlier case, the air force also denied the sightings were valid, saying they were of conventional objects that had been misinterpreted. In that earlier case, the believers also accused the air force of lying, slandering witnesses, and covering up. Much time was spent examining official statements for any inconsistencies. When a university study supported the air force position, it was also accused of being part of the cover-up—the air force cover-up of unidentified flying objects.

AURORA DOES NOT EXIST, ELVIS IS DEAD—ACCEPT IT

Although many people have said that they have seen or heard Aurora in flight, there is only one person who has publicly claimed to have seen Aurora close up on the ground. This person is named Robert S. Lazar. On May 1, 1993, Lazar said about Aurora:

> The Aurora I did see once on the way out there, and the only reason I say it's Aurora is, I was told by Dennis in the bus. And it makes an unbelievably loud sound, and I think when I heard it I said, it sounds like the sky is tearing. From what I understand, it operates on a liquid methane powered engine. A lot of this information has gotten out in *Aviation Week* and *Popular Science*. If this in fact was Aurora, it was certainly a strange aircraft. It looks like, if you know what the old X-15 looked like—a very long slender craft with short wings on it—and a square exhaust that had little vanes in it . . . it's quite large. It's a really overgrown thing.[61]

Robert Lazar also claims to have seen nine captured alien flying saucers.[62] This is not an exception—the whole Aurora story has been pushed by a tight circle of Black airplane buffs, aerospace writers, and believers in various far-out UFO-conspiracy theories.[63] One person stated, "The Aurora tales came straight out of UFO groups, and a lot of [the published] material . . . does too."[64] Some of those involved in spreading the "AUFOrora" stories are known to believe that the air force has "reverse engineered" the captured UFOs in order to build Aurora.[65] They also believe that the air force has perfected optical invisibility (a Romulan cloaking device), antigravity, and time dilation.

The original 1990 articles, including the "flattened diamond" Aurora, were based on the stories of an individual with no connection to the aerospace industry, but who claims that he was hired to design several disk-shaped, antigravity-powered flying saucers for a shadowy military-industrial group. He claims that he was once taken out to a test site to see the diamond-shaped Aurora. He said it had steel wheels, like a Bonneville salt flats racer, because rubber tires could not withstand the very fast takeoff speeds it required. He also stated that a smell of burned carbon comes from the aircraft. This individual also claims to have found the cure for cancer.[66]

The April 20, 1992, *NBC Nightly News* story on Aurora had, as one of its sources, John Lear.[67] He also took the 1978 photo at Groom Lake which showed a MiG 21. John Lear is the son of Learjet designer William Lear. He has also flown some 160 different aircraft and holds seventeen speed records set in the Learjet. John Lear also believed that in 1972–74, a huge underground base was built at Groom Lake. This was part of a secret treaty between the U.S. government and alien beings: in exchange for advanced alien technology, the U.S. government would allow the aliens to operate freely on the earth. Lear also believes that the aliens' digestive system has atrophied and does not function. To survive, the aliens take organs from humans and cattle, mix them with hydrogen peroxide, then spread the mixture on their bodies. He also stated, "It became obvious that some, not all, but some of the nation's missing children had been used for secretions and other parts required by the aliens."[68]

The "light-in-the-sky" shown in the NBC report did not demonstrate remarkable maneuverability "like a flying saucer," as the report put it. It just seemed to hang in the night sky. It was the landing light of one of the 737s used to bring workers to Groom Lake each morning. It arrives at 4:45 A.M. every weekday. The UFO believers call it "Old Faithful" and consider it a captured flying saucer, flown by a human pilot.

The XB-70 Aurora sightings also have problems. The landing at the Helendale Airport is impossible for so large an aircraft. An airport directory lists it as having two runways, one 3,800 feet long, while the other is 5,300 feet long. Both of them are dirt runways. These runways are too short and, being dirt, would not be able to support so large an aircraft. The field also has no lights, tower, navigation aids, or instrument landing systems. The only fuel is Shell 80, 100LL, and Autogas. Liquid methane is not listed.[69]

The drawing of the XB-70 forward fuselage being loaded on a C-5 is also questionable. The witness did not get a good look at it, as the object was obscured by bright lights and obstacles. No cockpit canopy was seen. The drawing made it look much more airplanelike than the evidence would suggest. The air force officially stated that it was a radar cross section test article.[70]

The other XB-70 sightings can be put in perspective by an experience of the author. On January 12, 1993, the author saw the XB-70 flying at low altitude, heading west. It was in a steep bank, so its delta wing and long nose could be seen. The author was quite surprised, as he had evidence by this time that the XB-70 Aurora stories were not to be relied upon. At second glance, the plane proved to be an F-14 coming into NAS Miramar. Its wings were swept back, giving it a long nose-delta shape. The Helendale sighting could have been a home-built Long E-Z light aircraft.

Belief in Aurora itself has generated sightings—the T-38 trainers at Beale Air Force Base, used for chase work and pilot proficiency flights, were repainted in an all-black finish. One pilot said: "One twilight, I was flying in close formation with my wingman as we returned to base after a mission. We had our landing lights on as we turned over the town on final and the base's switchboard lit up with calls from people inquiring if the 'Aurora' was arriving at Beale!"[71]

The photos of the doughnut-on-a-rope contrail has several internal inconsistencies. As the size of the lens (410mm) and the frequency of pulses (three hertz) are known, it is a simple matter to calculate the distance between each "puff" for a given altitude. This gives the speed of the aircraft. When the calculations are done, they indicate that if the aircraft was moving at supersonic speed, the puffs would be shooting out several hundred feet from the central contrail. They each would be the size of a football field. This represents the loss of a huge amount of energy. Blowups of the photos also show the regularity of the pulses breaks down. The contrail itself shows zigzags that are inconsistent with a very-high-speed object. This indicates it was from a much lower and slower aircraft.[72]

The circular contrails visible in the weather satellite photos were identified by several letter writers as being made by an E-3 AWAC aircraft flying a racetrack course.[73] It is an example of how any strange sight or sound would be credited to Aurora.

The North Sea sighting was of a KC-135, two F-111s, and a black triangle silhouetted against high clouds. The size of the Aurora was about that of the F-111s. The angle of an F-111 with its wings in the fully swept back position is 70 degrees. The undersides of F-111s are also painted black. In all probability, the Aurora was simply a third F-111 with its wings swept back. From far below, the nose and wings would blend together, and it would look like a black triangular flying wing.

READING BETWEEN THE LINES

This evidence of Aurora's nonexistence has been amassed from different sources over the past several years. Yet, even as the story was developing and spreading, it was possible to tell that belief in UFOs was closely connected with Aurora. The February 1988 *Gung-Ho* article on Aurora said, "Rumor has it some of these systems involve force-field technology, gravity-driven systems, and 'flying saucer' designs. Rumor further has it that these designs are not necessarily of Earth human origin—but of who might have designed them or helped us to do it, there is less talk." The author used the name Al Frickey; he has subsequently been identified by several sources as the author of books on stealth and Black airplanes. At the end of the article, the editor added: "The Air Force has had a unit at Nellis for several years; its name: The Alien Technology Center. The first question is, do you think they are studying Mexicans? The center is rumored to have obtained alien (not Earth) equipment and, at times, personnel to help develop our new aircraft star wars weaponry."[74]

An article on the "skyquakes" observed that earthquakes, meteors, or aircraft had all been eliminated as the source. It ended on a tongue-in-cheek note: "That leaves us with only one explanation to describe the particulars of this case—the shaking was caused by aliens, with a thing for Thursdays, whose spaceship is buzzing southern California."[75]

An October 1992 issue of *Aviation News* noted in an article on Aurora, "Indeed, within the last year, established figures of U.S. government agencies have indicated that captured UFOs are also here [Groom Lake], but that is another story!"[76]

Television programs on Aurora also showed UFO influences—as already noted, the NBC film of a light in the night sky above Groom Lake was described as maneuvering "like a flying saucer." A Fox network special titled *Sightings: UFO Report* showed a segment on Groom Lake. (An "unidentified" object was probably a car headlight on the Groom Lake Road.) On October 16, 1992, the Fox program *Sightings* had a segment on Aurora—it had been preceded by a "report" on UFO abductions (in which persons claim they were taken aboard a UFO and their sperm or ova removed).

Local television programs have also stressed the Aurora-UFO link. The three Las Vegas stations, Channels 3, 8, and 13, have all done a number of stories on Area 51, Aurora, and UFO claims. It was Channel 8 that first interviewed Lazar in November 1989. WFAA-TV of Dallas, Texas, carried a two-part program titled, *The Aurora Project*. The first part was on Aurora and Groom Lake, while part two was on the UFO subculture in the area. This is typical of how television programs deal with the story.[77]

The Testor Corporation, which released models of the SR-75 and XR-7, followed up with another kit in the late summer of 1994. It was a model of a UFO, as described by Robert Lazar. The order form reads: "This particular disc, nicknamed the 'Sports Model' was one of nine different discs being 'back engineered' at a secret U.S. Installation known as 'S4' on the Nellis Air Force range in the Groom Lake area of Nevada."[78]

The most obvious example of the link between Aurora and the UFO subculture is computer bulletin boards (BBS). The America Online BBS has two sections (called "folders") titled "Area 51" dealing with Black airplanes and "Above Top Secret" on UFOs. Using screen names such as "FR8-Driver," "Steve 1957," "BlackSky," "GrahamP," "Stealth C," and "Velvt-Elvis," individuals can post messages and exchange information. These messages can be read by anyone with a personal computer and a phone modem. The author has a seventeen-foot printout of the messages from June 26 to October 15, 1993.[79] There are numerous references to UFOs. One message reads: "The crafts seen could be man made craft that might utilize an anti-matter/anti-gravity propulsion system designed around flying saucer shapes. As I look at all the information I have, I feel that there are some vehicles of extraterrestrial origin being seen flying above and around Area 51."[80]

It reached the point that one person observed, "The Area 51 folder seems to have degenerated into a discussion of UFO's."[81]

THE AURORA MYTH

The myth of Aurora can serve as an example of how a belief system can develop and spread. It began with vague stories in the late 1970s and early 1980s. They did not take solid form until the name Aurora was publicized in 1985. Soon, the name was attached to the rumors of a superfast airplane. By the late 1980s, the first sightings were being made and Aurora had become part of the UFO subculture. Media access, on the part of the UFO believers, brought the story to national attention, which created more reports and more sightings. Each time the story was repeated, it grew and was embellished. Rumor became fact and then proof. Yet few of the people reading the stories or watching the television reports knew they were based on fringe UFO stories.

A close examination shows the difference between the Aurora myth and research into stories of real Black airplanes. In historical research, the separate bits of information are assembled into a complete picture. It is much like the pieces of a puzzle; individually they may seem meaningless, but together the pattern becomes clear.

With the Aurora myth, it is different. These tales of darkness and shadows are without substance or coherence. Each sighting exists separately, without connection to a whole. Each new tale replaces the ones that came before. For example, the "flattened diamond" stories, which ushered in the first series of reports, are now passed off by believers. The intercepted radio messages between Gaspipe and Edwards Air Force Base are now thought to be from security guards playing with their radios.[82]

The ebb and flow of the Aurora myth also followed the earlier pattern of the flying saucer myth. As with Aurora, the flying saucer myth had been developing for several years before the first widely publicized sightings in June and July 1947. In the years to follow, the number of UFO sightings went up sharply during times of public unease. These included presidential election years, the launch of *Sputnik 1,* and the 1960s.[83] The Aurora articles followed a similar development—the first major articles were published in late 1990, as the United States prepared for war with Iraq. The number of reports then dropped off following the victory. This was followed by a major upturn in 1992, which was an election year and a period of economic uncertainty.

In 1954, the noted psychologist Carl Jung gave an interview in which he expressed a degree of skepticism about UFOs. In 1958 the interview was republished, but in a distorted form which made Jung appear to be a UFO believer. He issued a statement giving a true version of his beliefs, "but," as he said later, "nobody, so far as I know took any notice of it." Jung concluded:

> The moral of this story is rather interesting. As the behavior of the press is a sort of Gallup test with reference to world opinion, one must draw the conclusion that news affirming the existence of UFOs is welcome, but that skepticism seems to be undesirable. To believe that UFOs are real suits the general opinion, whereas disbelief is to be discouraged. This creates the impression that there is a tendency all over the world to believe in saucers and to want them to be real, unconsciously helped along by a press that otherwise has no sympathy with the phenomenon.
>
> This remarkable fact in itself surely merits the psychologist's interest. Why should it be more desirable for saucers to exist than not?[84]

This was written thirty years before the Aurora stories were published; yet one need only substitute "Aurora" for "UFOs" and "saucers," and the statement still stands.

REVISIONIST AURORA

On July 1, 1994, believers in Aurora suffered a stunning blow. On that date the Senate Appropriations Committee added $100 million in funding to bring three SR-71s back into operation.[85] This was an outgrowth of the problems in reconnaissance revealed by the Gulf War. If Aurora did exist, why would the SR-71s be taken out of storage?

While the believers debated this question, a second blow came at the end of the year, with publication of Ben Rich's book *Skunk Works*. He identified the Aurora funding item as being for the competition between Lockheed and Northrop for the B-2 contract. Both companies came up with flying-wing designs and built one-quarter scale models for ground testing on a radar range. The Lockheed design was smaller and had a fin. Northrop's design was larger and had a heavier payload and a longer range. Rich wrote that "the rumor surfaced that it was a top secret project assigned to the Skunk Works—to build America's first hypersonic plane . . . there is no code name for the hypersonic plane, because it simply does not exist."[86]

Although some believers whispered about secret crashes and technical problems or cast doubts on Rich's word, it was becoming clear a hypersonic Aurora was becoming untenable. The Aurora story began to undergo a shift.

The first example of this "revisionist Aurora" was an article in the January 1995 *Popular Science* on the A-17—a Northrop stealth attack aircraft. It was described as a cross between the F-111 and the YF-23 prototype. The article alleged the aircraft had a two-man crew, swing wings, and the ability to reach supersonic speeds without an afterburner. The A-12's future home base was to be Cannon Air Force Base in New Mexico. An A-17 sighting was made in September 1994, above Amarillo, Texas. The article concluded by saying, "Here's a legend-shattering notion: Was the triangular craft seen refueling over Britain's North Sea actually an A-17—and not the rumored Aurora hypersonic spyplane?"[87]

The Black airplane community greeted the A-17 with less than enthusiasm. One suspects this was due to the challenge it posed to the "orthodox" Aurora. A more significant problem was that it was inconsistent with the policy governing post–F-117A stealth aircraft. Although the F-117A remained secret, the existence of the B-2, YF-22, and YF-23 programs were public knowledge. Shortly before the aircraft were rolled out, artists conceptions were released.

Another negative factor is the sheer number of supposedly still secret Black aircraft. From the published accounts, there are at least eleven such planes—two Shamus (the B-2 prototype and the "funny-looking" one), at least three differing Auroras (the diamond, North Sea, and XB-70), the TR-3, the large delta wings, the very large deltas, the F-121, the Artichoke, and, finally, the A-17. How many Black airplanes are there? At best, the only one with any credibility is the "funny-looking" Shamu.

The situation is similar to the early 1950s, when numerous reports and blurry photos were published of "new" Soviet aircraft. In this case too, they later proved to be false.

LOST IN DREAMLAND

The tales of Aurora and captured flying saucers brought a degree of attention to Groom Lake that no real Black airplane ever had. Soon, people were going to sites near the restricted area to watch the UFOs. One popular site was "The Black Mailbox," a rancher's mailbox that stands by the side of a highway. The site allows a view of the airspace above Groom Lake. Wednesdays are said to be the best night for seeing the saucers.

For the would-be saucer watcher, the skies above Groom Lake offer many sources of UFO reports. These include parachute flares, infrared decoy flares, aircraft lights and afterburners, planets, satellites, meteors (a brief film of Aurora was actually a bright meteor), and the Old Faithful airliner used to fly workers to Groom Lake.

Two hills, dubbed White Sides Mountain and Freedom Ridge, actually overlook Groom Lake. These were the only sites from which a person could have a close look at the facility itself from public land. They soon became a mecca for Black airplane buffs, UFO believers, radio intercept hobbyists, television reporters, and pilgrims.

The visitors caused problems with Groom Lake operations. Whenever someone was on White Sides or Freedom Ridge, altitude and route changes had to be made to avoid their seeing the aircraft (i.e., the MiGs). In some cases, operations had to be delayed or canceled.[88]

Signs were posted announcing, "Warning, There Is a Restricted Military Installation to the West." At the Groom Lake border, signs read, "Restricted Area, No Trespassing Beyond This Point, Photography Is Prohibited." The small print on yet another sign read, "Use of Deadly Force Authorized." Security guards, (originally in tan Broncos or Blazers, then later in white Cherokees and HUMVEES) watched the visitors. When they observed a group climbing White Sides, the local sheriff was called. The group was called down off the mountain, IDs were checked, warnings were issued about entering the restricted area or taking pictures, then the visitors were sent on their way. Video cameras also were installed to watch visitors on White Sides Mountain. The few foolish enough to enter the restricted area were arrested on the spot. They were required to sign a notice that they were being removed for trespassing on a military reservation, a form listing home address, social security number, and so on and a secrecy agreement related to "intentional or accidental exposure to classified information." The fine was typically $600, but could run as high as $5,000 or a year in jail. It was also illegal "to make any photograph, film, map, sketch, picture, drawing, graphic representation of this area or equipment at or flying over this installation." It was also illegal to sell, publish, or give away any of the above. In either case, the fine was $1,000 and/or a year in jail.

Throughout 1993, there was a standoff, which grew increasingly tense, between the viewers and the security guards. Film was confiscated, and hikers were often buzzed by low-flying helicopters. In other cases, the visitors were followed by guards. Rumors of a land grab by the air force of White Sides Mountain and Freedom Ridge began circulating in the spring of 1993. As support for this, when Freedom Ridge was discovered, there were already survey stakes in the ground. It was claimed that congressional personnel had already been warned that the seizure would be made in several months. Three congressional groups were (reportedly) shown Groom Lake and came away "dazzled" by what they had seen.[89]

The White Sides Defense Committee was organized to bring political pressure in the event of a land seizure. In their handouts, they have depicted as absurd the idea of a secret base that anyone can see. They have castigated Groom Lake as a Cold War relic and demanded that the land seized in 1984 be returned to public access.[90] On October 16–17, 1993, the committee organized a camp out at White Sides. About thirty people showed up and had "a grand old time."[91]

The following day, the Federal Register carried a notice of the withdrawal of 3,972 acres of public land in Lincoln County, Nevada. Air Force Secretary Sheila Widnell said the land was "necessary to the safe and secure operation of the activities on the Nellis Range." The withdrawal would be until the year 2001, and the land would be closed to public access, mining, or any other government agency. The land was, of course, White Sides Mountain and Freedom Ridge.[92]

THE GROOM LAKE LAND SEIZURE

The land seizure attracted both protests and media attention. The White Sides Defense Committee asked, "Is the military realistically trying to hide critical national secrets from foreign enemies, or is it trying to inhibit watchdog groups, interested taxpayers, and other domestic critics who could challenge the wisdom of continued billion dollar projects?" The Federation of American Scientists called the action "profoundly offensive . . . this move should be put on hold."[93] A Black airplane buff–UFO believer said, "What the hell is going on at Groom Lake? Why all this extreme secrecy after the Cold War is over? We think someone should shine a bright light on this place."[94]

Newsweek carried a brief story that artfully evaded the UFO aspects. It talked about "reporters and aviation buffs" climbing the two hills, and "a craft that, judging from its lights, has extraordinary maneuverability." This article ignored the fact that the main group was UFO believers, while the "craft" was actually Old Faithful, the 737 airliner UFO believers believed was a human-flown UFO.[95]

The controversy also drew the attention of the counterculture. During early 1994, magazines such as *Wired, Spin,* and *Nose* carried articles on Groom Lake. They featured a trendy nihilism as well as extravagant rhetoric and accusations, (e.g., "militarists who view [Nevada] as nothing more than featureless sandbox to bomb," "horribly irradiated," "a gung-ho think tank," "elite Air Force colonels," and "they move in like the Mafia)."[96] A list of projects at Groom Lake was called "Satan's Shopping List." This included not only Aurora, the "flattened diamond," the XB-70, the TR-3A, and UAVs, but also weapons with "microwave, electromagnetic pulse and psychtronic (mind-control) payloads." Groom Lake was described as equipped with defensive systems to protect against air and missile attack, including antiballistic missile launchers. The floor of the large hangar was described as descending "22 stories to a cavernous facility below" (an apparent reference to the tales of "secret alien underground bases" at Groom Lake). The budget for Groom Lake projects was claimed to be in the "tens of billions per year."

It was also suggested that any visitors make sure they have no "wants and warrants," as "your name is entered into a national crime and counterintelligence data base, your life, friends and family researched, your phone tapped, your mail monitored, and the movies you rent screened."[97]

Throughout the end of 1993, the controversy over the Groom Lake land seizure built, with accusations that the air force's request was too vague, that illegal chemical waste burning had taken place at the site, and that civilian contractors had cheated on their property taxes for their facilities at the site. The goal was to focus attention on Groom Lake, to make the land seizure more difficult, and to force the Bureau of Land Management to hold public hearings. In this effort, they were successful.

The final public hearing was held on March 2, 1994, at Las Vegas. One person there commented that "entertainment like this can't be bought."[98] Only about one-third of the speakers actually talked about the land seizure. The remainder were UFO believers, conspiracy buffs, paranoids, and other "colorful characters." A radio talk show host from Alaska said,

> God knows what they're testing out there. Do they have genetic engineering programs? Do they have bacteriological warfare programs? Did they not, in fact, create a thing called AIDS? Do they want to reduce the population of the planet by 25% by the year 2000? You cannot allow them to take your property. This is YOUR land. It is not their spread.
>
> Adolf Hitler wrote the book, *The New World Order,* and he used the Big Lie. He could not allow the people to go in and take a look at the camps at Nordhaus, Auschwitz, and Dachau. There are missing children, a hundred thousand plus across this nation. Where are they being taken to? Are they being used for medical experiments? Are there anti-gravitational disks being flown over there that were first developed under Adolf Hitler? Yes, yes, yes. You cannot, you must not allow it.[99]

The high point of the evening's "entertainment" was an individual with a Mideastern accent and a very loud voice who quoted from the Koran (chapter 51, of course). He announced, "All aliens! All aliens! . . . We want to see the freedom of those captured aliens. . . . Freedom, freedom of captured aliens! We are here to save the good from the bad!" He pointed accusingly at the Bureau of Land Management officers and said, "Fear in your God!"[100] This was not the typical land use meeting.

In the desert, the situation was getting increasingly out of hand. On March 22, 1994, a group of visitors, including a reporter and photographer from the *New York Times Sunday Magazine,* were near the restricted area. The reporter wanted to interview one of the private security guards (nicknamed "Cammo Dudes"). This had proven difficult, as, when approached, the guards would quickly cross back into the restricted area to prevent being identified. When one of the white Cherokees passed their three-vehicle convoy, the vehicles turned diagonally across the road, trapping the Cherokee between them. The reporter then walked over and interviewed the guard. Showing remarkable restraint, the guard only said, "No comment," and, "Don't ask me any questions."

The following day, the group was on Freedom Ridge when they discovered they were being watched by a telescopic camera. When they tuned a scanner to the sheriff's radio frequency, they discovered that search warrants were being issued. When the sheriff's deputy found them, he bluntly told them to surrender their film or be held until search warrants were obtained. Two rolls of film were surrendered. Never before had it been taken to this point. (The implication was that the new get-tough policy was in response to the "ambush journalism" of the day before.)[101]

On April 8, 1994, an ABC news crew was stopped, searched, and detained for two hours. A video camera, sound-mixing equipment, tape recorders, microphones, batteries, cables, a tripod, radio scanners, walkie-talkies, and audio and videotapes were seized. The total value was estimated at $65,000. This was the first time that a search warrant had been served.[102]

Accounts of the seizure were carried in local newspapers and by *Aviation Week and Space Technology.*[103] The equipment and tapes were returned six days later, and the report was aired on April 19. It included shots of the crew being questioned, a Russian satellite photo of the base on the XR-7 instruction sheet, and "sound bites" of enraged citizens at the public hearings. Other than the satellite photo, there were no shots of the Groom Lake facility itself. (It was accusations that the crew had filmed the site that had led to the search warrants being issued.)[104]

This situation continued for another year. Then, during the weekend of April 8–9, 1995, the warning signs were put up. White Sides and Freedom Ridge were closed off. Groom Lake, and its secrets, were again hidden.

* * *

RACHEL, NEVADA

The center of the Aurora (and UFO) watching is Rachel, Nevada. A wide spot in the road, its population is about 100 people. The town consists of a Quik Pik gas station, RV park, thrift store, and the world famous "Little A-Le-Inn" (Little Alien) bar/meeting place/restaurant/hotel/UFO research center. Originally called the Rachel Bar and Grill, the name was changed in 1990 when the UFO watchers started showing up. Inside, the walls are lined with photos of UFOs and personalities. UFO books, T-shirts, bumper stickers, and souvenirs are for sale. An extensive UFO reference library contains numerous books, magazines, maps, and videotapes. The food is described as excellent.[105]

On the surface, Rachel resembles the small towns (and their eccentric inhabitants) of fiction. But it is not another *Lake Wobegon,* it is more akin to *Twin Peaks.* Like the Black projects conducted at Groom Lake, Rachel has its "dark side."[106]

Although the sign says "Earthlings Welcome," this does not extend to liberals. This political category is defined rather broadly in Rachel. During the standoff with the Branch Davidians in Waco, Texas, during early 1993, opinion in the town was solidly behind David Koresh. President Clinton and the federal government are vehemently cursed and despised. (Local federal employees [and their money], however, are "loved.")[107] During the Los Angeles riots, one person was heard to say, "If those damn [rioters] come near here we'll be ready."[108] The town's inhabitants think environmentalists "taste as good cooked as roast spotted owl."[109]

Rachel was also the site for annual UFO conferences. These had "a no-holds-barred Bible-thumping and conspiracy" slant. There was talk about "Frankenstein experiments" being done to humans at the secret alien underground bases.

In 1993, the audience of about two hundred met in an old tent. One observer thought this was appropriate, likening the atmosphere to an "evangelical flying saucer camp meeting" where the speakers' "every utterance is taken as the gospel truth." John Lear talked about the secret bases, the exchange program, the eighty alien races visiting the earth, and some forty UFO crashes over the years.

Robert Lazar also appeared and was mobbed by the faithful everywhere he went. Lazar spent two and a half hours answering questions from the eager audience. These covered such areas as how the saucers worked, antimatter generators, gravity waves, and, of course, Aurora. (Many Aurora believers also accept Lazar's claims and view Aurora as a test that would prove them—if Aurora was true, so must his captured saucer stories.)

One of the conference moderators was Gary Schultz. He runs a group called Secret Saucer Base Expeditions, which has tours to the area. Schultz described a dark and sinister web of conspiracy, run by a "shadow government." UFOs represent only one strand of this web, which includes the death of David Koresh and the Branch Davidians, the B-2, the Council of Foreign Relations, Dreamland, and the local sheriff. This was backed up by quotes from "the only authorized version of the Bible."

One observer, who believed UFOs are alien spaceships, wondered, "Does UFOlogy give rise to paranoia or vice versa?"[110]

It is only fitting that in a time of delusion, the final word on Aurora, the nonexistent Dark Eagle, should be given by a nonexistent person. In late 1990, at the time the Aurora stories were published, there was a fad on the U.S. east coast for T-shirts with a black Bart Simpson. One read:

"It's A Black Thing, You Wouldn't Understand."

Chapter 13

Dark Eagles in a Changing World

War is a matter of vital importance to the State; the province of life or death; the road to survival or ruin. It is mandatory that it be thoroughly studied.

Sun Tzu
ca. 400 B.C.

The Dark Eagles have made an impact on history far greater that their numbers would indicate. The XP-59A introduced jet engine technology to the United States and created the concept of the Black airplane. The U-2 revolutionized intelligence gathering by showing how much could be learned from high-altitude photos alone. This led directly to reconnaissance satellites and a true Open Skies. The A-12 was aviation's greatest achievement and led to nearly a quarter century of SR-71 operations. At the other end of the spectrum, the drones gathered intelligence that manned aircraft could not provide. The Have Blue and F-117A changed the nature of airpower, while the MiGs restored the lost art of the dogfight. The HALSOL has the potential of a new dimension in atmospheric studies, while the GNAT-750 ushered in a new facet in aerial reconnaissance.

But the world is now a very different place than it was when the U-2 first lifted off from Groom Lake. In 1955, the picture was clear—Them versus Us—in a nuclear standoff. A Black airplane was needed to probe the darkness of the Soviet Union. Now, the post–Cold War world is less clear. A long war has ended, and an empire has fallen. As in the past, the aftermath is at best confused, at worst chaotic. The image that springs to mind is not so much "the end of history," as that old Chinese curse, "May You Live in Interesting Times."

Perhaps the most difficult aspect of foreseeing the shape of post–Cold War events is their sheer unpredictability. In 1985, the headline "Israel and PLO Sign Peace Treaty" would have been imaginable only in a science-fiction novel. Today, no technothriller can match the morning paper. One hundred years from now, as the twenty-first century nears its close, historians may look back and ask themselves if people today actually realized the enormity of the events taking place around them. It will be up to those future historians to decide if the world that arose from the ashes of the Cold War was better or if it was worse. We, at least, know that it will be different.

Are the Dark Eagles an endangered species in this changing world? The answer depends on two factors—military and political.

TOMORROW'S ENEMIES

In the post–Cold War world, the U.S. military is geared to fight "major regional contingencies." Although not the scale of a war between NATO and the Warsaw Pact, these represent conflicts as large as any faced by the United States in the years since the end of World War II. The most significant potential area of future conflict is North Korea. Its military forces have a total of 1.2 million men. They are equipped with 3,500 light and medium tanks, 4,000 armored personnel carriers, 10,000 artillery pieces, 2,000 short-range rocket launchers, and Scud missiles. Its home-built No Dong missile has sufficient range to hit Japan. North Korea's ongoing nuclear weapons program has been a source of friction during the early 1990s.

It is estimated that if the North Koreans should decide to attack, the warning time would be very short, possibly a matter of hours or minutes. It is further estimated that by D day plus seven, the North Koreans would either have been beaten or have reached Pusan. Another possibility is that the battle will become a prolonged war of attrition, like the first Korean War. In either case, even if the attack is defeated, the casualties among U.S., South Korean, and Allied forces would be massive.

North Korea also faces a desperate internal situation—the country endures extreme economic hardships, with reports of food shortages, starvation, and gasoline shortages. The subsidies from the Soviet Union and Communist China, which once supported the regime, are gone. The death of the "Great Leader," Kim Il Sung brings the potential of a succession crisis. Historically, nations faced with internal and dynastic problems have often lashed out. The shooting down of a U.S. Army helicopter in December 1994 was only one in a long record of attacks and border incidents. It shows North Korea is not the most rational of nations.

In the longer term, two other nations pose a major risk of sparking a regional conflict. Iraq is one of them. The Gulf War left Iraq a shadow of its former military power. The Iraqi army is half its 1990 size and is tied up with internal security duties against the Kurds, the Shiite insurgents, and coup attempts. Weapons, ammunition, and spare parts are in very short supply. United Nations inspectors also keep watch for any Iraqi efforts to develop weapons of mass destruction. While the Iraqis have attempted to interfere with the UN inspectors, the controls have, so far, proven effective. The danger is that, like Germany in the years following World War I, the inspection effort will fade. The Iraqis, like Hitler, will then feel they can openly renounce the agreements. In 1990 the Iraqis had chemical and biological weapons and were within perhaps a year of having their first nuclear weapon. There is no evidence that the Iraqis have given up efforts to build such weapons, nor have they abandoned the vision of a restored Babylonia.[1]

The other trouble spot is Iran. The loser in the Iran-Iraq War, it has been actively rebuilding its forces. Iran has used its oil to barter for weapons from Russia, the Ukraine, the Czech Republic, China, and North Korea. As with Iraq, Iran has chemical weapons and is actively seeking nuclear weapons. It has ordered some 150 No Dong missiles from North Korea. They would enable Iran to strike targets throughout the Mideast. The Iranians have also bought Tu-22M Backfire strategic bombers from Russia, along with antishipping and cruise missiles.

A major Iranian goal is to reassert control of the Strait of Hormuz, through which most of the world's oil flows. The Tu-22M bombers, as well as the three Kilo-class diesel submarines also bought from Russia, could be used to restrict access to and from the straits in a military confrontation. Another area of Iranian interest is the predominantly Moslem republics of the former Soviet Central Asia. The mixture of revolutionary fervor, imperial ambition, and religious fanaticism make Iran unpredictable.[2]

All these potential future enemies share factors in common with old enemies such as Nazi Germany and the Stalinist Soviet Union. All are totalitarian countries built around the cult of the leader and an ideology of violence, revenge, and hatred. They define their stature in terms of large military forces to carry out their leader's will. All have secret police to crush any internal enemies, real or imagined, and any idea outside the closed ideology the leader has created.[3] They are simply the latest chapter in the eternal struggle between the rights of the individual and the dominance of the state. In the end, it is not that different from the struggle between Athens and Sparta.

The weapons of the post–Cold War era are very different from those faced by the United States and its allies in regional conflicts of the past. During the Cold War, only the major powers had strategic bombers, ballistic missiles, or chemical, biological, and nuclear weapons. Today, several Third-World countries have, or are seeking, such weapons.

It does not seem that anything can prevent countries from getting such weapons. "Rogue nations" seem immune to economic sanctions or political pressures. Any country with a chemical plant can produce chemical weapons. A modern nuclear weapons program would take ten years and $10 billion. Nor does it require cutting-edge technology to build the bomb. The Iraqi, North Korean, and Iranian nuclear programs were all based on technology viewed as obsolete in the West.[4] The aircraft and missiles needed to deliver the weapons are freely available on the open market. The mere fact that a potential enemy has nuclear weapons both raises the stakes in a crisis and makes the options more limited.

Any future regional war would be as different from the Gulf War as it was from Vietnam or Korea. Nonetheless, the core of U.S. strategy remains the same. In the Gulf War, the spearhead of the Coalition air strikes was the F-117As. They attacked the means by which the Iraqi military was controlled, leaving it disorganized, fragmented, and only able to respond on a piecemeal basis.

In any future conflict, the United States would have to use its superior technology, intelligence, and training to prevent weapons of mass destruction from being used. Their hiding places, the support facilities needed to operate them, and the communications links used to control them would all have to be destroyed or disrupted. Historically, it has been Black airplanes, such as the U-2, A-12, Model 147 drones, and the F-117A, that have given the U.S. military this superiority against an enemy.

POLITICS OF THE DARK EAGLES

The size, nature, and goals of a nation's military forces are not defined solely by the threats facing it. Rather, these are largely political decisions. The sudden and painful downsizing of the U.S. military following the end of the Cold War is understandable, given American history. The huge military forces built up and maintained by the United States during the forty years of the Cold War were the exception. Traditionally, Americans have been suspicious of large standing armies. The military's role was limited to a small group of experienced officers and enlisted men that would serve as the core of an expanded army in time of war. After the war was over, the U.S. military was sharply cut back. This was the pattern following World Wars I and II.

As part of the current downsizing, Black airplanes are under attack from both Congress and the slopes of Freedom Ridge. The liturgy of "Waste, Fraud, and Abuse" is repeated at every opportunity. Black projects are described as having "pornographic cost and endemic mismanagement," as being a betrayal of democracy, and as beyond the control of even the most powerful congressmen.[5]

Reality is not so black and white. Lockheed returned $2 million on the $20 million U-2 contract. Three decades later, the F-117A came in $30 million under budget. Black airplane programs were able to achieve these cost savings and short development timescales due to a small, tightly organized team that works closely with the customer, toward a common goal. This requires that the contractor be given authority to make decisions. Hand in hand with this is a responsibility to give the customer straight answers about the status of a project.[6]

Although this pattern has continued from the XP-59A to the CIA's GNAT-750 project, the details have changed over the years. Kelly Johnson was given almost complete authority over the U-2, A-12, and D-21 projects. Today, a contractor would never be given that kind of flexibility. The F-117A underwent a top-to-bottom management review every six weeks. The GNAT-750 was subjected to constant Pentagon and congressional carping. Black programs also require costly security measures. Everyone connected with the project must undergo extensive security checks. Top secret documents cost money to store and inventory. The most visible expense is the daily commuter flights to and from Groom Lake. All these are problems that less-secret programs do not face. Despite this, Black programs are at least as efficient as White development was in the 1950s. In contrast, current White programs are subjected to a maze of reviews, reorientation, and second-guessing.[7]

The critics of Black airplane programs have their own "credibility gap." On the eve of the Gulf War, they gave their predictions about the outcome. The Center for Defense Information said the war could last for months and cost up to 40,000 U.S. casualties. Former Navy Secretary James Webb predicted the U.S. Army would be "bled dry" within three weeks. The British American Security Information Council predicted U.S. troops would "panic and run" all the way back to Saudi Arabia.[8]

While disaster was predicted for U.S. forces, Iraqi military forces were viewed as requiring a "sledgehammer" to defeat. They were described as having "high institutional self-esteem" and "remarkable bravery," and the ability to "fight with great tenacity." Iraqi officers were described as showing "remarkable dash and flexibility."[9]

Others were not so sure. One person said, well before the ground war started, "The Iraqis are not terribly proficient, not all that well trained. . . . I am confident as hell that when we meet [the Iraqis] on the ground, man to man, tank to tank, we're going to clobber them." The speaker was Tom Clancy.[10]

One of the lessons of the Gulf War is that it is easier to be an expert or a critic than it is to be right.

Another complaint is that Black projects lack oversight—that they are beyond the control of even the most powerful congressmen. This, it is claimed, means that there is no accountability. In the 1950s information about the U-2 overflights was limited to only four people in the White House and four hundred in the entire country. Today, the full defense and intelligence committees review the GNAT-750.

Questions about accountability cut both ways. The 1984 Groom Lake land seizure stands in contrast to the accomplishments of the Have Blue prototypes. While the Have Blues were changing the nature of airpower, the land seizure was being used for political posturing, power plays, petty attempts to settle scores over land policy, and as the means to satisfy the greed of special-interest groups. It saw years of inaction, followed by last-minute extensions.

As the U.S. military budgets shrink, congressmen are increasingly using their power to keep military projects in their districts. In the 1993 budget, some $8 billion out of the $260 billion total was for projects the Department of Defense did not want. No money was added for these projects, so the funding had to come from real defense needs, such as pay, fuel, training, or weapons. In reality, many now-useless defense projects continue, not because of "power-mad generals," but, rather, due to congressional "pork."[11] The $8 billion of pork is equivalent to over half the $14.3 billion total for Black projects. Some forms of "Waste, Fraud, and Abuse" are more politically correct than others.

REFLECTIONS OF SOCIETY

If a nation's military is a reflection of its political system, then its politics is a reflection of the larger society. As the twentieth century nears its end, American society has become increasingly nihilistic, bitter, resentful, and hostile to any positive thought or accomplishment. The media, at all levels, increasingly slips into a tabloid mentality. On the network news and "news magazines," there is a steady diet of scandal, gossip, and "shocking disclosures" about government conspiracies and cover-ups. These stories are often presented as entertainment. In such an atmosphere, the "story" becomes more important than the facts, and truth becomes irrelevant.[12]

Political debate has been replaced by sound bites that reduce problems to eight-second slogans. Issues are replaced by crusades, in which no quarter is given and compromise is evil. On the right, talk-show hosts play to the resentments and fears of their audiences. On the left, academics hold conferences titled, "Rage! Across the Disciplines."[13] The critic is judged superior, all efforts are judged failures from the start, and a blind will to destroy is created.

The result is a hatred of government or any kind of authority save the authority of the talk-show hosts and the academics. The belief grows that government and authority are evil, that there is no hope, and that it is useless to try to solve problems. An employee of a left-wing radio station said of its audience, "They distrust the American government. Actually, they distrust *all* government."[14] A fan of the show *The X-Files* observed, "There are a lot of people of my generation who have reason to question what they've been told by the government or by their parents."[15] In such an environment, belief in plots to establish a "New World Order" blends with belief that Aurora is based on captured flying saucer technology. Groom Lake no longer is a flight test center, but a New Age (un)holy place.

DARK FRONTIERS

In the end, the story of the Dark Eagles is about vision. It requires vision to realize it is possible to build a plane without a propeller, to fly higher and faster than any airplane has ever done before, to make an airplane invisible, or to remain aloft for months at a time. Such vision is not found in committees, it is not found in analysis, it is not found among critics and experts or among cynics. It is found among individuals who can see beyond the accepted, and have the courage to make it a reality.

If there are to be future Dark Eagles, certain things must be understood. First, that there are larger issues surrounding the Black airplanes than the slick cliches of "experts." It must also be understood that remarkable achievements cost money. When one is exploring the technological frontiers, setbacks and crashes must be expected. All too often, the timid turn back at the first sign of problems, while the politically ambitious begin the hunt for scapegoats.

The A-12 went through agony before the inlet and engine system could be made to work. Had the F-117A been a White project, it is doubtful it could have withstood the inevitable criticism over the loss of three aircraft and the technical problems that beset the stability system and the targeting and bombing equipment. One can almost hear the outraged speeches in Congress and the television news "exposes" of the plane. The congressmen would have gotten their momentary political advantage, and the news media would have their cover-up stories. And in January 1991, when it was needed, the F-117A would not have been in the dangerous skies of Baghdad.

Most of all, the Dark Eagles represent technological breakthroughs that will gain for the United States an advantage over an enemy in any potential future conflict. The secrecy surrounding these projects is necessary. It is not a joke or a game. They must be hidden, they must be guarded by extraordinary means, as would any treasure.

Out in the desert, at a place whose name is never spoken, the future of military aviation technology awaits.

Chapter Notes

CHAPTER 1

1. Daniel Ford, "Gentlemen, I Give You the Whittle Engine," *Air & Space* (October/November 1992): 88–98.

2. Dr. Jim Young, "Lighting The Flame: The Turbojet Comes To America," *Society of Experimental Test Pilots 1992 Report to the Aerospace Profession,* 247, 248.

3. Don Middleton, *Test Pilots: The Story of British Test Flying 1903–1984* (London: Willow Books, 1985), 125, 126.

4. Ford, "Whittle Engine," 88–94.

5. David M. Carpenter, *Flame Powered: The Bell XP-59A Airacomet and the General Electric I-A Engine* (Jet Pioneers of America, 1992), 8–13.

6. Ford, "Whittle Engine," 95–97.

7. Carpenter, *Flame Powered,* 14, 15, 19.

8. Ibid, 19–22.

9. Ford, "Whittle Engine," 97. It was General Electric's earlier work with superchargers that caused the British to recommend they produce the Whittle engine.

10. Carpenter, *Flame Powered,* 16, 17, 27.

11. John Ball Jr., *Edwards: Flight Test Center of the USAF* (New York: Duell, Sloan and Pearce, 1962), 12, 14–16, 42, 43.

12. Young, "Turbojet," 254. At the time of the XP-59A test flights, North Base was known as the Materiel Command Flight Test Base or Muroc II. Similarly, press accounts of the mid-1960s often refer to the "Skonk Works." Rachel, Nevada, was known until the 1980s as "Sand Springs." (It was renamed in memory of the first child born in the town, Rachel Jones.) To avoid confusion, the later names of North Base, the Skonk Works, and Rachel will be used throughout.

13. Carpenter, *Flame Powered,* 27–31, 33–35.

14. Ford, "Whittle Engine," 88–90.

15. Young, "Turbojet," 254.

16. A. M. "Tex" Johnston and Charles Barton, *Tex Johnston, Jet-Age Test Pilot* (Washington, D.C.: Smithsonian Institution Press, 1991), 59.

17. Young, "Turbojet," 254. The open cockpit did reduce the maximum speed and altitude the first XP-59A could reach. (The observer had only a plastic windshield for protection from the slipstream.) Later, Brown Recorders were installed and an instrument panel was fitted into the aft fuselage. This panel was photographed in flight to provide a continuous record. The XP-59A did much to usher in the modern era of flight test instrumentation. Today, hundreds of separate readings are transmitted to the ground in real time. When the first XP-59A was given to the Smithsonian, the observer's cockpit was removed and the aircraft was restored to its October 1–2, 1942, appearance.

18. Carpenter, *Flame Powered,* 40, 41. It is often thought the fake prop was used as cover throughout the XP-59A program. It appears, from photographic evidence, that it was used only during the move and while at Harpers Lake. The photos of the plane with the fake prop have become a symbol of the secrecy that enveloped the project.

19. Johnston and Barton, *Tex Johnston, Jet-Age Test Pilot,* 61–64. It was Woolams who came up with the idea of the black derby hats. Tex Johnston tells of jumping a P-38 lining up for a gunnery pass; when the pilot saw him, Johnston tipped his hat and climbed away. It appears this was a rather common practice.

20. Young, "Turbojet," 256.

21. Carpenter, *Flame Powered,* 27, 40–48, 51, 55, 56.

22. Young, "Turbojet," 259. An aspect of jet flight that prop pilots had to get used to was the slow throttle response of jet engines. It took much longer for the turbine to spin up compared to a piston engine. On the positive side, the jet pilot did not have to deal with prop pitch, mixture, manifold pressure, and the roar and vibration of a propeller airplane.

23. Richard P. Hallion, *Test Pilots: The Frontiersmen of Flight* (Garden City: Doubleday, 1981), 172, 173.

24. Clarence L. "Kelly" Johnson with Maggie Smith, *Kelly: More Than My Share of It All* (Washington, D.C.: Smithsonian Institution Press, 1985), 96–98.

CHAPTER 2

1. Curtis Peebles, *Guardians* (Novato, Calif.: Presidio Press, 1987), chap. 1.

2. Paul F. Crickmore, *Lockheed SR-71 Blackbird* (Osceola, Wis.: Motorbooks, 1986), 9.

3. Curtis Peebles, *The Moby Dick Project* (Washington, D.C.: Smithsonian Institution Press, 1991), 99, 100, 119, 120.

4. Paul Lashmar, "Skulduggery at Sculthorpe," *Aeroplane Monthly,* (October 1994): 10–15.

5. Robert Jackson, *Canberra: The Operational Record* (Washington, D.C.: Smithsonian Institution Press, 1989), 60. These early overflights took advantage of the commitment of the Soviet air force to the Korean War. The first Soviet MiG 15s saw action in November 1950. Between then and the spring of 1953, when the Soviet "Honchos" were withdrawn, a full twelve air divisions had seen action. The result was to strip the western Soviet Union of air defenses. The story is told of an RB-45C that overflew *Moscow*.

6. Richard H. Kohn and Joseph P. Harahan, ed., *Strategic Air Warfare* (Washington, D.C.: Office of Air Force History, 1988), 95, 96. The mission was as much a show of force as it was for reconnaissance. Strategic Air Command chief Gen. Curtis E. LeMay later said that the loss rate of SAC bombers, had there been a war with the Soviet Union during the 1950s, would have been no higher than that of the peacetime accident rate.

7. Jay Miller, *Lockheed U-2* (Austin, Tex.: Aerofax, 1983), 10–12.

8. Jay Miller, *The X-Planes: X-1 to X-31* (New York: Orion, 1988), 131–33.

9. Miller, *Lockheed U-2,* 12, 15–18.

10. Central Intelligence Agency, Memorandum for Record, Subject: Special Aircraft for Penetration Photo Reconnaissance, (Washington, D.C.: May 12, 1954).

11. Dino A. Brugioni, *Eyeball to Eyeball* (New York: Random House, 1991), 12–15.

12. Chris Pocock, *Dragon Lady: The History of the U-2 Spyplane* (Shrewsbury, England: Airlife, 1989), 8.

13. Edwin H. Land, Memorandum for: Director of Central Intelligence, Subject: A Unique Opportunity for Comprehensive Intelligence (Central Intelligence Agency, Washington, D.C.: November 5, 1954).

14. A. J. Goodpaster, Memorandum of Conference with the President (Dwight D. Eisenhower Library, Abilene, Kans.: November 24, 1954).

15. Brugioni, *Eyeball to Eyeball,* 17, 22, 185.

16. Pocock, *Dragon Lady,* 10–14.

17. Miller, *Lockheed U-2,* 19, 20.

18. U.S. Air Force Oral History Interviews, Maj. Gen. Osmond J. Ritland, March 19–24, 1974, vol. 1, 142–44 (Edwards AFB History Office, Ritland Files).

19. Brackley Shaw, "Origins of the U-2: Interview with Richard M. Bissell Jr.," *Air Power History* (Winter 1989): 18.

20. Clarence L. "Kelly" Johnson and Maggie Smith, *Kelly: More Than My Share of It All* (Washington, D.C.: Smithsonian Institution Press, 1985), 122, 123.

21. Pocock, *Dragon Lady,* 14.

22. Miller, *Lockheed U-2,* 20.

23. Johnson and Smith, *Kelly,* 123–25.

24. Skip Holm, "Article Airborne," *Air Progress Aviation Review* (1986): 25–29. Different sources give contradictory dates for the U-2's first flight. This article is composed of the transcripts of the radio communications, notes, and LeVier's postflight reports.

25. Pocock, *Dragon Lady,* 15.

26. Miller, *Lockheed U-2,* 22, 23.

27. Pocock, *Dragon Lady,* 19, 20; and Kenneth W. Weir, "The U-2 Story," *Society of Experimental Test Pilots 1978 Report to the Aerospace Profession,* 186. *Officially* U-2 stood for "Utility," a cover for its reconnaissance role (much as the "X" was a cover for the X-16).

28. Shaw, "Interview with Bissell," 18.

29. Pocock, *Dragon Lady,* 16, 17, 20–23. The X-16 cancellation was also a case of history repeating itself—although Bell had invented the Black airplane with the XP-59A, it was Lockheed's P-80 that won the production contract.

30. Pocock, *Dragon Lady,* 22. Comparing the X-16 and the U-2 indicates just how farsighted Johnson was. The X-16 represented aeronautical conventional thinking, a conventional aircraft that attempted to perform an extraordinary mission. Johnson realized it would take an extraordinary aircraft. In contrast to the large X-16, Johnson used as his model the aerodynamic efficiency of a glider. One example of this was the two engines of the X-16. Conventional wisdom held this made the aircraft more reliable. However, if an engine was lost, altitude could not be maintained. In reality, two engines meant more weight and complexity. None of the U-2s lost over "denied areas"—the USSR, China, or Cuba—was due to flameouts.

31. Pocock, *Dragon Lady,* 18.

32. Francis Gary Powers and Curt Gentry, *Operation Overflight* (New York: Holt, Rinehart and Winston, 1970), 3–5.

33. Pocock, *Dragon Lady,* 18. These requirements were a mixture of security and the demands of flying the U-2. The need for a Top Secret clearance is obvious. As reserve officers, their resignations would not attract notice. (Many pilots left for airline jobs.) Because of the U-2's demanding nature, good "stick and rudder" men were needed. The F-84 had poor handling and an unreliable engine. (Thirteen pilots had been killed in crashes at Turner AFB in the previous year.) This was good preparation for the U-2.

34. Pocock, *Dragon Lady,* 23.

35. Powers and Gentry, *Operation Overflight,* 21, 33.

36. Ronald Rubin, "A Day at the Ranch," *Gung-Ho* (July 1983): 57.

37. Francis Gary Powers, Flight Log (Central Intelligence Agency, Washington, D.C.: May 1956–April 1960).

38. Shaw, "Interview with Bissell," 20.

39. Evaluation, July 6, 1956 (Francis Gary Powers File, Central Intelligence Agency: Washington, D.C.).

40. Miller, *Lockheed U-2,* 90, 91.

41. Pocock, *Dragon Lady,* 20–22.

42. Powers and Gentry, *Operation Overflight,* 30–33.

43. Shaw, "Interview with Bissell," 21. The U-2 could glide 250 nautical miles from 70,000 feet. The descent took 73 minutes. This was for "still air" conditions only, as one member of the first group learned in early 1956. He suffered a flameout over the Grand Canyon. As the plane descended, he pointed it toward the Ranch, less than 100 miles away. The U-2 entered a strong jet stream, which nearly pushed it backward. After several relight attempts, it looked as if he would have to land on a dirt strip in the canyon. Finally, he was able to restart the engine and return to Groom Lake.

44. Pocock, *Dragon Lady,* 22, 145.

45. *David Frost Show,* Transcript, April 28, 1970 (Francis Gary Powers File, Central Intelligence Agency: Washington, D.C.), 4.

46. Brugioni, *Eyeball to Eyeball,* 20.

47. Pocock, *Dragon Lady,* 145. Of the first seven U-2 crashes, only Ericson survived.

48. Powers and Gentry, *Operation Overflight,* 28, 33.

49. Pocock, *Dragon Lady,* 22, 23, 38.

50. Rubin, "A Day at the Ranch," 52–57, 74–76.

51. Brugioni, *Eyeball to Eyeball,* 27.

52. Peebles, *The Moby Dick Project,* chap. 6.

53. Johnson and Smith, *Kelly,* 119, 120.

54. Shaw, "Interview with Bissell," 21.

55. Brugioni, *Eyeball to Eyeball,* 18.

56. Powers and Gentry, *Operation Overflight,* 29, 49. Initially, Powers thought the eighteen-month contract was an unduly optimistic estimate of the U-2's lifetime. Accordingly, the U-2s were flown only the mimimum necessary for training and test flights, to avoid wearing them out. Powers later recalled that the pilots heard rumors that the original concept for the overflights was for each U-2 to make a single overflight, ending with a belly landing (an apparent reference to the original CL-282's takeoff cart and skid landing gear).

57. Brugioni, *Eyeball to Eyeball,* 30.

58. Shaw, "Interview with Bissell," 21.

59. Pocock, *Dragon Lady,* 26–29. Of the fifty-five early-model U-2s built, only ten still survived when the type was retired in 1980. One of these was Article 347, the first U-2 to overfly the Soviet Union. It is now on display, along with a B camera, at the National Air and Space Museum.

60. Brugioni, *Eyeball to Eyeball,* 30–32.

61. Pocock, *Dragon Lady,* chap. 3.

62. Steven J. Zaloga, *Soviet Air Defence Missiles* (Alexandria, Va.: Jane's Information Group, 1989), 37; and Shaw, "Interview with Bissell," 18, 19, 22.

63. Peebles, *The Moby Dick Project,* 199, 200.

64. Steven J. Zaloga, *Target America* (Novato, Calif.: Presidio Press, 1993), 255.

65. Peebles, *The Moby Dick Project,* 204–6.

66. Pocock, *Dragon Lady,* 63, 64, 157.

67. *The Chicago Show,* Transcript, April 30, 1970, WLS-TV (Francis Gary Powers File, Central Intelligence Agency: Washington, D.C.), 9.

68. Pocock, *Dragon Lady,* 30, 31, 43, 66. It was initially thought Sieker had loosened his faceplate to eat when the engine flamed out. Ground tests, however, indicated the clip had failed under pressure.

69. Ben R. Rich and Leo Janos, *Skunk Works* (New York: Little, Brown, 1994), 151–55.

70. A. J. Goodpaster, Memorandum for the Record (Dwight D. Eisenhower Library, Abilene, Kans.: February 8, 1960).

71. Pocock, *Dragon Lady,* 46.

72. Brugioni, *Eyeball to Eyeball,* 43. Sverdlovsk was a major Soviet military and industrial city. The Kyshtym 40 facility outside town included reactors that produced the nuclear material, the weapon assembly facilities, and waste storage tanks. Also in the area was a flight test center for trials of aircraft weapons systems. It was an SA-2 site guarding Kyshtym 40 that shot down Powers's U-2. The Su-9 and MiG 19s that tried to intercept the plane came from the test site.

73. Powers and Gentry, *Operation Overflight,* 79, 80.

74. Pocock, *Dragon Lady,* 41–44. Article 360 was the first U-2C. The aircraft had been modified with a more powerful J75 engine. This restored the altitude lost due to the weight of equipment added to the U-2, which included the midnight blue paint, ejector seat, and reconnaissance equipment. (The paint alone added 80 pounds.) The aircraft was also more demanding to fly; in some flight conditions only four knots separated the maximum and minimum speeds. Article 360 was sent to Detachment C in Japan, but was damaged on September 24, 1959, when it made a belly landing on a glider strip after running out of fuel. The plane was repaired, but it had a reputation for sporadic problems. Article 360 was not to have made the overflight, but the delays caused the U-2 originally picked to be grounded for maintenance.

75. Powers and Gentry, *Operation Overflight,* 81–84.

76. Kelly and Smith, *Kelly,* 128.

77. "Soviets Downed MiG along with U-2 in '60," *San Diego Tribune,* April 30, 1990, sec. A. The Soviets did not admit the loss of the MiG and its pilot, or the mass SA-2 firings, until 1990. Up to this point, they claimed only one SAM had been fired. "Scholarly" studies and textbooks through the 1980s continued to claim the U-2 had suffered a flameout, long after the true events were known. This was combined with claims the overflight was a plot by the CIA to sabotage the Paris Summit and prolong the Cold War. One of the author's professors said it was a possibility that both the U-2 and the shooting down of Korean Air Lines flight 007 were both due to this plot. It was not until 1992 that the Russian government admitted the Soviets had systematically lied about the shooting down of the airliner.

78. Pocock, *Dragon Lady,* 48–58.

79. Orin Humphries, "High Flight" *Wings* (June 1983): 10–31, 50–55.

80. Pocock, *Dragon Lady,* 68.

CHAPTER 3

1. Robert Hotz, "Editorial Laurels for 1962," *Aviation Week and Space Technology* (December 24, 1962): 11.

2. Ben Guenther and Jay Miller, *Bell X-1 Variants* (Arlington, Tex.: Aerofax, 1988), 25, 32.

3. John L. Sloop, *Liquid Hydrogen as a Propulsion Fuel 1945–1959* (Washington, D.C.: NASA SP-4404, 1980), 141–45.

4. Ibid., 147–49.

5. Ibid., 152–62. There seems to be a connection between Pratt and Whitney's 304 hydrogen-fueled engine and its design of a nuclear-powered jet engine for the WS-125A bomber. In the nuclear engine, a reactor heated water, turning it into pressurized steam. This went through a steam turbine, which powered the compressor fan via a reduction gear. The steam then flowed through a heat exchanger, which heated the compressed air to produce thrust. The water then flowed back to the reactor to begin the cycle all over again. The flow is the reverse of the 304 engine, but many of the technical features, such as the design of the heat exchanger, are identical. The WS-125A bomber could cruise indefinitely on nuclear power alone. To provide added thrust, such as for takeoff and the Mach 3 dash to the target, boron would be sprayed through an afterburner. The boron fuel also acted as a radiation shield. The air force canceled Pratt and Whitney's nuclear engine in August 1957, the month before the first 304 engine runs started.

6. Clarence L. "Kelly" Johnson and Maggie Smith, *Kelly: More Than My Share of It All* (Washington, D.C.: Smithsonian Institution Press, 1985), 137, 138.

7. Sloop, *Liquid Hydrogen,* 163–67. Ironically, given Sun Tan's secrecy, a number of aircraft projects during the 1950s had configurations similar to the CL-400. These include the Soviet M-50 jet bomber and MN-1 reconnaissance aircraft, the British Avro 730, the Bristol T.188, the Armstrong Whitworth AW.166 and AW.169, and the French SO 9000.

8. Thomas P. McIninch, "The Oxcart Story," *Studies in Intelligence* (Summer 1982): 26. This document was declassified when the surviving A-12s were given to museums. A copy was sent to each museum as an "owner's manual" for the airplane. It was soon being passed around like samizdat. (Some nth generation copies were unreadable.) "Thomas P. McIninch" is a pseudonym for John P. Parangoski, the Oxcart project officer. *Studies in Intelligence* is an internal, classified CIA magazine.

9. Jay Miller, *The X-Planes: X-1 to X-31* (New York: Orion, 1988), 43.

10. A-12/SR-71 Lecture, San Diego Aerospace Museum, June 14, 1991.

11. Steven J. Zaloga, *Soviet Air Defence Missiles* (Alexandria, Va: Jane's Information Group, 1989) 75, 76. The SA-2's booster rocket burns for four to five seconds, while the second stage fires for twenty-two seconds. Thus, it would have to be launched well before the A-12 was within range. The closure speed would be over Mach 6. The computer and guidance radar would have to direct the SA-2 to a point in space ahead of the A-12. This proved impossible in some 1,000 launches against the A-12/SR-71.

12. McIninch, "The Oxcart Story," 26.

13. *"Kelly's Way": The Story of Kelly Johnson and the Lockheed Skunk Works* (Edwards AFB: Air Force Flight Test Museum, 1993), videotape.

14. Dino Brugioni, *Eyeball to Eyeball* (New York: Random House, 1991), 39.

15. Rich and Janos, *Skunk Works,* 197–99.

16. The President's Appointments, Monday July 20, 1959 (Dwight D. Eisenhower Library: Abilene, Kans.).

17. McIninch, "The Oxcart Story," 26, 27.

18. Clarence L. Johnson, "Development of the Lockheed SR-71 Blackbird," *Lockheed Horizons* (Winter 1981/82): 4.

19. McIninch, "The Oxcart Story," 27, 29.

20. A. J. Goodpaster, Memorandum for Record (Dwight D. Eisenhower Library: Abilene, Kans., June 2, 1960).

21. Brugioni, *Eyeball to Eyeball,* 53.

22. Johnson, "Development of the Lockheed SR-71 Blackbird," 3, 8.

23. Johnson and Smith, *Kelly,* 139, 140.

24. McIninch, "The Oxcart Story," 28.

25. Johnson and Smith, *Kelly,* 140–42.

26. A-12/SR-71 Lecture, San Diego Aerospace Museum, June 14, 1991.

27. Richard P. Hallion, *Designers and Test Pilots* (Alexandria, Va: Time-Life Books, 1983), 153.
28. A-12/SR-71 Lecture, San Diego Aerospace Museum, June 14, 1991.
29. Johnson, "Development of the Lockheed SR-71 Blackbird," 3.
30. McIninch, "The Oxcart Story," 27, 29.
31. William H. Brown, "J58/SR-71 Propulsion Integration or The Great Adventure into the Technical Unknown," *Lockheed Horizon* (Winter 1981–82): 6–9.
32. Clarence L. Johnson, "Some Development Aspects of the YF-12A Interceptor Aircraft," AIAA Paper No. 69-757 (July 14–16, 1969): 7.
33. McIninch, "The Oxcart Story," 29–31.
34. Johnson, "Some Development Aspects of the YF-12A Inteceptor Aircraft," 5.
35. A-12/SR-71 Lecture, San Diego Aerospace Museum, June 14, 1991.
36. *"Kelly's Way,"* videotape.
37. Paul F. Crickmore, *Lockheed SR-71 Blackbird* (Osceolo, Wis.: Motorbooks, 1986), 21, 22.
38. A-12/SR-71 Lecture, San Diego Aerospace Museum, June 14, 1991.
39. McIninch, "The Oxcart Story," 33. The two men who had done so much to bring both the U-2 and A-12 into existence, Allen Dulles and Richard Bissell, had both resigned following the Bay of Pigs disaster a year before.
40. Crickmore, *Lockheed SR-71 Blackbird,* 23.
41. McIninch, "The Oxcart Story," 33, 34.
42. Ibid., 30.
43. Lt. Col. Steve Stowe, *USAF Test Pilot School 1944–1989* (Edwards AFB: USAF Test Pilot School, 1991), 42. Two of Skliar's classmates were Mercury astronauts L. Gordon Cooper and Virgil I. "Gus" Grissom.
44. SR-71/A-12/YF-12 Flights/Checkout (Edwards AFB History Office, June 14, 1991).
45. Crickmore, *Lockheed SR-71 Blackbird,* 21, 23, 26, 29.
46. Private source. The author has extensive contacts within the Black airplane community, both personal and through publications. This provides an insight into both the latest information and the beliefs behind it.
47. Milton O. Thompson, *At the Edge of Space* (Washington, D.C.: Smithsonian Institution Press, 1992), 55, 58–60. Groom Lake may have had a secret role in the X-15 program. In the event of an engine failure during the climb, the X-15 would have to land on dry lake beds scattered across the desert. If the X-15 suffered an engine failure forty to forty-six seconds after ignition, it would "land at an unnamed lake bed in a highly classified restricted area." Groom Lake meets this description, and it is along the line

between Delamar Lake and Edwards AFB. Groom Lake was also larger than many of the emergency lake beds. The high mountains to the west and east would have made the unpowered approach difficult.

48. Private source

49. Private sources.

50. McIninch, "The Oxcart Story," 35; and Crickmore, *Lockheed SR-71 Blackbird,* 26–28.

51. Jules Bergman, "Our Watchdog at the Edge of Space," *Readers Digest* (December 1964): 59.

52. Crickmore, *Lockheed SR-71 Blackbird,* 25, 26.

53. Untitled comments by Kelly Johnson, *Lockheed Horizons,* (Winter 1981/82): 16.

54. A-12/SR-71 Lecture, San Diego Aerospace Museum, June 14, 1991.

55. Crickmore, *Lockheed SR-71 Blackbird,* 24–26.

56. McIninch, "The Oxcart Story," 34, 35.

57. A-12/SR-71 Lecture, San Diego Aerospace Museum, June 14, 1991.

58. Untitled comments by Kelly Johnson, *Lockheed Horizons,* (Winter 1981/82): 15, 16. Sometimes even these efforts were not enough. On March 30, 1965, Article 126's right engine sucked off its manufacturer's nameplate during a ground run.

59. McIninch, "The Oxcart Story," 37.

60. Crickmore, *Lockheed SR-71 Blackbird,* 25.

61. A-12/SR-71 Lecture, San Diego Aerospace Museum, June 14, 1991.

62. Steve Pace, *Lockheed Skunk Works* (Osceolo, Wis.: Motorbooks, 1992), 183–87.

63. Crickmore, *Lockheed SR-71 Blackbird,* 37–41.

64. Paul F. Crickmore, *Lockheed SR-71 Blackbird: The Secret Missions Exposed* (London: Osprey Aerospace, 1993), 68–70.

65. Crickmore, *Lockheed SR-71 Blackbird,* 40, 41.

66. McIninch, "The Oxcart Story," 35.

67. Memorandum for the President: Agenda for Luncheon with Secretaries Rusk and McNamara (Lyndon B. Johnson Library: Austin, Tex., February 28, 1964).

68. McIninch, "The Oxcart Story," 36.

69. Crickmore, *Lockheed SR-71 Blackbird,* 47.

70. A-12/SR-71 Lecture, San Diego Aerospace Museum, June 14, 1991.

71. Crickmore, *Lockheed SR-71 Blackbird,* 65–70.

72. McIninch, "The Oxcart Story," 38.

73. Dean Rusk, Memorandum for the President, Subject: Warning to Cubans and Soviets against Interference with our Aerial Surveillance of Cuba (Lyndon B. Johnson Library: Austin, Tex., March 15, 1964).

74. Memorandum for the President, Re: NSC Agenda, Tuesday, May 5, 1964 (Lyndon B. Johnson Library: Austin, Tex.). There was no intention of using the A-12 over the Soviet Union, except in a national emergency. Satellites could provide routine coverage of the USSR with zero political risk. However, in the 1960s, the number of satellites was too small to allow them to cover targets in Cuba, North Vietnam, or North Korea.

75. McIninch, "The Oxcart Story," 38, 39.

76. Jay Miller, *Lockheed SR-71 (A-12/YF-12/D-21)* (Austin, Tex.: Aerofax, 1984), 6.

77. McIninch, "The Oxcart Story," 38.

78. Jay Miller, *Lockheed U-2* (Austin, Tex.: Aerofax, 1983), 21; and *Electronic Spies* (Alexandria, Va.: Time-Life Books, 1991), 36.

79. SR-71/A-12/YF-12 Flights/Checkout, Edwards AFB History Office, June 14, 1991.

80. Crickmore, *Lockheed SR-71: The Secret Missions Exposed,* 16, 41.

81. McIninch, "The Oxcart Story," 38, 39

82. David A. Anderton, *North American F-100 Super Sabre* (Osceola, Wis.: Motorbooks, 1987), 108–11.

83. Crickmore, *Lockheed SR-71: The Secret Missions Exposed,* 63, 64.

84. Crickmore, *Lockheed SR-71 Blackbird,* 36.

85. McIninch, "The Oxcart Story," 38.

86. Ibid., 41, 42.

87. Crickmore, *Lockheed SR-71 Blackbird,* 36.

88. McIninch, "The Oxcart Story," 42.

89. *Astronautics and Aeronautics, 1967* (Washington, D.C.: NASA, 1968), 5.

90. Stowe, *USAF Test Pilot School 1944–1989,* 61, 62. Among Scott's classmates were NASA astronauts David Scott, James B. Irwin, and Theodore Freeman, as well as X-15 astronaut Mike Adams.

91. McIninch, "The Oxcart Story," 41.

92. Crickmore, *Lockheed SR-71: The Secret Missions Exposed,* 25–31. Sullivan's A-12 was the only U.S. aircraft over Hanoi at the time. Thus, the SA-2 batteries had their best opportunity.

93. McIninch, "The Oxcart Story," 41–44.

94. Crickmore, *Lockheed SR-71 Blackbird,* 171.

95. *SR-71 Blackbird: The Secret Vigil* (New York: Aviation Week video, 1990).

96. Crickmore, *Lockheed SR-71: The Secret Missions Exposed,* 31–33.

97. McIninch, "The Oxcart Story," 45.

98. Ibid., 46–48.

99. Crickman, *Lockheed SR-71: The Secret Missions Exposed,* 1–8, 34, 41.

100. McIninch, "The Oxcart Story," 49, 50. Years later, the SR-71 operations building at Kadena was still known as the "Oxcart" building. After nearly a quarter century in storage, JP-7 fuel was still found in the A-12 tanks when they were disassembled for transfer to museums. The fuel had not evaporated even after all those years.

CHAPTER 4

1. William Wagner, *Lightning Bugs and Other Reconnaissance Drones* (Fallbrook, Calif.: Aero Publishers, 1982), 6–14.

2. Ibid., 15–22.

3. Ibid., 23–32.

4. Ibid., 35–41.

5. Dino A. Brugioni, *Eyeball to Eyeball* (New York: Random House, 1991), 104, 105, 116, 117.

6. Chris Pocock, *Dragon Lady: A History of the U-2 Spyplane* (Shrewsbury, England: Airlife, 1989), 93, 196.

7. Brugioni, *Eyeball to Eyeball,* 133, 135–40.

8. Ibid., 51, 153, 157, 159, 164, 166, 181–86.

9. Ibid., 187–217.

10. Pocock, *Dragon Lady,* 78–80.

11. Brugioni, *Eyeball to Eyeball,* 217, 254, 276, 277, 296, 363, 364, 452, 453.

12. "One Minute to Midnight: The Real Story of the Cuban Missile Crisis" (NBC News special, October 24, 1992).

13. Pocock, *Dragon Lady,* 80.

14. Orin Humphries, "High Flight," *Wings* (June 1983): 50.

15. Brugioni, *Eyeball to Eyeball,* 454–82.

16. James G. Bright and David A. Welch, *On the Brink* (New York: Hill and Wang, 1989), 311, 327, 369. Although only forty-five at the time of the missile crisis, Statsenko was retired soon afterward, apparently because of the shooting down of Anderson's U-2. There is some evidence he tried to blame the Cubans for the firing.

17. "One Minute to Midnight; The Real Story of the Cuban Missile Crisis."

18. Brugioni, *Eyeball to Eyeball,* 483–89.

19. Pocock, *Dragon Lady,* 82–85. Ten air force U-2 pilots, Majors Richard Heyser, Buddy Brown, Ed Emerling, Gerald McIlmoyle, Robert Primrose, and Jim Qualls, and Captains George Bull, Roger Herman, Charles Kern, and Dan Schmarr received Distinguished Flying Crosses for their Cuban overflights between October 14 and 28. Major Rudolph Anderson was posthumously awarded the Distinguished Service Medal, the highest peacetime decoration. Another six U-2 pilots who made Cuban overflights

after October 28 received no awards, despite the uncertainty in the days following the resumption of U-2 missions.

20. Wagner, *Lightning Bugs and Other Reconnaissance Drones,* 42.

21. Ibid., 44, 46–49, 51.

22. Dean Rusk, Memorandum for the President, Subject: Warning to Cubans and Soviets against Interference with Our Aerial Surveillance of Cuba (Lyndon B. Johnson Library: Austin, Tex., March 15, 1964).

23. Memorandum for the President, Re: NSC Agenda, Tuesday, May 5, 1964 (Lyndon B. Johnson Library: Austin, Tex.). Interestingly, the text of the original memo was changed after it was typed. It originally read, "we have drones which could do this job . . ." The word "could" was crossed out and "might" was written in with an underline. Another line originally read "which would enable them to continue indefinitely even if Castro tried to bring them down." The word "indefinitely" was crossed out. Clearly, at some point, second thoughts began to grow about the drones.

24. "U.S. Studies Drones for Use Over Cuba," *San Diego Union,* May 6, 1964, sec. A. The "F-104" should be "RF-101"—no reconnaissance version of the F-104 was ever used over Cuba. Also, the article implied that low-level flights were still being made in 1964. In fact, they had ended when the missiles were removed.

25. *Electronic Spies* (Alexandria, Va.: Time-Life Books, 1991), 100–5.

26. Wagner, *Lightning Bugs and Other Reconnaissance Drones,* 53–56, 62, 65.

27. Pocock, *Dragon Lady,* 96, 194–96.

28. Wagner, *Lightning Bugs and Other Reconnaissance Drones,* 57, 64–66, 99.

29. Ibid., 67, 69, 70, 71.

30. Ibid., 74.

31. "Pilotless U.S. Plane Downed, China Says," *New York Times,* November 17, 1964, 1. All citations for the *New York Times* are from the microfilm edition.

32. "Report Baffles Officers," *New York Times,* November 17, 1964, 14.

33. Wagner, *Lightning Bugs and Other Reconnaissance Drones,* 75.

34. "Silent on Peking Report," *New York Times,* November 19, 1964, 2.

35. Wagner, *Lightning Bugs and Other Reconnaissance Drones,* 77, 125.

36. "Red China Says It Downed U.S. Drone," *New York Times,* January 3, 1965, 3.

37. "Drone Downed, Peking Says," *New York Times,* April 1, 1965, 2.

38. "U.S. Drone Shown, Peking Announces," *New York Times,* April 3, 1965, 2.

39. "U.S. Plane Down, China Says," *New York Times,* April 4, 1965, 9.

40. "Red China Downs U.S. Robot Plane," *San Diego Union,* April 19, 1965, sec. A.

41. Robert Zimmerman, "Ryan Receives Contract for More Drones," *San Diego Union,* April 21, 1965, sec. A.

42. "China Reports Downing Plane," *New York Times,* August 22, 1965, 2.

43. Wagner, *Lightning Bugs and Other Reconnaissance Drones,* 93, 95.

44. "The View from the Top," *Warplane,* (Vol. 6, Issue 69, 1987): 1362–64.

45. "Hanoi and Peking Report 3 U.S. Planes Downed," *New York Times,* February 8, 1966, 14.

46. "China Reports Downing Plane," *New York Times,* March 6, 1966, 54.

47. Jeffery L. Levinson, *Alpha Strike Vietnam* (New York: Pocket, 1989), 33.

48. Wagner, *Lightning Bugs and Other Reconnaissance Drones,* 93–97, 103, 104, 108, 109, 118.

49. Ibid., 125.

50. Ibid., 101, 102.

51. Ibid., 102.

52. Ibid., 46, 110, 111.

53. Ibid., 111, 112.

54. Ibid., 105, 111, 118, 138, 139.

55. Levinson, *Alpha Strike Vietnam,* 116, 117.

56. Wagner, *Lightning Bugs and Other Reconnaissance Drones,* 125.

57. Ibid., 118, 122–29.

58. Ibid., 118–20. During the testing of the Rivet Bouncer system, an unusual incident occurred. Due to a family emergency, a technical representative working on Rivet Bouncer was allowed home leave. He booked the first available flight, which was on Air France. He did not check the route, so was very surprised to discover its first stop was *Hanoi*. He remained on the plane and his passport was not checked.

59. Ibid., 213.

60. Levinson, *Alpha Strike Vietnam,* 177–79, 214, 215, 223, 250, 251.

61. "Chinese Pilots Guided by the Thoughts of Mao," *New York Times,* January 15, 1968, 3. The "Heroic Sea Eagle Regiment" of the Chinese Navy was described as being inspired "to completely wipe out the bourgeois military line and smash the illogical, stereotyped foreign programs and manuals." The thoughts of Chairman Mao were described as better than compasses and radar. "With Mao Tse–tung's thoughts in our minds," one pilot was quoted as saying, "we can tell exactly where we are heading, even in clouds or dense fog."

62. "U.S. Plane Downed, China Says," *New York Times,* April 30, 1967, 5.

63. "U.S. Plane Downed, Says China," *New York Times,* June 13, 1967, 3.

64. Pocock, *Dragon Lady,* 113.

65. Wagner, *Lightning Bugs and Other Reconnaissance Drones,* 133, 134.
66. Ibid., 135, 136.
67. Ibid., 136, 137.
68. Ibid., 134.
69. "U.S. Plane Claimed by Hanoi," *New York Times,* June 9, 1968, 9.
70. Wagner, *Lightning Bugs and Other Reconnaissance Drones,* 191.
71. A. J. Wang, "The Air Force of the Dragon," *Air Combat* (September 1991): 19.
72. "Hanoi Will Demand Reconnaissance Flights' End," *New York Times,* May 13, 1968, 4.
73. Gene Roberts, "Enemy Increases Convoys in North," *New York Times,* November 16, 1968, 1.
74. Wagner, *Lightning Bugs and Other Reconnaissance Drones,* 139, 140–45.
75. Ibid., 172, 190, 191.
76. "Hanoi Claims U.S. Drone," *New York Times,* April 20, 1969, 61.
77. Wagner, *Lightning Bugs and Other Reconnaissance Drones,* 190, 191.
78. Ibid., 157–65.
79. Ibid., 166–71, 213.
80. Benjamin F. Schemmer, *The Raid* (New York: Harper and Row, 1976), 35, 36.
81. Ibid., 98–100, 173–80.
82. Earl H. Tilford Jr., *Search and Rescue in Southeast Asia 1961–1975* (Washington, D.C.: Office of Air Force History, 1980), 109–11; and Schemmer, *The Raid,* 200–10.
83. Wagner, *Lightning Bugs and Other Reconnaissance Drones,* 172, 208.
84. Ibid., 193, 194.
85. Ibid., 199. The 147SC/TV did have a major problem—a tendency to just fall out of the sky. The problem was traced to the fin-shaped antenna that transmitted the television signals to the DC-130. It was found that the fin caused the drone to become directionally unstable. Once it was replaced with a flush antenna, the problem disappeared.
86. Ibid., 198.
87. Ibid., 198–200. The 147 drones live on in China. In the 1970s, photos were published of a Chinese copy of reconnaissance drones very similar to the 147G/H high-altitude drones. The launch aircraft was a Tu-4 (a Soviet-built copy of the B-29) modified with turboprop engines.
88. "Drone Operations in Vietnam," *Warplane,* (Vol. 4, Issue 48, 1986), 946.
89. Wagner, *Lightning Bugs and Other Reconnaissance Drones,* 24, 25, 191, 192, 197, 199, 200, 209, 213. Much of the original work on the 147SD's systems was actually done for the Model 124I reconnaissance

drone built for Israel in the early 1970s. This was a ground-launched drone able to fly both high- and low-altitude missions.

CHAPTER 5

1. William Wagner and William P. Sloan, *Fireflies and other UAVs* (Arlington, Tex.: Aerofax, 1992), 36, 37.

2. Ibid., 38–40.

3. Ibid., 40, 41.

4. Ibid., 28–34, 41. Following a U-2 crash in the late 1950s, a small-town newspaper agreed not to publish a photo of the wreckage. By 1969, the press's attitude toward "security" had changed.

5. "Secret Plane Test Disclosed," *New York Times,* August 7, 1969, 24.

6. Wagner and Sloan, *Fireflies and other UAVs,* 29, 30.

7. Ibid., 42–45.

8. Ibid., 45–47.

CHAPTER 6

1. Jay Miller, *Lockheed's Skunk Works: The First Fifty Years* (Arlington, Tex.: Aerofax, 1993), 135.

2. Ben R. Rich and Leo Janos, *Skunk Works* (New York: Little Brown, 1994), 263, 264.

3. Interview with Keith Beswick, January 12, 1994.

4. Miller, *Lockheed's Skunk Works,* 135.

5. Rich and Janos, *Skunk Works,* 264, 265.

6. *"Kelly's Way": The Story of Kelly Johnson and the Lockheed Skunk Works* (Edwards AFB: Air Force Flight Test Museum, 1993), videotape.

7. Miller, *Lockheed's Skunk Works,* 136.

8. Steve Pace, *Lockheed Skunk Works* (Osceola, Wis.: Motorbooks, 1992), 179–82.

9. Miller, *Lockheed's Skunk Works,* 139, 141, 199.

10. Paul Crickmore, *Lockheed SR-71 Blackbird* (Osceola, Wis.: Motorbooks, 1986), 31, 32.

11. Paul F. Crickmore, *Lockheed SR-71: The Secret Missions Exposed* (London: Osprey Aerospace, 1993), 36, 38; and *"Kelly's Way,"* videotape. This included scenes inside the launch control officer's cockpit showing the control panel. There are also shots of the wind-tunnel tests, a D-21/M-21 on the ground, refueling, and two successful D–21 launches from the onboard camera.

12. Miller, *Lockheed's Skunk Works,* 136, 137; and Crickmore, *Lockheed SR-71: The Secret Missions Exposed,* 32, 35–38.

13. Miller, *Lockheed's Skunk Works,* 137, 138.

14. Beswick interview, January 14, 1994.

15. Crickmore, *Lockheed SR-71: The Secret Missions Exposed,* 38.

16. Beswick interview, January 12, 1994; and *"Kelly's Way,"* videotape. The video contains footage of two D-21/M-21 launches. The second one shown was actually the first launch. It is not identified in the film.

17. Miller, *Lockheed, Skunk Works,* 138.

18. Beswick interview, January 12, 1994.

19. Crickmore, *Lockheed SR-71: The Secret Missions Exposed,* 38.

20. Miller, *Lockheed's Skunk Works,* 138.

21. Ibid., 139, 140. To put the 24,000-pound weight of each D-21B/booster rocket in perspective, the Hound Dog missile weighed 10,140 pounds, the MK 15 nuclear weapon weighed 7,600 pounds, the MK 39 and MK 41 nuclear weapons each weighed 10,000 pounds.

22. Crickmore, *Lockheed SR-71 Blackbird,* 35.

23. Miller, *Lockheed's Skunk Works,* 139.

24. Ibid., 139, 141.

25. Ibid., 140; and private sources. In what may have been a sighting of the accidental D-21B drop of September 28, 1967, a student on his lunch break glanced up and noticed a large jet accompanied by four other, smaller aircraft. The planes were very high, 40,000 feet or more, he judged. He could barely see the shape of the large plane, while the smaller planes could be seen only because of their contrails. The group was flying west and seemed to be directly over Torrance, California, south of Los Angeles. He then noticed one of the large plane's contrails (the inner one, on the left side as he recalled later) began growing. Suddenly, the contrail "got huge" and another contrail shot out ahead of the large plane. He estimated it was going twice as fast as the other planes. The whole formation, new object included, began a gradual turn to the northeast. They were then lost from sight. The whole event had lasted only a minute. The student, who was interested in space, thought he had seen an X-15 launch. He had seen films of X-15 launches and knew several chase planes followed the B-52 drop plane. Unfortunately, he does not recall the date. According to one source, a D-21 launch had been made over the Los Angeles area. (This was before the Miller book was published.) The description is consistent with the large solid rocket used to boost the D-21B, but one problem is that he believes the sighting occurred during the 1966–67 school year. This is inconsistent with the known D-21B modification program. The Skunk Works is unwilling to say where the September 28, 1967, drop occurred. The lack of official confirmation, along with the witness's inability to remember the date, makes an exact identification impossible.

26. Miller, *Lockheed's Skunk Works,* 140.

27. Ibid., 140, 141, 199.
28. Rich and Janos, *Skunk Works,* 268, 269.
29. Miller, *Lockheed's Skunk Works,* 141.
30. Ibid., 141, 199; and Rich and Janos, *Skunk Works,* 269.
31. Miller, *Lockheed's Skunk Works,* 141.
32. Anthony M. Thornborough and Peter E. Davies, *Lockheed Blackbirds* (Osceola, Wis.: Motorbooks, 1988), 56, 57.
33. "Lockheed D-21," *Aerophile* (March/April 1977), 106, 107.
34. Lou Drendel, *SR-71 Blackbird in Action* (Carrollton, Tex.: Squadron/Signal, 1982), 21.
35. Crickmore, *Lockheed SR-71 Blackbird,* 35, 36.
36. Thornborough and Davies, *Lockheed Blackbirds,* 56, 57; and Bill Sweetman, *Aurora: The Pentagon's Secret Hypersonic Spyplane* (Osceola, Wis.: Motorbooks, 1993), 9.
37. Michelle Briggs, "Classified D-21 Mini-Blackbird On Display at Arizona Museum," *Pacific Flyer,* November 1993, A39.
38. "Top-Secret D-21 Drone Uncloaked," *Museum of Flight News* (January/February 1994), 4, 5.
39. Curtis Peebles, *Guardians* (Novato, Calif.: Presidio Press, 1987), chapters 7 and 8.
40. Miller, *Lockheed's Skunk Works,* 129.
41. Rich and Janos, *Skunk Works,* 270. The Soviet counterpart to the D-21 was the 123 DBR. Although it was a high-speed, high-altitude reconnaissance drone, the 123 DBR had a very different mission profile. It was 95.12 feet long, with a top speed of Mach 2.55, a range of 1,620 nautical miles, and an operating altitude of 69,000 to 72,000 feet. Flight time was about 90 minutes. It was launched from a large trailer by its own KR-15 jet engine and two rocket boosters that separated after liftoff. It carried three AFA-54 vertical cameras and one oblique camera in a detachable nose. As with the D-21, at the end of the mission the 123 DBR's nose would separate; it would then be recovered while the rest of the airframe would crash. Liftoff weight of the 123 DBR was over 63,800 pounds. It was built by the Tupolev Design Bureau and operated by the Soviet army. The 123 DBR first entered service in the late 1960s and was phased out in the late 1970s/early 1980s. As with the D-21, it would remain Black for a decade. It was not until early 1994 that photos of the 123 DBR were released by the Russians.

CHAPTER 7

1. Peter M. Grosz, "So, What's New about Stealth?," *Air International* (September 1986): 147–51. There is very little documentation about the German World War I "stealth" aircraft. Although this might suggest some

type of secrecy, it is more likely that it was not seen as a very practical project.

2. Alfred Price, *The History of U.S. Electronic Warfare,* vol. 1 (Privately Printed: Association of Old Crows, 1984), 260, 261.

3. Bill Gunston, "Back to Balloons and Gliders?," *Air International* (May 1986), 228.

4. Doug Richardson, *Stealth* (New York: Orion, 1989), 24–33. From time to time, some aircraft showed a reduced RCS by chance. The German Go 229, another flying wing design, was nearly invisible to radar. It was made of wood and had charcoal mixed with the glue to absorb radar signals. The wood construction also hid the radar return from the engines. The British Vulcan bomber was also hard to detect from some angles.

5. Alfred Price, *The History of U.S. Electronic Warfare,* vol. 2 (Privately Printed: Association of Old Crows, 1989), 199, 200. There may have been an attempt to reduce the large RCS of the XB-70. The wing and fuselage formed a corner reflector. The second XB-70 had 5 degrees of dihedral added to the wings. Although described as being done for aerodynamic reasons, it would also cause the radar echo to be dispersed by 10 degrees away from the radar. The (unbuilt) third prototype, the YB-70, was to have had the canard sweep changed from 31 degrees to 51 degrees, which would bring the echo of the canards and wings into line. Before this, the canards would have their own echoes. At best, these efforts had only a minimal effect.

6. *Air Combat* (Alexandria, Va.: Time-Life Books, 1990), 40, 41.

7. Steven J. Zaloga, *Soviet Air Defence Missiles* (Alexandria, Va.: Jane's Information Group, 1989), 222–25.

8. Bill Sweetman and James Goodall, *Lockheed F-117* (Osceola, Wis.: Motorbooks, 1990), 18, 19.

9. Robert and Melinda Macy, *Destination Baghdad* (Las Vegas: M&M Graphs, 1991), 41.

10. Jay Miller, *Lockheed F-117 Stealth Fighter* (Arlington, Tex.: Aerofax, 1990), 7.

11. "We Own the Night," *Lockheed Horizons* (May 1992), 6–9.

12. Ben R. Rich and Leo Janos, *Skunk Works* (New York: Little, Brown, 1994), 26–31. Ironically, at the time the "F-19" kit came out, the author made the comment, "This thing better not be disk-shaped, or the air force will never be able to explain it."

13. "Northrop's 1976 Stealth Fighter Proposal Featured Faceted Body with Overhead Inlet," *Aviation Week and Space Technology* (February 10, 1992), 23.

14. "Declassified Photos Show 'Have Blue' F-117 Predecessor," *Aviation Week and Space Technology* (April 22, 1991), 30.

15. Jay Miller, *Lockheed's Skunk Works: The First Fifty Years* (Arlington, Tex.: Aerofax, 1993), 161.
16. Rich and Janos, *Skunk Works,* 34–36.
17. Miller, *Lockheed's Skunk Works,* 159, 163.
18. Jim Cunningham, "Cracks in the Black Dike: Secrecy, the Media, and the F-117A," *Airpower Journal* (Fall 1991), 17.
19. "Lockheed California Co. Is Developing a Small Fighter Intended to Demonstrate Stealth, or Low Signature Technologies," *Aviation Week and Space Technology* (August 2, 1976), 11.
20. John W. R. Taylor, ed., *Jane's All the World's Aircraft* (London: Jane's Yearbooks, 1978), 326.
21. "First Flight of Lockheed's New Stealth Fighter Demonstrator Being Built by the Company's 'Skunk Works'," *Aviation Week and Space Technology* (June 20, 1977), 11.
22. Sweetman and Goodall, *Lockheed F-117,* 21.
23. Steven Pace, *Lockheed Skunk Works* (Osceola, Wis.: Motorbooks, 1992), 222–24.
24. Ralph Vartabedian, *Now It Can Be Told—He Has the Right Stuff, Los Angeles Times,* September 29, 1989.
25. Rich and Janos, *Skunk Works,* 50–53; and Miller, *Lockheed's Skunk Works,* 161.
26. "Declassified Photos," 30. The camouflage scheme used on the first Have Blue and F-117 is similar to the "dazzle" finish used on ships during World Wars I and II. The idea was not to "hide" the ships but, rather, make it difficult for a U-boat commander to determine the ship's speed and heading. On an airplane, particularly one with an odd shape, like the Have Blue and F-117, a similar effect could be expected.
27. Rich and Janos, *Skunk Works,* 54–57; and Pace, *Lockheed Skunk Works,* 224.
28. Ralph Vartabedian, *Now It Can Be Told, Los Angeles Times,* September 29, 1989.
29. Miller, *Lockheed's Skunk Works,* 162.
30. Pace, *Lockheed Skunk Works,* 224.
31. Rich and Janos, *Skunk Works,* 58–61.
32. Ralph Vartabedian, "Now It Can Be Told," *Los Angeles Times,* September 29, 1989.
33. "Plane Crash Shrouded in Mystery," *San Diego Union,* undated clipping, May 1978.
34. Miller, *Lockheed Skunk Works,* 162.
35. Rich and Janos, *Skunk Works,* 4.
36. Sweetman and Goodall, *Lockheed F-117,* 24, 25.
37. Rich and Janos, *Skunk Works,* 69.

38. Sweetman and Goodall, *Lockheed F-117,* 24, 25.

39. Pace, *Lockheed Skunk Works,* 225.

40. Ralph Vartabedian, "Now It Can Be Told," *Los Angeles Times,* September 29, 1989.

41. "Declassified Photos," 30.

42. Research File Groom Lake (Nevada), Area 51 and Project Red Light, (W. L. Moore Publications, Compiled 1987). The only date on the clipping is 1979, but the mention of a second crash suggests a date in the late summer of 1979. "Red Light" was claimed by a person named "Mike" to be the code name for the flight tests of a captured alien UFO at Groom Lake in the early 1960s. William Moore wrote a book in 1980 called *The Roswell Incident* in which he claimed that the army air forces had captured a flying saucer in July 1947, which has been covered up ever since. Several "UFOlogists" have claimed that stealth technology was actually developed from captured flying saucers. See the author's book *Watch the Skies!* for more details on the crashed saucer/stealth/Groom Lake stories.

43. Sweetman and Goodall, *Lockheed F-117,* 25, 26; and Cunningham, "Cracks in the Black Dike," 18, 19.

44. Richard P. Hallion, *Storm Over Iraq* (Washington, D.C.: Smithsonian Institution Press, 1992), 88.

45. Miller, *Lockheed F-117 Stealth Fighter,* 6.

46. Melinda Beck, William J. Cook, and John J. Lindsay, "Unveiling a Ghost Airplane," *Newsweek* (September 15, 1980), 23.

47. Cunningham, "Cracks in the Black Dike," 19, 20.

48. U.S. Air Force Oral History Interviews, Maj. Gen. Osmond J. Ritland, March 19–21, 1974, vol. 1 (Edwards AFB History Office: Ritland Files), 142.

49. Ed Vogel, "Military reportedly moving in to stake claim on Groom Mountains," *Las Vegas Review-Journal,* March 10, 1984, sec. B; and Ed Vogel, "Nevadans question Air Force seizure of land," *Las Vegas Review-Journal,* May 20, 1984, sec. B.

50. Mary O'Driscoll, "Reid backs USAF on land issue,"*Las Vegas Sun,* July 7, 1984, sec. B.

51. Ed Vogel, "Nevadans question Air Force seizure of land," *Las Vegas Review-Journal,* May 20, 1984, sec. B.

52. Ibid.

53. "Gov. Bryan fires at Air Force," *Las Vegas Sun,* August 7, 1984, sec. A.

54. Mary Manning, "Irate Nevadans 'march' on D.C. to protest federal land-grabbing," *Las Vegas Sun,* August 6, 1984, sec. B.

55. Timothy Good, *Alien Contact* (New York: William Morrow, 1993), 148, 149; and Chris Chrystal, "AF admits to illegality of NTS land grab," *Las Vegas Sun,* August 7, 1984, sec. A.

56. Kathleen Buckley, "Wilderness tradeoff proposed," *Las Vegas Review-Journal,* September 14, 1984, sec. B; and Chris Chrystal, "2-year limit in Groom Mtn. land grab bill," *Las Vegas Sun,* September 14, 1984, sec. C.
57. Kathleen Buckley, "Top secret maneuvers in jeopardy," *Las Vegas Review-Journal,* September 19, 1984, sec. B.
58. Christopher Bell, "Air Force will seek continued access restriction to secret base," *Las Vegas Review-Journal,* October 11, 1987, sec. B; and David Koenig, "House panel drops plan for wilderness," *Las Vegas Review-Journal,* November 19, 1987, sec. C.
59. David Koenig, "Air Force seeks extension of closed area on Groom Mountain," *Las Vegas Review-Journal,* March 18, 1988, sec. B.
60. David Koenig, "Congress extends fed takeover of Groom area," *Las Vegas Review-Journal,* April 2, 1988, sec. B.
61. "Group stakes claims during lapse," *Las Vegas Review-Journal,* June 17, 1988, sec. B.
62. David Koenig, "House separates Groom Mountain pact from wilderness issue," *Las Vegas Review-Journal,* June 15, 1988, sec. B.

CHAPTER 8

1. Steve Pace, *Lockheed Skunk Works* (Osceola, Wis.: Motorbooks, 1992), 227.
2. Robert and Melinda Macy, *Destination Baghdad* (Las Vegas: M&M Graphics, 1991), 42.
3. "Declassified Photos Show 'Have Blue' F-117A Predecessor," *Aviation Week and Space Technology* (April 22, 1991), 30.
4. Paul W. Martin, "Development of the F-117 Stealth Fighter," *Lockheed Horizons* (August 1992), 22.
5. Pace, *Lockheed Skunk Works,* 227–29. The first five aircraft are sometimes referred to as "YF-117." This is incorrect.
6. Jay Miller, *Lockheed's Skunk Works: The First Fifty Years* (Arlington, Tex.: Aerofax, 1993), 165.
7. Pace, *Lockheed Skunk Works,* 227.
8. William Scott, "F-117A Design Presented Avionics Challenges," *Aviation Week and Space Technology* (February 8, 1993), 43.
9. Harold C. Farley Jr. and Richard Abrams, "F-117A Flight Test Program," *Society of Experimental Test Pilots 1990 Report to the Aerospace Profession,* 144, 145.
10. Ben R. Rich and Leo Janos, *Skunk Works* (New York: Little, Brown, 1994), 83–85.
11. "We Own the Night," *Lockheed Horizons* (May 1992), 8, 9, 14, 15; and Miller, *Lockheed's Skunk Works,* 166.

12. "We Own the Night," 10–11.
13. Farley and Abrams, "F-117A Flight Test Program," 147, 148.
14. "We Own the Night," 6, 12, 13, 61.
15. Bill Sweetman and James Goodall, *Lockheed F-117A* (Osceola, Wis.: Motorbooks, 1990), 27.
16. "We Own the Night," 13. Ben Rich, who preferred gray, explained that, "The Skunk Works plays by the Golden Rule: he who has the gold sets the rules! If the General had wanted pink, we'd have painted them pink."
17. Farley and Abrams, "F-117A Flight Test Program," 148, 149.
18. Ibid., 150, 151.
19. Ibid., 153, 154.
20. Scott, "F-117A Design Presented Avionics Challenges," 43, 44.
21. Macy, *Destination Baghdad,* 27, 50, 53–55; and Robert Shelton Jr. and Randy Jolly, *Team Stealth F-117* (Stillwater, Minn.: Specialty Press, 1993), 29, 31.
22. "We Own the Night," 18, 19, 24.
23. Jim Goodall, *F-117 Stealth in Action,* (Carrollton, Tex.: Squadron/ Signal Publications, 1991), 16.
24. Pace, *Lockheed Skunk Works,* 233; and Sweetman and Goodall, *Lockheed F-117A,* 29, 30.
25. "We Own the Night," 15, 60, 61. "Bandit" was the radio call sign for the F-117 during flight tests. All the operational air force pilots received numbers counting from 150. The Lockheed and air force test pilots had numbers in the 100 range. Hal Farley, the first man to fly the plane, was given the special number "117."
26. D. M. Giangreco, *Stealth Fighter Pilot* (Osceola, Wis.: Motorbooks, 1993), 34, 35.
27. Goodall, *F-117 Stealth in Action,* 30.
28. "We Own the Night," 20.
29. Goodall, *F-117 Stealth in Action,* 31–39.
30. Macy, *Destination Baghdad,* 28.
31. Goodall, *F-117 Stealth in Action,* 31–33.
32. Macy, *Destination Baghdad,* 27, 67.
33. Interview with Capt. John Hesterman and Lt. Col. Bob Maher, Edwards AFB Air Show, October 9, 1990.
34. Goodall, *F-117 Stealth in Action,* 29.
35. "We Own the Night," 20, 21.
36. Goodall, *F-117 Stealth in Action,* 35. The effect of the phase of the moon on F-117A operations is not unique. During World War II, RAF bomber squadrons stood down during the "moon period" because it made the night sky too bright.

37. Michael A. Dornheim, "F-117A Provides New Freedom in Attacking Ground Tragets," *Aviation Week and Space Technology* (May 14, 1990), 106.
38. Goodall, *F-117 Stealth in Action,* 19–21, 30.
39. Richard P. Hallion, *Storm over Iraq* (Washington, D.C.: Smithsonian Institution Press, 1992), 99.
40. Goodall, *F-117 Stealth in Action,* 42.
41. "We Own the Night," 60.
42. *Electronic Spies* (Alexandria, Va.: Time-Life Books, 1991), chap. 1.
43. Goodall, *F-117 Stealth in Action,* 19, 42. This book contains two interviews with Steve Paulson, one of the original 4450th Tactical Group members. He stated, "Yes, let's just say we were within hours of going on two different occasions." He would not elaborate.
44. "We Own the Night," 14.
45. This information was obtained from a framed set of F-117A development patches shown at the Edwards AFB Air Show as part of the F-117 display.
46. James C. Goodall, "Research Data: Lockheed's F-117A Stealth Fighter," *FineScale Modeler* (July 1990), 48.
47. "We Own the Night," 24.
48. Sweetman and Goodall, *Lockheed F-117A,* 78, 79.
49. F-117A development patches.
50. Goodall, *F-117 Stealth in Action,* 23.
51. Jim Cunningham, "Cracks in the Black Dike: Secrecy, the Media, and the F-117A," *Airpower Journal* (Fall 1991), 22, 23.
52. Washington Roundup, *Aviation Week and Space Technology* (October 12, 1981), 17.
53. Sweetman and Goodall, *Lockheed F-117A,* 28. Accounts differed as to why there was no "F-19." One version has it that Northrop wanted "F-20" to symbolize a new generation of fighter. Another was that it was to avoid confusion with the MiG 19. This was not the only skipped F-number. There is no "F-13."
54. Benjamin F. Schemmer, "Stealth Goes Public," *Armed Forces Journal International* (October 1983), 23, 24.
55. Richard Barnard, "AF Abandons Stealth Reconnaissance Plane; Deploys Fighter Units," *Defense Week* (November 21, 1983), 1, 6.
56. Richard Dietrich, "Design of 'stealth' plane kit a model of intelligence work," *San Diego Tribune,* September 12, 1986, A-1.
57. Ron Labrecque, "Alone and without Stealth, John Andrews Makes Child's Play of the Pentagon's Most Secret Weapon," *People* (no date on clipping), 127, 128; and Cunningham, "Cracks in the Black Dike," 20, 24, 25.
58. Washington Roundup, *Aviation Week and Space Technology* (September 14, 1987), 21.

59. Cunningham, "Cracks in the Black Dike," 25. The document control problem occurred because of standard Skunk Works practice. When a document was no longer needed, it was destroyed. When the air force showed up to conduct a classified document audit, they discovered there were no records of their destruction. As there was no record, the documents were ruled "lost." Later, a desk-by-desk audit system was established.

60. *Weekly World News,* August 2, 1986.

61. Cunningham, "Cracks in the Black Dike," 25.

62. "Triangle over California," *The APRO Bulletin* (vol. 33, no. 6, 1986), 6.

63. William B. Scott, "F-117A Crash Reports Cite Pilot Fatigue, Disorientation," *Aviation Week and Space Technology* (May 15, 1989), 22; and Sweetman and Goodall, *Lockheed F-117A,* 81, 82.

64. "Pentagon seals plane crash site," *San Diego Union,* July 12, 1986, sec. A.

65. "Triangle over California," 6.

66. Eric Malnic and Ralph Vartabedian, " 'Stealth' jet said to crash in California," *Boston Globe,* July 12, 1986.

67. "Stealth jet fighter fleet is put at 50," *San Diego Union,* August 22, 1986, sec. A; and Cunningham, "Cracks in the Black Dike," 26. It has been claimed that the air force dug up a large area around the impact point to recover every scrap of debris, then scattered parts from an F-101 to deceive anyone entering the site. This is incorrect.

68. Scott, "F-117A Crash Reports Cite Pilot Fatigue, Disorientation," 22, 23.

69. Giangreco, *Stealth Fighter Pilot,* 47–49, 51–53.

70. Scott, "F-117A Crash Reports Cite Pilot Fatigue, Disorientation," 22, 23.

71. "AF pilot killed in Nevada crash said to be flying Stealth fighter," *San Diego Union,* October 16, 1987, sec. A.

72. Giangreco, *Stealth Fighter Pilot,* 50, 51; and Scott, "F-117A Crash Reports Cite Pilot Fatigue, Disorientation," 22.

73. "A-7D Crash Accents Different USAF, Navy Emergency Rules," *Aviation Week and Space Technology* (November 2, 1987), 31; and Giangreco, *Stealth Fighter Pilot,* 51.

74. Benjamin F. Schemmer, "Is Lockheed Building a Super-Stealth Replacement for USAF's Mach 3 SR-71?" *Armed Forces Journal* (January 1988).

75. Al Frickey [pseud.], "Stealth—and Beyond," *Gung-Ho* (February 1988), 38.

76. "Do It Your Stealth," *Aviation Week and Space Technology* (September 14, 1987), 21.

77. Doug Richardson, *Stealth* (New York: Orion, 1989), 122.

78. Edwards AFB Air Show, October 23, 1988. The invisible "F-19" was not the only stealth prank at the show. Around noon, the announcer came on the public address system and said that the B-2 had just taken off from Palmdale and would soon be arriving at Edwards. A few minutes later, it was announced that the B-2 would be making a flyby. One heard the engine noise, but, again, one did not actually "see" it. The punchline came at the 1993 Edwards Air Show—*three* B-2s (two in flight, one on the ground), an F-117A formation flyby, and the HALSOL-Pathfinder.

79. Giangreco, *Stealth Fighter Pilot,* 60, 61.

80. John D. Morrocco, "USAF Unveils Stealth Fighter; Black Weapons Probe Likely," *Aviation Week and Space Technology,* (November 14, 1988), 28.

81. "Air Force lifts veil on Stealth fighter squadron," *San Diego Union,* November 11, 1988, sec. A.

82. Giangreco, *Stealth Fighter Pilot,* 60.

83. *San Diego Union,* November 11, 1988, sec. A.

84. Cunningham, "Cracks in the Black Dike," 27.

85. Giangreco, *Stealth Fighter Pilot,* 62. A later report said the whine was no longer being "heard."

86. Mary Enges-Maas, "Stalking the Stealth Fighter," *Times-Advocate,* December 3, 1989, sec. C.

87. "We Own the Night," 24, 25; and Sweetman and Goodall, *Lockheed F-117A,* 89, 90.

88. *Sky Soldiers* (Alexandria, Va.: Time-Life Books, 1991), 128.

89. David F. Bond, "Six F-117As Flown in Panama Invasion; Air Force Broadens Daytime Operations," *Aviation Week and Space Technology* (March 5, 1990), 30.

90. Giangreco, *Stealth Fighter Pilot,* 65.

91. "Stealth error kept under wraps," *San Diego Union,* April 7, 1990, sec. A. In all the adverse comments following the Panama mission, one aspect has been overlooked. The PDF had three hours warning that an attack was about to begin. Even had the bombs fallen where planned, the United States had already lost the element of surprise.

92. "General didn't report Stealth flaws in Panama," *San Diego Tribune,* July 2, 1990, sec. A.

93. John D. Morrocco, "F-117A Fighter Used in Combat for First Time in Panama," *Aviation Week and Space Technology* (January 1, 1990), 32.

94. Letters, *Newsweek* (July 16, 1990), 12.

95. Kenneth Freed, "Panama Tries to Bury Rumors of Mass Graves," *Los Angeles Times,* October 27, 1990, sec. A.

96. "Bombing Run on Congress," *Time* (January 8, 1990), 43. F-117A pilots have spent considerable time denying the plane was ever called the

"Wobbly Goblin." Many have called it the best-handling plane they have ever flown. The fact that the press continued to use the term into 1992 says more about their "accuracy" than that of the plane.

97. "Stealth error kept under wraps," *San Diego Union,* April 7, 1990, sec. A; and "General didn't report Stealth flaws in Panama," *San Diego Tribune,* July 2, 1990, sec. A. The final word on the F-117A's first combat mission came from a *Newsweek* press pool member. The reporter told a Department of Defense public relations officer that he did not think the F-117A attack was that significant, as no one could hear it coming. The officer laughed in the reporter's face.

98. Charles Krauthammer, "Don't Cash the Peace Dividend," *Time* (March 26, 1990), 88.

99. Bruce Van Voorst, "Who Needs the Marines? From the Halls of Montezuma to the Shores of Redundancy," *Time* (May 21, 1990), 28; and Bill Turque and Douglas Waller, "Warriors without War," *Newsweek* (March 19, 1990), 18–21.

100. Wade Greene, "An Idea Whose Time Is Fading," *Time* (May 28, 1990), 90.

101. "Three early Iraqi incursions are revealed," *San Diego Union,* October 7, 1990, sec. A.

102. Giangreco, *Stealth Fighter Pilot,* 66–73.

103. Macy, *Destination Baghdad,* 19, 20; and Giangreco, *Stealth Fighter Pilot,* 86.

104. Jolly and Shelton, *Team Stealth F-117,* 52, 54, 56.

105. Giangreco, *Stealth Fighter Pilot,* 82–84.

106. Hallion, *Storm over Iraq,* 2, 159, 162. Such press criticism sometimes backfired— a "leading journalist" on a Washington, D.C., news show announced the B-2 would be used in the Gulf, while another commented that it would have to "do better than it did in Panama"!

107. *Gulf War Air Power Survey,* vol. II, part I (Washington, D.C.: Government Printing Office, 1993), 113.

108. Hallion, Storm over Iraq, 2, 3.

109. Joe Hughes, "500 protest war threat at Balboa Park," *San Diego Tribune,* December 24, 1990, sec. B.

110. Giangreco, *Stealth Fighter Pilot,* 87.

111. *Gulf War Air Power Survey,* vol. II, part I, 50, 51, 77–79.

112. Giangreco, *Stealth Fighter Pilot,* 91.

113. *Gulf War Air Power Survey,* vol. IV, 126, 127; and vol. II, part I, 36.

114. Giangreco, *Stealth Fighter Pilot,* 87–91.

115. Thomas B. Allen, F. Clifton Berry, and Norman Polmar, *CNN War in the Gulf* (Atlanta: Turner Publishing, 1991), 119.

116. Giangreco, *Stealth Fighter Pilot,* 90–96.

117. Hallion, *Storm over Iraq,* 170.

118. Giangreco, *Stealth Fighter Pilot,* 96, 97.

119. *Gulf War Air Power Survey,* vol. II, part I, 124, 126–33.

120. Macy, *Destination Baghdad,* 11, 12.

121. Giangreco, *Stealth Fighter Pilot,* 98, 99.

122. *Gulf War Air Power Survey,* vol. II, part I, 134.

123. Giangreco, *Stealth Fighter Pilot,* 100.

124. Martin Middlebrook, *The Berlin Raids* (New York: Viking, 1988).

125. *Gulf War Air Power Survey,* vol. II, part I, 137, 147, 337, 338. Saddam Hussein responded in the usual manner of dictators to the night of thunder. He claimed over a hundred Coalition aircraft had been shot down, and he had Iraqi Air Force General Muzahim Saab Hassan arrested and executed.

126. Macy, *Destination Baghdad,* 17.

127. Giangreco, *Stealth Fighter Pilot,* 100, 101; and *Gulf War Air Power Survey,* vol. II, part I, 147, 148.

128. Giangreco, *Stealth Fighter Pilot,* 100, 101; and *Gulf War Air Power Survey,* vol. II. part I, 152.

129. Giangreco, *Stealth Fighter Pilot,* 101.

130. *Gulf War Air Power Survey,* vol. II, part I, 157, 225.

131. Giangreco, *Stealth Fighter Pilot,* 18, 101. Even if the Package Q strike had used LGB, the F-16 force would only have been cut to sixteen aircraft, and the tankers to eleven.

132. *Gulf War Air Power Survey,* vol. II, part I, 201, 207, 223.

133. Ibid., 177.

134. Hallion, *Storm over Iraq,* 194.

135. Private source.

136. *Gulf War Air Power Survey,* vol. II, part I, 339.

137. Giangreco, *Stealth Fighter Pilot,* 102–4.

138. Jolly and Shelton, *Team Stealth F-117,* 62, 67.

139. Stan Morse, *Gulf Air War Debrief* (London: Aerospace Publishing, 1991), 84, 85, 106.

140. Giangreco, *Stealth Fighter Pilot,* 105–10.

141. *Gulf War Air Power Survey,* vol. II, part I, 204, 205.

142. *Gulf War Air Power Survey,* vol. II, part II, 388, 389.

143. Giangreco, *Stealth Fighter Pilot,* 116, 117.

144. Macy, *Destination Baghdad,* 13.

145. Giangreco, *Stealth Fighter Pilot,* 111, 112.

146. "Bunker: If military use known, why wasn't civilian role detected?," *San Diego Union,* February 14, 1991, sec. A.

147. *Gulf War Air Power Survey,* vol. II, part I, 206, 207, 220, 221, 388, 389.

148. Allen, Berry and Polmar, *War in the Gulf,* 137, 140. The 1907 Hague Conventions, the 1949 Geneva Conventions, and the 1977 Protocols state: "In sieges and bombardments, all necessary measures must be taken to spare, as far as possible, buildings dedicated to . . . charitable purposes . . . hospitals, and places where the sick and wounded are collected, *providing they are not being used at the time for military purposes* (emphasis added). It is the duty of the besieged to indicate the presence of such buildings or places by distinctive and visible signs, which shall be notified to the enemy beforehand." As the bunker was being used as a military installation at the time, it would be considered a legitimate target. And, as one letter writer noted, "At least four internationally recognized emblems other than the red cross/red crescent can . . . be placed on civilian shelters to indicate protected status. Camouflage paint is not one of them."

149. *Gulf War Air Power Survey,* vol. II, part I, 220, 221, 242.

150. Ibid., 224–28.

151. Ibid., 229–31.

152. Giangreco, *Stealth Fighter Pilot,* 115, 116.

153. Macy, *Destination Baghdad,* 29; and Giangreco, *Stealth Fighter Pilot,* 113–15.

154. *Gulf War Air Power Survey,* vol. II, part I, 247; and Giangreco, *Stealth Fighter Pilot,* 118, 119.

155. Jolly and Shelton, *Team Stealth F-117,* 93.

156. *Gulf War Air Power Survey,* vol. II, part I, 322–25.

157. Private source. An army sergeant recalled his first sight of Iraqi troops. He was driving his fuel truck across the desert when he saw two Iraqis beside an overturned jeep. As he drove past, they waved. He thought, "This is the enemy?" Then, at a refueling stop, two columns marched in and sat down. They were either very young or old men, ragged and hungry. He never saw an Iraqi officer or NCO; they had fled, abandoning their troops in the desert.

158. "'Peace with Honor'—and at Great Price," *Life,* (March 18, 1991), 58, 59.

159. Macy, *Destination Baghdad,* 68, 69, 74, 75.

160. Giangreco, *Stealth Fighter Pilot,* 122; and private source.

161. Giangreco, *Stealth Fighter Pilot,* 120–23.

162. Timothy R. Gaffney, "Secrets of the F-117A," *Popular Science* (September 1993), 73–77, 105, 106.

163. David A. Fulghum, "Planners Seek to Exploit U.S. Technology Lead," *Aviation Week and Space Technology* (January 17, 1994), 52.

164. John D. Morrocco, "Lockheed Returns to Navy with New F-117N Design," *Aviation Week and Space Technology* (March 7, 1994); and Miller, *Lockheed's Skunk Works,* 182–84. The F-117N proposal highlights the

decline of naval aviation. While the air force is deploying a second generation of stealth aircraft, the B-2 Spirit and F-22 Lightning II, the navy has yet to build even a first-generation stealth aircraft. When the last A-6 Intruder is retired in 1999, the navy will be without a long-range, all-weather attack aircraft for a decade.

165. Fulghum, "Planners Seek to Exploit U.S. Technology Lead," 52.

CHAPTER 9

1. Don Dwiggins, *Man-Powered Aircraft* (Blue Ridge Summit, Pa.: Tab, 1979).

2. AeroVironment Inc., "Backgrounder—Pathfinder Solar-Electrical Aircraft" (Simi Valley, Calif.: AeroVironment Inc., 1993).

3. Ibid.

4. Stuart F. Brown, "The Eternal Airplane," *Popular Science* (April 1994), 73.

5. Michael A. Dornheim, "Raptor/Pathfinder to Test High-Efficiency Propulsion," *Aviation Week and Space Technology* (October 11, 1993), 60; and Brown, "The Eternal Airplane," 75.

6. Dornheim, "Raptor/Pathfinder to Test High-Efficiency Propulsion," 60. The overall program to develop UAVs to carry our ballistic missile detection-interception systems was called "RAPTOR." This stood for "Responsive Aircraft Program for Theater OpeRations."

7. Ballistic Missile Defense Organization, " 'Pathfinder' Solar-Electric Unmanned Air Vehicle Data Sheet" (Washington, D.C.: BMDO, 1993).

8. Author's examination of the HALSOL-Pathfinder aircraft at the 1993 Edwards AFB Air Show.

9. Letter: Ballistic Missile Defense Organization to Curtis Peebles, reference FOIA Request 93-F-2470, February 16, 1994.

10. Brown, "The Eternal Airplane," 75; and David A. Fulghum, "Solar-Powered UAV to Fly at Edwards," *Aviation Week and Space Technology* (October 4, 1993), 27.

11. AeroVironment Inc., "Backgrounder—Pathfinder Solar-Electrical Aircraft."

12. Ballistic Missile Defense Organization, "Pathfinder Questions and Answers" (Washington, D.C.: BMDO, 1993).

13. AeroVironment Inc., "Backgrounder—Pathfinder Solar-Electrical Aircraft."

14. Ibid.; and " 'Pathfinder' Solar-Electric Unmanned Air Vehicle Data Sheet."

15. Dornheim, "Raptor/Pathfinder to Test High-Efficiency Propulsion," 60.

16. United States Air Force Press Release "BMDO to Display HALE" (Washington, D.C.: BMDO, October 1, 1993); and "Pathfinder Questions and Answers."

17. Ballistic Missile Defense Organization, "Solar-Powered Pathfinder Soars on First Test Flight" (Washington, D.C.: BMDO, 1993); and NASA videotape of first Pathfinder flight, 1993 Edwards AFB Air Show.

18. Author's examination of the HALSOL/Pathfinder at the 1993 Edwards AFB Air Show.

19. Fulghum, "Solar-Powered UAV to Fly at Edwards," 27; and Brown, "The Eternal Airplane," 100.

20. Interview with Lt. Col. Dale Tietz, director of the Raptor/Talon program, April 5, 1994.

21. Interview with Bob Curtin of AeroVironment Inc., April 5, 1994.

22. David A. Fulghum, "Scud-Killing UAVs Now Science Tool," *Aviation Week and Space Technology* (October 31, 1994), 27.

23. Curtin interview, April 5, 1994.

24. AeroVironment Inc., "Backgrounder—Pathfinder Solar-Electric Aircraft."

25. Brown, "The Eternal Airplane," 100.

26. Lt. Col. Dale Tietz, "Raptor/Talon Briefing," Ballistic Missile Defense Organization, 1993; and Dornheim, "Raptor/Pathfinder to Test High-Efficiency Propulsion," 60.

27. Brown, "The Eternal Airplane," 74; and AeroVironment Inc., "Backgrounder—Pathfinder Solar-Electric Aircraft."

28. AeroVironment Inc., "Backgrounder—Pathfinder Solar-Electric Aircraft,"

29. Conversations with NASA personnel at the Pathfinder display, 1993 Edwards AFB Air Show; and "Solar-Powered Spy," *Air and Space* (February/March 1994), 11, 12.

30. Tietz interview, April 5, 1994.

31. Ballistic Missile Defense Organization, "'Pathfinder' Solar-Electric Unmanned Air Vehicle Data Sheet," 1993.

32. Ballistic Missile Defense Organization, "Pathfinder Questions and Answers," 1993.

33. Louis C. Gerken, *UAV—Unmanned Aerial Vehicles* (Chula Vista, Calif.: American Scientific Corporation, 1991), 110. The "Teal" prefix in Teal Rain commonly refers to sensor systems. (Recall that "Senior" is used for strategic reconnaissance such as the D-21, while "Have" was used for aircraft technology programs like Have Blue.)

34. Ronald D. Murphy, "AMBER for long endurance," *Aerospace America* (February 1989), 23.

35. Private source.

36. Leading Systems, Inc. "Demonstrated 35 hrs endurance; large payload; 27,800 ft" (Irvine, Calif.: Leading Systems Inc., 1988); and John D. Morrocco, "Navy Plans Operational Trials for Amber RPV in 1989," *Aviation Week and Space Technology* (December 14, 1987), 25, 26.

37. Murphy, "AMBER for long endurance," 33, 34.

38. Leading Systems, Inc. "Demonstrated," 1988.

39. Morrocco, "Navy Plans Operational Trials for Amber RPV in 1989," 26.

40. Forecast International, Amber/GNAT Orientation, 1990.

41. Murphy, "AMBER for long endurance," 32.

42. William Wagner and William P. Sloan, *Fireflies and Other UAVs* (Arlington, Tex.: Aerofax, 1992), 18.

43. Leading Systems, Inc., "The Leading Systems Story" (Irvine, Calif.: Leading Systems Inc., July 1989).

44. General Atomics, "Advanced Technology Amber I" (San Diego, Calif.: General Atomics, n.d.).

45. Murphy, "AMBER for long endurance," 33; and Leading Systems, "Amber Fact Sheet" (Irvine, Calif.: Leading Systems, n.d.).

46. Murphy, "AMBER for long endurance," 34.

47. Private source.

48. General Atomics, "Long-endurance Tactical Surveillance and Support Systems" (San Diego, Calif.: General Atomics, n.d.).

49. Leading Systems, Inc., "The Leading Systems Story," July 1989.

50. *Gulf War Air Power Survey,* vol. II, part I, (Washington, D.C.: Government Printing Office, 1993), 262, 263. The most extreme example of this was by army intelligence officers in the field. They only counted A-10 strikes in producing damage assessments. The 300 sorties by F-16s and 24 B-52 sorties made *per day* were not included. The result was that as late as January 31, 1991, they estimated the Republican Guard units were still at 99 percent effectiveness. Part of the problem was the army belief that airpower could not possibly be effective against dug-in troops and armor.

51. Richard P. Hallion, *Storm over Iraq* (Washington, D.C.: Smithsonian Institution Press, 1992), 204–9, 218; and *Gulf War Air Power Survey,* vol. II, part I, 262–64. The goal of the Coalition air campaign did not lend itself to easy numerical measurements. As the official history noted: "Much of the bean counting entirely missed the point. The number of tanks, vehicles, trucks, and artillery pieces destroyed did not determine whether the Iraqi Army would fight or even how well it would fight. Its battlefield effectiveness would depend on the state of mind of Iraqi soldiers and their officers. Consequently, the impact of the air war depended, to a great extent, on psy-

chological imponderables, and such uncertainties are not congenial to staff officers or to those statistical managers that have so bedeviled American military and intelligence agencies over the past twenty years."

52. Gerkin, *UAV—Unmanned Aerial Vehicles,* 35, 36, 75, 76, 88–91, 169–72.

53. Forecast International, Amber/GNAT Orientation, 1990.

54. General Atomics, "Long-endurance Tactical Surveillance and Support Systems."

55. Forecast International, Amber/GNAT Orientation, 1990.

56. General Atomics Press Release, January 21, 1994. The GNAT-750 was actually part of a family of UAVs with similar designs. The GNAT-BT (for "Basic Trainer") was designed to train the controllers without risking the bigger UAV. It was 8 feet long and had a 13.1-foot wingspan. It has made some 404 flights, involving 359 hours of flight time and 1,169 landings. On the low end of the design spectrum is the Prowler. This is half the size of the GNAT-750 and is rail-launched for short-range reconnaissance. On the upper end is the GNAT-750-45, also called the 750-TE Predator. It had a 41.7-foot wingspan and was 26.7 feet long. Its payload was increased to 400 to 500 pounds. The equipment was housed in a bullet-shaped nose. The Predator would carry a synthetic aperture radar with a resolution of one foot that would cover an 8,000-foot-wide swath from 25,000 feet. A turret under the nose would also house three electro-optical infrared sensors with a resolution of six feet. The data would be relayed to ground stations via a 30-inch-diameter satellite dish.

57. David A. Fulghum, "USAF Stresses UAVs for Recon," *Aviation Week and Space Technology* (September 27, 1993), 44.

58. John D. Morrocco, "Pentagon-CIA UAV Gains New Significance," *Aviation Week and Space Technology* (November 8, 1993), 28.

59. General Atomics, "Long-endurance Tactical Surveillance and Support Systems."

60. Fulghum, "USAF Stresses UAVs for Recon," 44; and David A. Fulghum and John D. Morrocco, "CIA to Deploy UAVs in Albania," *Aviation Week and Space Technology* (January 31, 1994), 20, 21. The CIA GNAT-750 effort was Tier 1 of a three-part UAV development program. Tier 2 would use the Predator. Ten Predators would be built in a $31.7 million program, to be fully operational in 30 months. Initially, Tier 3 was to be a large and costly program to develop a stealth UAV. These plans were soon scrapped. Reports likened the Tier 3, in a broad sense, to an unmanned B-2 in terms of size, complexity, and cost. The technology was considered so sensitive that should one have crashed, the wreckage would have to be bombed to ensure it was destroyed. The cost would have been so great that

only two to four could have been built. Replacing Tier 3 is "Tier 2–plus" and "Tier 3–minus." The first involved a UAV for broad area coverage for a major regional crisis like the Gulf War. It would fly at an altitude of 60,000 to 65,000 feet and carry a payload of 1,500 pounds. The sensor payload would include such equipment as ELINT and multispectral sensors. The other is a flying wing UAV with the stealth properties of the original Tier 3, but at a much lower cost.

61. Fulghum and Morrocco, "CIA to Deploy UAVs in Albania," 21.

62. Washington Outlook, *Aviation Week and Space Technology* (September 13, 1993), 19.

63. Fulghum and Morrocco, "CIA to Deploy UAVs in Albania," 20, 21; and Morrocco, "Pentagon-CIA UAV Gains New Significance," 28.

64. Fulghum and Morrocco, "CIA to Deploy UAVs in Albania," 21, 22.

65. News Breaks, *Aviation Week and Space Technology* (December 13–20, 1993), 21.

66. Fulghum and Morrocco, "CIA to Deploy UAVs in Albania," 20.

67. CBS Radio News, January 31, 1994.

68. Fulghum and Morrocco, "CIA to Deploy UAVs in Albania," 20.

69. Nicholas Bethell, *Betrayed* (New York: Times Books, 1984).

70. Washington Outlook, *Aviation Week and Space Technology* (February 14, 1994), 19.

71. David A. Fulghum, "CIA to Fly Missions from Inside Croatia," *Aviation Week and Space Technology* (July 11, 1994), 20, 21.

72. Washington Outlook, *Aviation Week and Space Technology* (June 6, 1994), 23.

CHAPTER 10

1. Robert F. Dorr, "Black Yak—USAF Mystery Jet," *Air International* (December 1994), 342.

2. Robert K. Wilcox, *Scream of Eagles* (New York: John Wiley and Sons, 1990), 4, 5, 10, 14, 32, 45, 102, 109, 210.

3. Mike Spick, *Jet Fighter Performance: Korea to Vietnam* (London: Ian Allen, 1986), 147, 148.

4. Wilcox, *Scream of Eagles,* 100.

5. Spick, *Jet Fighter Performance,* 149.

6. Wilcox, *Scream of Eagles,* 102–9; and Spick, *Jet Fighter Performance,* 149.

7. Wilcox, *Scream of Eagles,* 75. The "Doughnut" refers to the round gunsight used by fighters.

8. Spick, *Jet Fighter Performance,* 150. Several sources have claimed that the MiG 21F has a center of gravity problem. As the fuel burned off, it

was claimed, the center of gravity would move aft. With over 25 percent of the fuel still remaining, the safety limit would be exceeded. The plane would then pitch up, which the horizontal stabilizers could not counter. It is now clear that this is not correct.

9. Wilcox, *Scream of Eagles,* 76, 77. The use of captured enemy aircraft did not start with Have Doughnut and Have Drill. During World War II, the flight line at Wright Field was home to Zeros, ME 109s, FW 190s, and Ju 88s. The British had a special unit to test-fly captured German planes, while the Germans had examples of nearly every Allied fighter and bomber. The German's "Rosarious Flying Circus" took the captured planes to operational units for dogfight training. As with Have Drill, selected German pilots were allowed to fly the P-47s and P-51s. The Japanese had examples of U.S. aircraft, including an early model B-17, a P-40, a P-51, and an F4U Corsair. The Soviet air force actually used captured German planes in combat. During the Korean War, the U.S. actively tried to get a Soviet MiG 15 pilot to defect, offering $100,000. On September 23, 1953, North Korean air force Capt. Ro Kum Suk flew a MiG 15 to Kimpo Air Base. (He had not heard of the offer, but was given the $100,000.) The MiG 15 was flown in simulated attacks on B-36, B-47, F-86D, and F-84F aircraft. The difference between these earlier wartime examples and the Yak 23, Have Doughnut, Have Drill, and subsequent MiG operations is that the planes were "acquired" by secret means, in violation of the export agreement between the Soviet Union and client states. If it was done without the knowledge of the client state's government, then the United States would not want this publicized; if it was done with the government's knowledge, then the client state would not want this known. Intelligence arrangements (both official and unofficial) between governments (even with allies such as Israel) have always been considered particularly sensitive.

10. James Feron, "Iraqi Pilot in MiG 21 Defects to Israel," *New York Times,* August 17, 1966, 9.

11. Don Linn and Don Spering, *MiG 21 Fishbed in Action* (Carrollton, Tex.: Squadron/Signal Publications, 1993), 7; and James Feron, "Iraqi Pilot in MiG 21 Defects to Israel," *New York Times,* August 17, 1966, 9.

12. Ehud Yonay, *No Margin for Error* (New York: Pantheon, 1993), 220.

13. F. Clifton Berry Jr. and Benjamin F. Schemmer, "Soviet Jets in USAF Use: The Secret MiG Squadron," *Armed Forces Journal International* (September 1977), 26.

14. Yonay, *No Margin for Error,* 220; and Piotr Butowski with Jay Miller, *OKB MiG: A History of the Design Bureau and Its Aircraft,* (Stillwater, Minn.: Specialty Press, 1991), 96, 97.

15. James Feron, "Israel Asked Not to Let West See MiG," *New York Times,* August 19, 1966, 6.

16. Butowski with Miller, *OKB MiG,* 33, 219. The Israeli air force destroyed 363 planes and 8 helicopters on the ground. Sources differ on the total number of Israeli air-to-air losses in 1967. It may have been around 10—this would reduce the kill rate to 7.2 to 1. Even with this, both the "reduced" kill rate and the speed of the war are a stunning contrast to Vietnam.

17. Butowski with Miller, *OKB MiG,* 96, 97.

18. Terence Smith, "2 Syrian MiG 17's Flown to Israel," *New York Times,* August 13, 1968, 1.

19. Hans-Heiri Stapfer, *MiG 17 Fresco in Action* (Carrollton, Tex.: Squadron/Signal, 1992), 26–28.

20. Industry Observer, *Aviation Week and Space Technology* (February 17, 1969), 13. This item highlights the problems in reconstructing the history of the U.S. MiG effort. It says the MiG 21 (singular) was brought to the United States in the spring of 1968. *Scream of Eagles* says it was in 1967. It is also possible this is a reference to delivery of 007, and the dates were confused.

21. Wilcox, *Scream of Eagles,* 134–36.

22. Ibid., 136–38; and Lou Drendel, . . . *And Kill MiGs* (Warren, Mich.: Squadron/Signal, 1974), 41.

23. Wilcox, *Scream of Eagles,* 139, 140.

24. Ibid., 137, 154, 187.

25. Ibid., 186–88, 197.

26. Research File "Groom Lake (Nevada), Area 51 and Project Red Light" (W. L. Moore Publications, Compiled 1987).

27. Wilcox, *Scream of Eagles,* 187.

28. Ibid., 180, 181.

29. Maj. Gen. Marion E. Carl, USMC (Ret), with Barrett Tillman, *Pushing the Envelope* (Annapolis, Md.: Naval Institute Press, 1994), 112.

30. John T. Smith, "Day of the 'Top Guns'," *Air Enthusiast* 45 (1992), 21; and "'Fox 2' Part 3," *Take Off* #43, 1186, 1187.

31. Wilcox, *Scream of Eagles,* 269, 270; and "'Fox 2' Part 3," 1187.

32. Don Hollway, "Showdown between Two Top Guns," *Aviation* (March 1994), 43; and "'Fox 2' Part 3," 1188.

33. Wilcox, *Scream of Eagles,* 275, 276; and Hollway, "Showdown between Two Top Guns," 43.

34. Wilcox, *Scream of Eagles,* 276, 277; and "'Fox 2' Part 3," 1188, 1189.

35. Don Hollway, "Showdown between Two Top Guns," 44.

36. "'Fox 2' Part 3," 1189, 1190; and Hollway, "Showdown between Two Top Guns," 44, 45.

37. "'Fox 2' Part 3," 1190, 1191.

38. Spike, *Jet Fighter Performance,* 148.

39. Wilcox, *Scream of Eagles,* 288–91.

40. Drendel, . . . *And Kill MiGs,* 50; and Smith, "Day of the 'Top Guns'," 23.

41. Jeffrey Ethell and Alfred Price, *One Day in a Long War* (New York: Berkley Books, 1991), 190–95. The North Vietnamese have complicated the search for Toon by being unwilling to give a true account of wartime events. They claimed that 16 U.S. planes were shot down on May 10, 1972. This was later "improved" to 18. They have yet to admit the loss of any MiGs. Even when there is a rough match between accounts, North Vietnamese claims have major discrepancies as to location and timing.

42. Robert S. Hopkins III, "In Search of Colonel Toon," *Journal of Military Aviation* (July/August 1993), 22, 23. The Soviet pilots indicated that Pham Tuan was a gifted pilot. The Soviets traditionally looked down on Third-World pilots, but Tuan was described as an equal. Between 1973 and 1975, he was at the Mary Air Base in the Kara Kum Desert, which was the Soviet counterpart to Red Flag/Top Gun. In 1977, he attended the Soviet Air Force Academy. In 1979, he was selected for cosmonaut training.

43. Michael O'Connor, "Aces of the Yellow Star," *Air Combat* (January 1979), 20–22.

44. Ethell and Price, *One Day in a Long War,* 194, 195. That the epic Cunningham-Toon dogfight should have an element of controversy should come as no surprise. One need only recall the debate over who shot down the Red Baron or Admiral Yamamoto.

45. Wilcox, *Scream of Eagles,* 289.

46. Butowski with Miller, *OKB MiG,* 97.

47. Photos of MiG unit patches.

48. Butowski with Miller, *OKB MiG,* 178, 179.

49. Wilcox, *Scream of Eagles,* 138.

50. Berry and Schemmer, "Soviet Jets in USAF Use," 26, 27.

51. Research File "Groom Lake (Nevada), Area 51 and Project Red Light," (W. L. Moore Productions, Compiled 1987).

52. Butowski with Miller, *OKB MiG,* 150, 170, 195.

53. Michael Skinner, *Red Flag: Combat Training in Today's Air Force* (New York: Berkley Books, 1989), 28, 29.

54. Letters, *Aviation Week and Space Technology* (December 12, 1988), 146; and Talkback, *Air International* (March 1993), 157.

55. MiG unit patch.

56. Air Force Safety Agency, Bond accident report, "[deleted] Aircrew Training, November 1, 1979"; and "Phase I Ground Training Program" (May 17, 1984). The low utilization rate of the U.S. MiGs is shown by the

history of Bond's aircraft. In the three years before the crash, it had logged only 98.2 hours of flight time. Between its final inspection on March 15, 1984, and the crash, the plane had flown only 7.2 hours—a little more than one hour per week.

57. James C. Goodall, *America's Stealth Fighters and Bombers: B-2, F-117, YF-22 and YF-23* (Osceola, Wis.: Motorbooks, 1992), 21; and Don Ecker, "The Saucers and the Scientist," *UFO,* vol. 5, no. 6 (1990), 17.

58. Research File "Groom Lake (Nevada), Area 51 and Project Red Light," (W. L. Moore Publications: Compiled 1987). If it was a later–model MiG 21, it would indicate that the Egyptian MiGs were being delivered as early as the summer of 1978.

59. Butowski with Miller, *OKB MiG,* 178, 179.

60. Peter W. Merlin, "Dreamland—the Air Force's remote test site," *Aerotech News and Review* (April 1, 1994), 1.

61. Butowski with Miller, *OKB MiG,* 96, 97, 178.

62. Phil Pattee, "General killed in plane crash," *Las Vegas Review-Journal,* April 27, 1984, sec. A.

63. "Nellis AFB test plane crash kills general," *San Diego Union,* April 27, 1984, sec. A.

64. "'Secret' plane crash said to be MiG 23," *San Diego Union,* May 3, 1984, sec. A.

65. Butowski with Miller, *OKB MiG,* 102.

66. Air Force Safety Agency, Bond accident report (May 17, 1984). The type and serial number of both the aircraft and its engine were deleted from the released text of the report. Any specific details that might identify what type of plane was involved in the accident were also deleted. This allowed the report to be released, even with the extreme secrecy surrounding the MiGs.

67. Butowski with Miller, *OKB MiG,* 97.

68. James Lawrence, "The Ghost from MiG Alley," *Plane and Pilot* (March 1988), 26–33.

69. Arnold F. Swanberg, "MiGs Galore!," *Air Enthusiast* 35 (1988), 22.

70. Conversation with student test pilot, Edwards AFB Air Show, October 18, 1992.

71. Visit to USAF Test Pilot School, April 25, 1994.

72. Stapfer, *MiG 17 in Action,* 27, 28.

73. Wilcox, *Scream of Eagles,* 288–90.

74. "MiG Mystery: Why is Soviet plane at Offutt? Nobody's saying," Nebraska newspaper clipping, March 1992 (San Diego: San Diego Aerospace Museum MiG 21 file).

75. "A MiG for the Museum," *Air and Space* (August/September 1991), 16.

76. Bill Taylor, "USAF Armament Museum," *Air Enthusiast* 50 (1994), 60, 62.
77. Talkback, *Air International* (January 1993), 52.
78. Airscene Headlines, *Air International* (September 1992), 116.
79. Letter, Department of the Air Force to Curtis Peebles, February 10, 1993.
80. Stuart Brown, "Searching for the Secrets of Groom Lake," *Popular Science* (March 1994), 56, 58
81. "U.S. acquisitions of Russian weapons confirmed in U.N. document," *Aerospace Daily,* October 28, 1994.
82. "OTSA," *Flightline* (Winter 94–95), 20–26.
83. Kelth Rogers, "Budget for hypersonic spy plane rivals Nevada Test Site," *Las Vegas Review-Journal,* December 5, 1993, sec. A.
84. "Secret plane crash said to be MiG 23," *San Diego Union,* May 3, 1984, sec. A.
85. Kelth Rogers, "Groom Lake toxic burning alleged," *Las Vegas Review-Journal,* March 20, 1994, sec. A.
86. Private source, recovered debris, and documents.

CHAPTER 11

1. Clarence A. Robertson Jr., "Bomber Choices Near," *Aviation Week and Space Technology* (June 1, 1981), 18, 22.
2. Richard P. Hallion, *A Synopsis of Flying Wing Development* (Edwards AFB: History Office, January 9, 1986), 15, 16, 39.
3. "Multiple Sightings of Secret Aircraft Hint at New Propulsion, Airframe Designs," *Aviation Week and Space Technology* (October 1, 1990), 22.
4. Steve Douglass, "Project Black: The hunt for secret stealth aircraft," *Intercepts Newsletter* (December/January 1992/1993), 2.
5. Al Frickey [pseud.], "Stealth—and Beyond," *Gung-Ho* (February 1988), 41. The article makes a number of dubious claims—that the July 1986 crash near Bakersfield did not involve a stealth fighter, but rather a plane "that was more conventional, but just as black," or that the stealth fighter would be deployed in flight from a C-5. The rear doors would open, the stealth fighter would be extended out into the airflow on a hook, the wings would unfold, and it would be released. Once the mission was completed, another C-5 would make a midair pickup. This article also claimed the D-21 had a top speed of Mach 5. It also referred to Aurora and the possibility of captured flying saucers at Groom Lake.
6. Bill Sweetman and James Goodall, *Lockheed F-117* (Osceola, Wis.: Motorbooks, 1990), 27, 28.

7. Private sources.

8. John W. R. Taylor, ed., *Jane's All The World's Aircraft* (London: Jane's Publishing Co., 1984), 439.

9. Douglass, "Project Black," 2.

10. "Multiple Sightings of Secret Aircraft," 22.

11. "TR-3A Evolved from Classified Prototypes, Based on Tactical Penetrator Concept," *Aviation Week and Space Technology* (June 10, 1991), 20.

12. Gregory T. Pope, "America's Secret Aircraft," *Popular Mechanics* (December 1991), 34.

13. "Multiple Sightings of Secret Aircraft," 22; and Pope, "America's Secret Aircraft," 34.

14. Pope, "America's Secret Aircraft," 34, 109.

15. Private source. The author was given a drawing and a description of the "F-121."

16. Steve Douglass, "Special Report: New Sightings and Evidence Reveal Existence of New Black Aircraft," (n.p., 1994).

17. Private sources.

18. Peter Grier, "The (Tacit) Blue Whale," *Air Force Magazine,* (August 1996): 51–55.

19. Michael A. Dornheim, "Testbed For Stealth," *Aviation Week & Space Technology* (May 6, 1996): 20,21.

20. David A. Fulghum, "Secret Flights In 1980s Tested Stealth Reconnaissance," *Aviation Week & Space Technology* (May 6, 1996): 20,21, and "Air Force answers questions on 'Tacit Blue' program," *Northrop Grumman Newsbriefs* (May 13, 1996): 2–4.

21. Private Sources. The "F-121's" antigravity motor was claimed to have come from captured UFO technology. The "F-121" was featured in an early episode of *The X-Files.* It features a Groom Lake-like base (in the mid-west!).

CHAPTER 12

1. "Mach 4, 200,000-Ft.-Altitude Aircraft Defined," *Aviation Week and Space Technology* (January 29, 1979), 141.

2. Craig Covault, "Advanced Bomber, Missile in Definition," *Aviation Week and Space Technology* (January 29, 1979), 113, 121.

3. Rene Francillon, *Lockheed Aircraft Since 1913* (London: Putnam, 1982).

4. "'Unclassified' report brings no comment," *San Diego Union,* February 10, 1985, sec. A; and Michael G. Crutch, "Project Aurora: The Evidence So Far," *Aviation News* (9–22 October 1992), 496.

5. "Swift, high-flying new Stealth jet being developed," *San Diego Union,* January 10, 1988, sec. A.

6. Benjamin F. Schemmer, "Is Lockheed Building a Super-Stealth Replacement for USAF's Mach 3 SR-71?" *Armed Forces Journal* (January 1988). This article was the first to use the term "F-117" for the stealth fighter.

7. Al Frickey [pseud.], "Stealth—and Beyond," *Gung-Ho* (February 1988), 37–43. The article claimed that aircraft had been developed that could change color to match their background and self-correct battle damage in flight. It also claimed that the F-117 did not have an inflight refueling receptacle, as such a fixture could not be made stealthy.

8. Timothy Good, *Alien Contact* (New York: William Morrow, 1993), 255.

9. Elaine de Man, "Shooting the Stealth," *Air and Space* (August/September 1990), 93.

10. "Multiple Sightings of Secret Aircraft Hint at New Propulsion, Airframe Designs," *Aviation Week and Space Technology* (October 1, 1990), 23.

11. William B. Scott, "Black Programs Must Balance Cost, Time Savings with Public Oversight," *Aviation Week and Space Technology* (December 18, 1989), 42, 43.

12. "Secret Advanced Vehicles Demonstrate Technologies for Future Military Use," *Aviation Week and Space Technology* (October 1, 1990), 20, 21; and "Multiple Sightings of Secret Aircraft Hint at New Propulsion, Airframe Designs," 22, 23.

13. J. Antonio Huneeus, "Were Recent UFO Flaps Caused by Secret Military Aircraft?," *New York Tribune,* November 22, 1990; Bill Sweetman, "Air Force may be testing spy plane capable of cruising at 4,000 mph," *San Diego Tribune,* December 7, 1990, sec. A; and Editorial, "Aurora, C-17, ATF, B-2 may boost job count," *Antelope Valley Press,* December 9, 1990, sec. B.

14. William B. Scott, "Scientists' and Engineers' Dreams Taking to Skies as 'Black' Aircraft," *Aviation Week and Space Technology* (December 24, 1990), 41, 42.

15. Gregory T. Pope, "America's New Secret Aircraft," *Popular Science* (December 1991), 33–35, 109.

16. Curtis Peebles, *Watch the Skies! A Chronicle of the Flying Saucer Myth* (Washington, D.C.: Smithsonian Institution Press, 1994).

17. Betsy Woodford, "Phantom of the air tracked by Caltech's seismographic network," *Caltech News* (April 1992), 9. The author understands it is a common practice at some air force and navy bases to deny any knowledge when a sonic boom complaint comes in.

18. Edmund Newton, "Secret Is Out on 'Quakes': It's a Spy Plane," *Los Angeles Times,* April 17, 1992, sec. B.

19. *NBC Nightly News,* "In the 90s," KNSD Channel 39, April 20, 1992.

20. Russ Britt, "New Dawn for Aurora?" *Daily News,* May 17, 1992, Business sec.

21. Bill Sweetman, "Mystery contact may be Aurora," *Jane's Defence Weekly* (February 29, 1992), 333.

22. Christy Campbell, "Secret US spy plane is Kintyre's dark visitor," *Sunday Telegraph,* July 26, 1992. Infrared radar?

23. Bill Sweetman, "Clues hint of phantom spy plane," *Antelope Valley Press,* March 6, 1992, sec. B.

24. Editorial: "Aurora may be going 'boom' in the night," *Antelope Valley Press,* March 12, 1992.

25. Janice Castro, "Grapevine—R.A.F. to U.S.A.F.: Gotcha!," *Time* (May 25, 1992), 15.

26. "Possible 'Black' Aircraft Seen Flying in Formation with F-117s, KC-135s," *Aviation Week and Space Technology* (March 9, 1992), 66, 67.

27. William B. Scott, "New Evidence Bolsters Reports of Secret, High-Speed Aircraft," *Aviation Week and Space Technology* (May 11, 1992), 63.

28. Ibid., 62.

29. Private source.

30. Computer Message, Subject: Contrails, From: Steve 1957, America Online, 93-09-19, 17:39:11 EDT.

31. Industry Observer, "Combined Cycle Powerplant," *Aviation Week and Space Technology* (July 20, 1992), 13. "Impulse Motor" as in "Give me full impulse, Scotty!"

32. Scott, "New Evidence Bolsters Reports of Secret, High-Speed Aircraft," 62, 63.

33. Letters, *Flight International* (December 22, 1993/January 4, 1994), 39; and (January 26–February 1, 1994), 47.

34. William B. Scott, "Recent Sightings of XB-70-Like Aircraft Reinforce 1990 Reports from Edwards Area," *Aviation Week and Space Technology* (August 24, 1992), 23, 24.

35. William B. Scott, "Secret Aircraft Encompasses Qualities Of High-Speed Launcher for Spacecraft," *Aviation Week and Space Technology* (August 24, 1992), 25.

36. Peter W. Merlin, "Dreamland—the Air Force's remote test site," *Aerotech News and Review* (April 1, 1994), 9.

37. "Groom Lake's secret revealed?," *International Defense Review* (September 1993), 706. The new hangar is reportedly known as "Hangar 18." This is an inside joke; "Hangar 18" is the supposed name of the facility

at Wright-Patterson AFB where the Air Force keeps its captured flying saucers and dead aliens.

38. Michael A. Dornheim, "United 747 Crew Reports Near-Collision with Mysterious Supersonic Aircraft," *Aviation Week and Space Technology* (August 24, 1992), 24.

39. Steve Douglass, "Project Black: The hunt for secret stealth aircraft," *Intercept Newsletter* (December 1992/January 1993), 5.

40. Partial transcript of Dr. Rice's Media Availability, National Contract Management Association Conference, July 23, 1992, Los Angeles.

41. Nigel Moll, "Logbook: Aurora's Secret," *Flying* (March 1993), 100.

42. William B. Scott, "High Demand Stretches NRO Intelligence Assets" *Aviation Week and Space Technology* (February 1, 1993), 52. The NRO itself was founded in August 1960, but its existence was not officially acknowledged until the fall of 1992.

43. Steve Douglass, "Federal File: Aurora Doesn't Exist . . . ," *Monitoring Times* (March 1993), 42. The claim that a sonic boom cannot be heard at long range is incorrect. On several occasions in 1985 the author heard the double sonic boom of the space shuttle over Edwards AFB from Long Beach, California. This was 100 miles or more away.

44. Steven Aftergood and John E. Pike, "The High Cost of Secrecy," *Air and Space* (October/November 1992), 46, 47.

45. Russ Britt, "New Dawn for Aurora?," *Daily News*, May 17, 1992, Business sec.

46. Douglass, "Project Black," 5.

47. Moll, "Aurora's Secret," 100, 101.

48. Bill Sweetman, *Aurora: The Pentagon's Secret Hypersonic Spyplane* (Osceola, Wis.: Motorbooks, 1993), 12–15, 88, 89.

49. Bill Sweetman, "Hypersonic Aurora: A secret dawning?," *Jane's Defence Weekly* (December 12, 1992), 14.

50. Bill Sweetman, "Out of the Black: Secret Mach 6 Spy Plane," *Popular Science* (March 1993), 56–63, 98, 100, 101. A January 13, 1993, article in *Aerospace Daily* indicated that the Skunk Works had studied a Mach 4-5 replacement for the SR-71 during the 1980s, but abandoned it about 1986 as impractical. Only drawings and small models were produced. The aircraft would have been about the size of the B-1B, with a long, tapered fuselage, and would have had an intercontinental range. Believers in Aurora dismissed the story as government "disinformation." One said, "This article was a classic 'debunking' of a hypersonic Aurora in the complete spirit of UFO debunking from the 50's on."

51. Douglass, "Project Black," 3, 4.

52. Douglass, "Aurora Doesn't Exist . . . ," 42, 43.

53. Peter Roberson, "Mystery plane: Model depicts spy plane, but Air Force denies it exists," *The Bakersfield Californian,* November 29, 1993, sec. A; and Michael Sweeney, "If you want to see the SR-75, you must settle for the model," *General Aviation News and Flyer,* 1 November 1993, A-30.

54. Transcript of *CBS Evening News,* CBS Network (November 11, 1993).

55. Gary A. Warner, "Model maker's latest guessing game involves Air Force, secret spy planes," *San Diego Union-Tribune,* December 25, 1993, sec. A.

56. Sweetman, *Aurora,* 94.

57. Shawn Pogatchnik, "Magazine says U.S. is flying new spy plane," *Las Vegas Review-Journal,* December 5, 1993.

58. Fred Abatemarco, "Editor's Note," *Popular Science* (March 1994), 4.

59. Stuart F. Brown, "Searching for the Secrets of Groom Lake," *Popular Science* (March 1994), 54.

60. Sweetman, *Aurora,* 94.

61. *The Gospel According to Bob, Book V: The Ultimate UFO Seminar, May 1, 1993,* trans. by Glenn Campbell (Rachel, Nev.: Psycho Spy Productions, 1993), 13.

62. Bill Hamilton, "Aliens in Dreamland," *UFO Universe* (July 1990), 9, 10; and Grant R. Cameron, T. Scott Crain, and Chris Rutkowski, "In the land of dreams," *International UFO Review* (September/October 1990), 5, 6.

63. Private source.

64. Computer Message, Subject: AUFOrora, From: JoelC23358, America Online, 93-09-11, 17:13:21 EDT.

65. Private source; and Computer Message, Subject: UFOs at Groom, From: Stealth C, America Online, 93-08-13, 23:08:22 EDT. The complete message reads: "Some UFOlogists theorize that the debris from an alien spacecraft that crashed near Roswell, New Mexico, in 1947 is what they are working on at Area 51 (Unit S-4). Supposedly the USAF (in 1947 the AAF) recovered the debris and the bodies of aliens near Corona, New Mexico, in July 1947. It is possible that the USAF are trying to 'reverse engineer' the Roswell spacecraft and what people are seeing are examples of man-made copies of alien craft. However it is also possible what they are seeing is conventional (unconventional looking) secret stealth prototype aircrafts of earthly origin. Any comments?"

66. Private source. The author had received copies of this person's saucer designs long before they were published.

67. *Saucer Smear,* private newsletter (December 5, 1992), 6.

68. Statement released by John Lear, December 29, 1987, 2–4.

69. *APOA's Aviation USA* (Frederick, Md.: Aircraft Owners and Pilots Association, 1993), 78.

70. Private source.
71. "Beale's New Black Birds," *Air Combat* (December 1993), 42.
72. Private source.
73. Letters, *Flight International* (January 26–February 1, 1994), 47.
74. Al Frickey [pseud.], "Stealth—and Beyond," *Gung-Ho* (February 1988), 43.
75. Betsy Woodford, "Phantom of the air tracked by Caltech's seismograph network," *Caltech News* (April 1992), 9.
76. Micheal G. Crunch, "Project Aurora: The Evidence So Far," *Aviation News* (October 9–22, 1992), 496.
77. Glenn Campbell, *"Area 51" Viewer's Guide*, Edition 2.0 (Rachel, Nev.: Psycho Spy Productions, September 8, 1993), 98, 99.
78. "UFO Spacecraft," Testors New Model Releases, November 1, 1993.
79. Author's printout of computer messages, America Online, June 26 to October 15, 1993. There was quite a controversy over whether it was a good idea to discuss Black projects, like Aurora, in so public a forum. Rereading the printout indicates the material came from published accounts and Black buff rumors. One example of this was a September 19 message that claimed people at the site wear "foggles" to prevent them from seeing anything other than the project they are working on. Foggles are standard safety goggles with the plastic scratched so you can see only your feet. An escort leads you around. One doubts the KGB or GRU would care.
80. Computer Message, Subject: "Alien" Craft at Area 51, From: BlackSky, America Online, 93-06-29, 00:43:30 EDT.
81. Computer Message, Subject: Good Idea, From: DougBewar, America Online, 93-08-15, 15:55:09 EDT. The author should note that the vast majority of Aurora sightings (like the vast majority of UFO sightings) were made in good faith. I have talked personally with four people who heard the "Aurora roar," and I have no doubt they are honestly describing what they experienced. The problem, as with UFOs, is that the sky has an infinite number of tricks of light and shadow. (The author's "XB-70" sighting, for example.) It is ironic that in 1983, as the super-high-speed airplane stories were spreading, the *real* Black airplane was the HALSOL—a plane made of clear plastic and styrofoam, that could be outrun by an electric golf cart.
82. Private source.
83. Otto Billib, *Flying Saucers: Magic in the Skies* (Cambridge Mass.: Schenkmann Publishing, 1982).
84. C. G. Jung, Preface to the English edition, *Flying Saucers: A Modern Myth of Things Seen in the Skies*, (New York: Harcourt, Brace, 1959).
85. Tim Weiner, "Spy Plane That Came In from Cold Just Will Not Go Away in the Senate," *New York Times,* July 4, 1994, 8.

86. Ben R. Rich and Leo Janos, *Skunk Works* (New York: Little, Brown, 1994), 309–11.

87. Stuart F. Brown and Steve Douglass, "Swing Wing Stealth Attack Plane," *Popular Science* (January 1995), 54–56, 86.

88. Statement by Col. Warren Bennett, Commander 554th Range Squadron, Caliente Land Withdrawal Hearing, January 31, 1994.

89. Campbell, *"Area 51" Viewer's Guide,* 42, 43, 61–64, 88.

90. "The Government Screwed Up!," White Sides Defense Committee, May 1993.

91. "White Sides Camp-Out," White Sides Defense Committee, October 1993.

92. "Air Force tries to plug 37-year-old leak with Groom Lake land grab," *Aerospace Daily* (October 21, 1993), 131A.

93. "USAF Seeks To Keep Unwanted Eyes from Watching Secret Nevada Base," *Inside the Air Force* (October 29, 1993), 16.

94. Private source.

95. "The Mystery at Groom Lake," *Newsweek* (November 1, 1993), 4.

96. Micheal DiGregorio, "Reality Check," *Spin* (April 1994), 61–64, 103.

97. Agent X [pseud.], "Oh . . . THAT Secret Base!," *Nose* (April 1994), 26–28.

98. Private source.

99. *The Groom Lake Desert Rat,* no.4, March 6, 1994, 3.

100. Private source; and *The Groom Lake Desert Rat,* no.4, March 6, 1994, 3.

101. *The Groom Lake Desert Rat,* no.6, April 6, 1994, 1–3.

102. Ibid., no.7, April 10, 1994, 1.

103. Industrial Outlook, *Aviation Week and Space Technology* (April 18, 1994), 13.

104. *ABC News* Groom Lake transcript, quoted in *The Groom Lake Desert Rat,* no.7A, April 20, 1994. The broadcast did not include the statement about the "New World Order" or the reading from the Koran. The resulting impression of the hearings was less colorful than the actual event. Again, the involvement of fringe UFO believers in the controversy was glossed over.

105. Campbell, *"Area 51" Viewer's Guide,* 50–54.

106. Private source.

107. Campbell, *"Area 51" Viewer's Guide,* 51.

108. Private source. The exact quote used a derogatory term for black people. The source called this a "Rachel slur."

109. Roy J. Harris Jr., "'Earthlings Welcome' in Tiny Nevada Town

Where Mysterious Aircraft Often Fly Overhead," *The Wall Street Journal,* December 28, 1993.

110. Dennis Stacy, "The Ultimate UFO Seminar," *MUFON UFO Journal* (June 1993), 3–8. It is worth stressing that not all UFO believers accept the stories of Lear and Lazar. There are degrees of belief—the *MUFON UFO Journal* is a believer's publication, but it is much more critical than the magazines that carry the conspiracy tales. For a more complete account of the origins and development of the flying saucer belief systems, see the author's book *Watch the Skies!*.

CHAPTER 13

1. David Silverberg, "Old Demons, New Demons," *Armed Forces Journal* (November 1993), 14–16.

2. Francis Tusa, "Iran Bides Its Time," *Armed Forces Journal* (November 1993), 19.

3. Thomas L. Friedman, "Saddam's brazen behavior is no surprise to experts," *San Diego Union,* September 28, 1991, sec. A.

4. Mark Thompson, "Fear grows over ability of 'rogue nations' to build nuclear arms," *San Diego Union,* November 3, 1991, sec A.

5. Agent X [pseud.], "Oh . . . THAT Secret Base!," *Nose* (April 1994), 28; Russ Britt, "New Dawn for Aurora," *Daily News,* May 17, 1992, Business sec.; and Fred Abatemarco, "Editor's Note," *Popular Science* (March 1994), 4.

6. William E. Burrows, "How the Skunk Works Works," *Air and Space* (April/May 1994), 30–41.

7. Private source.

8. James Steinberg, "Zbig news on the prize fool awards front," *San Diego Union,* April 13, 1991, sec. C; and Bruce V. Bigelow, "Expert airs doubts about Tomahawks," *San Diego Tribune,* February 19, 1991, sec. A. A month into the war, a "civilian cruise missile expert" claimed the success rate for Tomahawk missiles was "far below" the 95 percent claimed by the navy. He said, "Before the war serious doubts existed about the Tomahawk's guidance system, especially about its performance at night." He added that he had heard the problems were not solvable. In issuing the statement, he explained that, "What I want to do . . . is instill in the press a sense of scientific skepticism about the performance of so-called smart weapons."

9. "Army raises ground war death toll estimates," *San Diego Union,* January 16, 1991, sec. A.

10. "Best-selling author Clancy's not surprised to see war unfold like a novel of his," *San Diego Union,* January 21, 1991, sec. A.

11. Mark Thompson, "Maine senator keeps dubious home-state radar from closing," *San Diego Union-Tribune,* December 10, 1993, sec. A.

12. Robert P. Laurence, "Punditry an illusion on talk shows," *San Diego Union-Tribune,* February 27, 1994, Night and Day sec. An example: a San Diego television station broadcast a two-part "report" on Groom Lake, Aurora, UFOs, Lazar, etc. on its 11:00 P.M. news. It opened with spectacular footage of a series of lights in the night sky near Groom Lake. They were curved, like lights on the rim of a disk. After the "report," the author called the station to inform them that the lights were infrared decoy flares. The next morning, the author called back, giving more information and leaving his phone number. They never called back or gave any hint that the "object" was not extraordinary.

13. Jim Gogek, "Talk radio catches static for cynicism in U.S.," *San Diego Union-Tribune,* January 3, 1994, sec. A; and Linda Seebach, "Voices of rage in academia's closed tower," *San Diego Union-Tribune,* June 23, 1993, sec. B.

14. Chris Jenkins, "Peace Freqs," *San Diego Union,* January 30, 1991, sec. E.

15. Kinney Littlefield, "'X'-static fans power bizarre show's takeoff," *San Diego Union-Tribune,* May 13, 1994, sec. E.

Index